Evolution-Revolution

Originally published in 1971 *Evolution – Revolution* is an interdisciplinary volume examining inquiry around the central topic of evolution and revolution. Containing contributions from a number of eminent academics of the time, the book addresses the meaning and application of evolution and revolution in the context, not of what things are, or even how they behave, but how they become. The broad interdisciplinary range of essays explores this concept through the idea of development and change and argues that both change, and development must be measured against concepts of flux and that which endures. The editors of the book suggest that these are the 'invariants' which contemporary thinkers are beginning to accept as the process-counterparts of Platonic 'immutables'. Thus this volume examines the two 'immutables' of evolution and revolution. The book covers the concept through essays in science, philosophic concepts of rationalism and existentialism, art and religion.

Evolution-Revolution

Patterns of Development in Nature Society, Man and Knowledge

Edited by Rubin Goetsky and Ervin Laszlo

Routledge
Taylor & Francis Group

First published in 1971
by Gordon and Breach, Science Publishers, Inc.

This edition first published in 2019 by Routledge
2 Park Square, Milton Park, Abingdon, Oxon, OX14 4RN
and by Routledge
52 Vanderbilt Avenue, New York, NY 10017

Routledge is an imprint of the Taylor & Francis Group, an informa business

Publisher's Note
The publisher has gone to great lengths to ensure the quality of this reprint but points out that some imperfections in the original copies may be apparent.

Disclaimer
The publisher has made every effort to trace copyright holders and welcomes correspondence from those they have been unable to contact.

A Library of Congress record exists under LCCN: 74160019

ISBN 13: 978-0-367-34358-3 (hbk)
ISBN 13: 978-0-429-32535-9 (ebk)
ISBN 13: 978-0-367-34371-2 (pbk)

Evolution–Revolution

Evolution–Revolution

Patterns of Development in Nature
Society, Man and Knowledge

Edited by
RUBIN GOTESKY and ERVIN LASZLO

GORDON & BREACH, SCIENCE PUBLISHERS
New York . London . Paris

Editors' Preface

The editors seek with this volume to present an interdisciplinary inquiry around a central topic: the meaning and application of the concepts 'evolution' and 'revolution'. We wish to justify the selection of this topic and its treatment.

With the rise of modern science, inquiries into patterns of development have won preference over inquiries into the 'nature' or 'essence' of things. The 'metaphysical mode of thinking', which dominated ancient and medieval philosophy, sought the nature of changing actualities in terms of their Being, Form, Idea, or Order reached when Becoming and development had done their work, or when the informed intellect could see through them to their immutable characteristics. By and large, the search for immutable essences has given way before the onslaught of facts and findings testifying not only to change, but to long-term development from common sources. With Darwin fell the doctrine of the immutable essence of distinct types of living things; with Hegel, Marx and Spencer that of the categorically distinct, preordained types of human societies and institutions, and with Einstein and Heisenberg the theory of immutable forms of matter and energy. The contemporary world picture is that of a world in process, where the significant actualities are not things but patterns of development. Yet if this world is to be comprehended, we must find more than constant and unco-ordinated change; change must be measured against logos: flux against the enduring. Hence we now seek constancy in change itself, more exactly, in the patterns of change and development. These are the 'invariants' which contemporary thinkers are beginning to accept as the process-counterparts of Platonic immutables. Two of these invariants constitute the theme of this volume: *evolution* and *revolution*.

Evolution and revolution are not concepts of what things are, or even of how they behave; basically, they refer to how things *become*. The 'things' we are talking about are not simply sense objects possessing 'thing-like qualities'; they are also 'events', provided we can define their characteristics and, however vaguely, their boundaries. These days 'things' tend to be thought of as 'systems', and 'systems' tend to include such widely divergent entities as the cell, the syndrome of human personality, and the organization—educational, economic or political. These are all 'things' (systems) which not only *are* and *behave*, but also *become*. If we are to make sense of their patterns of becoming, we must use concepts which themselves do not 'become' but remain constant. These are the theoretical invariants which include, among many others (for example, entropy, mass, inertia, equilibrium, growth, differentiation, integration) *evolution* and *revolution*. The fact that a dozen thinkers, approaching these concepts in the light of their own conceptual frameworks can use them to clarify and organize other concepts and theories and, through them, empirical experience itself, testifies to their scientific, philosophical and social importance.

There is a considerable diversity of views presented here. *Contextualism* (a form of process-pragmatism) is the point of view of Hahn; *general systems theory*, that of Taylor and Puligandla; *conceptual analysis* in the accepted Anglo-American tradition, that of Morrison; *phenomenology* and *rationalism,* creatively applied, that of Kockelmans; *existentialism,* that of Byrne, Maziarz and Schrader; and *epistemological constructivism* extended into a contemporary theory of communication, that of Thayer. Here we see how, across the great range of these diverse conceptual orientations, the central question of constancy in change is treated in an overlapping, yet never redundancy-creating fashion. 'Interdisciplinary inquiry', in this case, means the coming to bear of different theoretical disciplines (or their branches) upon common concerns.

It is perhaps inevitable that, given such breadth of interest, non-communication should sometimes occur. It is due here, above all, to our contributors' basically different attitudes. Some concern themselves with 'explanation' as typified in science and any from of rationalism; others with 'understanding', characteristic of existentialism, art and religion. Consequently, those concerned with explanation take the problem to be the grasp of the typical parameters of objective processes; the others take the problem to center on the subjectivity of man. Thus scientific detachment is opposed to existential commitment. The connection between the two, as they bear on patterns of development, must be supplied by the reader. Nevertheless, such contrasting attitudes bring out the contours of each and permit us to ascertain not only their accomplishments, but also their limitations.

This volume provides living material for further reflection rather than a finished research report to be accepted or rejected. But it is such constructive dialogue, not only between different individuals but between divergent schools of thought, that provides the value of, and the justification for, contemporary interdisciplinary inquiry.

Rubin Gotesky
Ervin Laszlo

Contents

PART THREE

PART FOUR

PART ONE

CONTEXTUALISM AND COSMIC EVOLUTION-REVOLUTION

LEWIS E. HAHN
Southern Illinois University

NINETEENTH-CENTURY developments in biology and geology helped trigger a series of remarkable developments in other sciences and helped make for a radically new way of seeing our world and our place in it, giving rise to a host of philosophical interpretations. Central in the biological developments was the Darwinian notion of evolution, and a cluster of pragmatic views constituted one of the most interesting of the philosophical outlooks. In this essay I should like to outline some features of the revolutionary changes in world view and relate them to contextualism as typifying some of the major pragmatic emphases. In this connection, it seems to me, contextualism is both an expression of evolutionary theory and a way of interpreting it. Accordingly, I shall begin with a brief account of contextualism, follow this with a discussion of the philosophical significance of the debate over evolution, next treat contextualism as an expression of the changed outlook stemming from this debate, and conclude with some comments concerning a contextualistic interpretation of evolution.

1. CONTEXTUALISM

In treating any form of pragmatism questions are fairly sure to be raised concerning both the name "pragmatism" and whether what it refers to has sufficient unity for the designation to be illuminating unless otherwise qualified. A full answer to these questions, moreover, might well take far more time than we have at our disposal; but some brief comments may be helpful. As to the name, most of the proponents, living and dead, of what we may call pragmatism apparently prefer to characterize their views by some other name.

3

For example, William James, who perhaps did most to bring it to the attention of the philosophical world and who helped frame the issues which divided it from other positions, spoke (in the Preface to *Pragmatism*) of not liking "pragmatism" as a name for the movement or collection of tendencies but thought that it was probably too late (1907) to change it. Although the term does not occur in Charles Sanders Peirce's famous essay of 1878 in the *Popular Science Monthly on* "How to Make Our Ideas Clear," that essay is usually said to be the first clear-cut formulation in print of the view; and years later he explained how he took over from Kant the adjective *pragmatic* to stress the relation of thought and knowledge to definite human purposes.[1] But he later decided to use the label "pragmaticism" as one ugly enough to protect his views from unwanted associations going with pragmatism.[2] John Dewey, the one who gave the view its most nearly definitive formulation, for many years preferred to call his view instrumentalism or perhaps experimentalism.

With reference to the matter of unity, few would be willing to argue that pragmatism as "a new name for some old ways of thinking" designates a single well-unified movement from the early 1870's to the present rather than a cluster of tendencies and movements. Indeed, it seems clear that a movement which at one time or another included such diverse thinkers as Peirce, James, Dewey, Mead, F. C. S. Schiller, Chauncey Wright, John Fiske, O. W. Holmes, Jr., J. H. Tufts, E. S. Ames, A. W. Moore, C. I. Lewis, George Boas, H. C. Brown, D. A. Piatt, Josiah Royce, Charles W. Morris, Sidney Hook, E. A. Burtt, Irwin Edman, Van Meter Ames, Max Otto, George Geiger, John L. Childs, and George Counts, to mention only a few, is not likely to be reducible to a single doctrine or even a simple set of views. Except possibly for periods like the one from 1894 to 1904 for the Chicago School few would be willing to claim that pragmatism was a unified movement, and even then Peirce and James were holding forth along rather different lines. Perhaps we should not expect too much unity for an outlook in which such a critic as A. O. Lovejoy in 1908 rather impatiently professed to find thirteen logically independent varieties, not all of which the pragmatists themselves could recognize.

At any rate, in point of fact pragmatism is a name which has referred to many varieties of doctrine, as one might infer from the reservations many leaders of the movement had about calling themselves pragmatists lest it commit them before the philosophic public to various doctrines they could not accept. And I shall not attempt to discuss all these variations. But I should like to direct attention to one important unifying factor in this body of doctrine since James, at least : namely, the attempt to take time seriously—time as passage, as ongoing process, as felt duration. Most of the great pragmatists since James have been concerned with things in time, things in process, changing realities, patterned events. With James, moreover, they have insisted that the truly empirical philosopher is concerned not with eternal or static realities but with things in the making or in the process of becoming. In Dewey's words, for the pragmatist "every existence is an event"—a history.[3] Although some things change at a slower rate and thus give a relative stability, no concrete thing exists apart from temporal process. The pragmatist argues that the things that seem to "exclude movement and change" are only "*phases* of things," perhaps legitimate abstractions but not concrete things.[4] Things in process, histories, changing events, with their mixture of contingency and stability, are thus basic for any adequate metaphysics or world view. These patterned events have within them movements from and toward other events, and they always occur in certain contexts and have reference to other events within these contexts.

No two pragmatists, it is true, treat these basic facts in quite the same way, and some prefer other realms of discourse than the metaphysical, but between the accounts of any two of them there is likely to be a large measure of continuity, frequently accompanied by major differences. Consider, for example, the approaches of Dewey, James, Schiller, and Stephen C. Pepper, remembering that James credits Peirce with much of his view. As I put it in another study of pragmatism some years ago :[5]

> Dewey, coming at events, histories, occurrences, primarily from the point of view of the experimental scientist interested in effecting change, noticed that certain events could be used to direct the occurrence of others. They could be used as instruments. If any particular event could be brought within the context of certain other events, control of consequences was effected. Here was a natural means of rendering good things more stable and bad ones less persistent. Experimentation was necessary

to determine just what kind of events were valuable for controlling any special type of existent; and until a possible means of control was tried out, it was impossible to say whether or not it would work. The ultimate criterion, however, was whether or not the intended consequences turned up. And fortunately enough, this process of bringing a particular set of events into a control context seemed to be a way of solving any empirical problem including those of philosophy. Hence Dewey set about hammering home the importance of setting up such relations between complexes of events and thereby making such a process an instrument for bringing about desired consequences. Because of his emphasis upon this aspect of pragmatism, his view came to be called Instrumentalism.

James before Dewey, of course, had stressed the temporal character of reality and the importance of consequences. He discerned the vital significance of the fact that events may lead to or refer to something beyond themselves, and about this fact he built up his pragmatic theory of meaning and truth. An idea means what it leads to, and the test of its truth is whether or not it leads as was expected. Specific consequences provide the test. If there is no difference in the specific consequences, James maintained, there is no practical difference in meaning.

Schiller was profoundly impressed by the shift in emphasis produced by such a view. Instead of speaking in terms of metaphysical entities often far removed from the empirical level, instead of stressing what seemed to him impossible idealistic goals somehow concerned with Absolutes rather than with human affairs, this view brought philosophical discussions back to the realm of human endeavor. He called the view Humanism.

Interested primarily in what pragmatism had to offer for aesthetic theory, Pepper was struck by the significance of context for aesthetic quality. Histories, patterned events, are not isolated affairs. They occur in certain contexts of other events; and differences in context may make important differences in quality. To control the quality of an event, one does not isolate it, but rather surrounds it with a suitable context; and given the proper context, practically any event may have the enhanced quality of beauty. Nor, Pepper was convinced, is context significant merely for aesthetic matters. It occupies an equally important place in other fields. Hence for that form of pragmatism which stresses this fact he coined the name Contextualism.

In terms of my own commitments and reservations the form of pragmatism I am most interested in is contextualism, an outlook which finds much of value in the other approaches listed above. I think of contextualism as a form of pragmatic naturalism which takes as its basic fact or root metaphor patterned events, things in process, or historical events.[6] The main traits of such events constitute the fundamental categories of the view I hold, and may be used to characterize or explain any set of problematic facts. Though there are various alternative statements of these categorial features,[7] one convenient grouping divides them into (1) a set of filling or textural traits which indicate the nature or "stuff" of an event, and (2) a group of contextual or environmental traits which denote the place of the event in relation to other events. The textural categories include *texture, strand, quality, fusion, and reference* (direction-distance values), whereas the most important contextual ones are perhaps *environment, initiations, means* (or *instruments*), *consummations,* and *frustrations* (blocking).

The patterned event or affair with which the contextualist starts is not a discrete atomic unit but rather a complex interrelationship of tendencies all interwoven into an integral whole with its own individual character or quality. As the term texture (borrowed from the weaver's art) suggests, each historical event is a web or network of happenings (strands), a focal center into which features of other histories somehow enter. Textures may be analyzed into constituent strands, and texture, strand, and context are relative to each other. What is strand or detail in a larger context may become texture in another. As a detail of a texture a strand reaches out into the context and brings some of the quality of the latter into the texture. Since the character of a texture is a fusion of the qualities of its strands and the latter are partly from its context, analysis of a texture takes us into the texture of other events.

References are both part of the character or nature of an event and links with its context. Strands reach out or refer to other textures. They move from initiations through means-objects or instruments to frustrations or consummations, and to control the direction of affairs we must direct our attention to the means.

We shall want to elaborate on certain of these categorial features later; but perhaps already enough has been said to suggest that the language and categories of the contextualist reflect in various ways developments in evolutionary theory; and I hope that a survey, to which we now turn, of some of the issues in the debate over evolution will make this plainer.

2.   DEBATE OVER EVOLUTION

In some ways the debate over evolution after the publication of Darwin's *The Origin of Species* in 1859 was livelier and more heated among theologians, philosophers, scientists from fields other than biology, and popular writers than among biologists; and we may wish to distinguish with A. O. Lovejoy and Philip Wiener between evolution as a scientific theory in a special field and evolutionism as a generalization invading every field from biology and cosmology to sociology and philosophy of history.[8] At any rate, writers in many diverse fields, some of them remote from biology, found or thought they found that Darwinian theory had implications for their field; and the end in this process has not yet been reached. The theory seemed to raise basic questions as to the nature of reality, law and design in nature, man and his place in the cosmos, human experience, knowledge and the methods of knowing, morals, and religion. Darwin was apparently headed for a career as a clergyman when his appointment as naturalist for the famous Beagle expedition provided experiences which changed his outlook and his vocation; but he himself avoided the wider implications of his view. In spite of his reluctance to speak out on some of these matters, debate over the scientific issues, however, seemed to make the broader implications stand out all the more clearly for others, and it has become obvious that a revolution has taken place in our ways of thinking. Indeed, it has become increasingly clear that whatever the reservations of his fellow biologists concerning his interpretation of natural selection, sexual selection, and the agencies of evolution, the fact or principle of evolution is well established by a host of different lines of evidence from many fields: for example, paleontology, embryology, comparative anatomy, analysis of the blood and other fluids of various species, vestigial organs, geographical distribution of plants and animals, and plant and animal breeding.

But why, it may be asked, should the general public get so excited over a scientific principle, even if it be one which ranges from stardust to men? Perhaps a sampling of some of the issues will help make clear the reasons for the heat and liveliness of the debate. The issues ranged from whether man is a brother to the apes or a Son of God and whether the species are products of special divine creation or of natural selection through whether the Bible or science affords the more trustworthy account of creation, whether human reason is a relatively perfect mirror of reality or an instrument of adjustment, and whether analysis is a matter of reducing a complex to permanent elements or of tracing a pattern of changes, to broad questions of naturalism versus supernaturalism and of change or permanence as the touchstone of the real, with a host of associated issues.

The issue of whether man is a brother to the apes or a Son of God is one as to whether man is a part of the animal kingdom and the natural order or a member of a distinct and higher spiritual order. The traditional Christian outlook had assumed that man was a creature apart, a being set over against his world. One set of principles applied to his world and the animal kingdom, another to man. To be sure, man was set down for a time in this vale of tears to enact his drama of salvation; but it was clear that this natural order was an alien realm and that man was destined for something higher. Thus the Darwinian notion that man was a part of the animal kingdom and the natural order seemed to many an affront to the dignity of man. The more the Darwinians traced continuities within the animal kingdom and the more they spelled out of man's place in nature, the more insulted many became. It was as if one had a choice of one's ancestors and kinship groups, God and the angels, on the one hand, or on the other, the apes; and faced with this choice, the Darwinians, it seemed to many devout individuals, deliberately and perversely chose to align themselves with a base, brutal realm rather than with a glorious spiritual order. A favorite ploy was for a speaker to retort that he did not know about the relatives of Darwin and the Darwinians but that he was quite sure that there were no monkeys among his own kinfolks. Lest one think that this issue faded out more than a century ago, perhaps a reminder of the famous Tennessee Scopes trial of 1925, with Clarence Darrow and William Jennings Bryan as opposing lawyers, is in order. Their debate, as contemporary newspaper accounts show, echoed the line of argument sketched above and took up various of the other issues I have mentioned.

A closely related issue was whether man and other living creatures were products of a special divine creation or of natural selection. It was sometimes formulated as a question of divine selection or design as opposed to natural selection and chance. Could any species of living things, much less man, have come about as a result of small chance variations appearing spontaneously in the individuals of a species struggling to survive in a world in which there always seemed to be too little food and too many offspring? Granted that in some instances small variations taken in conjunction with just the right external conditions might mean the difference between a species making an appropriate adjustment and surviving or failing to adjust and perishing, is the origin of species to be explained in terms of this sort of natural selection? Can the instances of a species having just the right conditions of climate and food supply for its maintenance be explained in terms of an unplanned, fortuitous natural process? Could a giant crab, say, with just the right pincers for stripping coconut husk from the three-eye holes of the fallen nut and the right secondary pincers for extracting the meat have come about by any unplanned process? Or is this not rather an illustration of the admirable provision of Infinite Wisdom by which each created thing is adapted to the place for which it was intended? Is not speaking of natural selection another way of saying that no intelligence made the selection, or, indeed, that no selection occurred at all? And are there not too many regularities in nature for one not to seek some design or plan? How could people who were brought up in a tradition which accepted the Bible and held to divine providence find the doctrine of natural selection credible? Faced with a choice between divine foresight and a blind, brutal, natural process, the choice in favor of the former was easily made by many. To decide otherwise seemed to them to rob the world of meaning and strip man of his focal place in the universe. The evolutionary explanation seemed to threaten Christianity and a religious view of man and his world.

This second issue was closely intertwined with a third as to whether the Bible or science affords the more trustworthy account of creation. Since many believed the Bible to be divine revelation, it seemed clearly more certain than the admittedly fallible, hypothetical account of the scientists. Why settle for a fallible version when infallible truth was at hand? Some thought, moreover, that the Biblical account interpreted in terms of Bishop Ussher's chronology was a great deal more plausible than the scientific estimate of the age of the earth. According to the former, the world was only about six thousand years old, but the changes the scientists spoke of would have required millions of years, and the great time periods seemed incredibly long. There was the further fact that the Bible spoke of a special creation which gave man a central place in the scheme of things whereas on the other interpretation man was a tiny detail, one instance among an indefinitely large number selected blindly by natural process, in a vast evolving universe.

These first three issues, however, were in a sense all subordinate to the great issue of naturalism versus supernaturalism. Is it possible to account for natural events wholly in terms of other natural events without going outside the system, or must we not bring in God and/or supernatural agents to account for what we find in our world? Some argued for a supernatural agent in terms of relevation and the authority of the Bible, but most philosophers have become convinced that the problems of determining whether a revelation is genuine and of justifying the choice of an authority drive us to a consideration in terms of the evidence available to human thinkers; and if we must use ordinary reflective thinking to decide these matters, why bring in revelation and religious authority?

Others argued for a supernatural intelligence as a first cause, holding that God is necessary to set the natural process in motion. After the system of nature is started off, perhaps it can take it from there; but how could it have started if there were no outside intelligent agent to get it going? The naturalists have replied by asking who caused or started the outside supernatural intelligence or God, and upon being told that God is self-causing and needs no further causative agent, they have inquired why we should not stop the regress one step sooner and treat nature as the self-sufficient system instead of moving to the deity and then stopping there.

Others have held that getting things started is not the only reason why God is needed to explain the evolutionary process, that there are also various gaps in the line of development from the inorganic to the organic realm, and that supernatural powers are needed to bridge them. The naturalists have conceded some gaps but have pointed to the impressive progress science has made in filling them in and have indicated that it seems reasonable to expect further progress along these lines. In the long run, they have argued, reliance on the working hypotheses of the scientists affords a more hopeful prospect than acceptance of religious dogmas of revelation.

Some other defenders of supernaturalism maintained that the admirable adaptation of organism to environment and the regularities in nature, the lawful character of nature, were clear indications of design and that the very evidence the evolutionists had piled up made design all the more evident. Men like Asa Gray, Fiske, and McCosh held that one could retain belief in God and combine it with evolution properly interpreted. Perhaps the creation was spread over a longer time than was previously thought and the whole evolutionary process of the Darwinians was simply God's way of carrying out His design for creation. Although the Biblical account was not to be taken literally, some noted, it suggested the general nature of the development described by the scientists; and the evidence from design could be used as the deists claimed to establish the existence and nature of God. One would not have to have a special revelation. All the key facts of design could be read in the great book of Nature. But the facts of evil in the world seemed to many to raise insuperable difficulties for any attempt to prove the existence and nature of God from His works. And if the deistic way was blocked, some theists thought the supernaturalists had to depend on some antecedent revelation or some other way of first establishing the deity so that it might then be possible to show how evil somehow fitted into the divine design.

The naturalists tended to argue that, on the one hand, the supernaturalistic special creation explained too much or too little and that, on the other hand, a prior intelligent causal agent to plan and preordain organic adaptations was unnecessary. So long as we stress the fact that nothing is beyond the power of the deity, any kind of universe or any kind of happening in it may be explained as stemming from the divine will, but surely this is to "explain" too much in that any and everything is undifferentiatedly attributed to the divine will. And if we insist on a differentiated account, some things being more an expression of the divine will than others or some being an expression in different ways from others, the naturalist questions whether the supernaturalist has explained enough. How illuminating it is to be told that God created the world, of course, depends partly on what kind of deity is the creator; and the naturalist wants something more than an alleged revelation to establish a benevolent deity. Once we accept the idea of an external supernatural power in terms of some revelation, we seem to be operating primarily within the domain of our ignorance rather than building on what we have learned through critical inquiry. Without extensive supplementation by what we have discovered through the sciences, the supernaturalistic account tells us very little about what kind of world has been or is being created, and the supernatural base and the scientific developments seem to require radically different assumptions as to how to get knowledge. If we accept the Darwinian account or some outgrowth of it, moreover, the supernatural base is not needed. Organic adaptations may be explained in terms of constant variations and the elimination of harmful variations in the struggle for existence engendered by excessive reproduction and limited food supply.

Another issue concerned knowledge and the role or character of reason or intelligence. Although the Calvinists had thought of human reason as basically depraved and sadly insufficient as a substitute for relevation, many pre-Darwinians, taking their cue from the Greeks, thought of reason as a mirror of the real. They adopted a spectator view of knowledge and thought of reason as surveying and simply noting what was real. To be sure, according to this outlook, prejudices and biases may distort one's vision of reality, but pure reason reflects fairly perfectly what is there in nature or the external world. With Darwin's account, however, came the suspicion that intelligence is a device for adjusting to a favorable or hostile environment rather than for providing a purely rational carbon copy of the real. Whatever pure reason might provide, what we have is something much less pure; and such being the case, context, or situation, and conditions of observation become significant in determining or interpreting what is perceived. In terms of the opposition between pure reason and intelligence as an agent of adaptation or adjustment, once more some felt that the evolutionists were denigrating man and conceiving of him as a brute rather than as a rational spirit.

In many ways, however, the issue with the greatest revolutionary impact was that of change or permanence as the touchstone of the real. For two thousand years in field after field permanence and perfection tended to be equated; and during this period, as John Dewey wrote in a famous essay on Darwin,[9] the ruling conceptions in the philosophy of nature and knowledge had rested on the assumption of the superiority of the fixed and final. Absolute permanency marked at one and the same time the height of reality and perfection;

and the species or forms typified fixity and perfection. Becoming, originating and passing away, change, these were signs of defect and unreality, at best the marks of a lower reality. Imagine, then, the impact of a soberly reasoned, massively detailed treatise with the title of Darwin's book, *The Origin of Species*! Here was an account of the coming to be and passing away of what had been taken to be fixed and inflexible. In the light of his account no longer can they be regarded as fixed forms of inflexible species; rather they become alterable characters,[10] and a last major fixity thereby comes to be expelled from nature. If even the species change, changing reality becomes something to be reckoned with; and some began to suspect that the fixed and permanent, far from being signs of perfection, are expressions of the dead and outmoded.

This is not to say, of course, that everyone agreed at once that change rather than permanence is the mark of the real, for many are still unconvinced. Old ways of thinking, especially ones grounded in numerous traditions, are hard to dislodge, and patterns of thinking have tended to change slowly. Nor is it to say that most people saw the full implications of the Darwinian revolution with reference to change and permanence. I suspect that we are still far from seeing all of them. What is perhaps most noteworthy about the Darwinian revolution is not that one scientist spoke of the origin of what had been assumed to be fixed and permanent, remarkable though that be. Still earlier Galileo and others had spoken of the earth as the setting for incessant alterations and generations and had shifted their interest and attention from the permanent to the changing. It is rather that most of his fellow scientists by now, whatever their reservations about his account of natural selection, assume his big point that species come into being and pass away and offer their version of the origin of species and evolutionary change.

The issue as to whether analysis is primarily a matter of reducing a complex to permanent or fixed elements or an affair of tracing a pattern of changes is obviously closely related to the one we have just been considering. The advocates of the latter interpretation of analysis charged the proponents of the former with reductionism and were answered by the charge that they were confusing logic and psychology. Is analysis an affair of timeless distinctions or one of distinguishing temporal patterns? Or does it depend on what one is attempting to analyze and why? At any rate, more and more explanations were being given in terms of origin and development, in terms of phases of a career or history, whether with reference to the stars and galactic systems, or to the constituents of matter, or to plants and animals, or to linguistics, or to economic institutions, or to religion and cultural patterns. And increasingly the genetic emphasis has entered into the analyses of investigators in such fields as experimental physiology, experimental psychology, ecology, physical and cultural anthropology, social psychology, sociology, comparative religion, institutional and historical economics, linguistics, and a host of new social sciences appearing in the late nineteenth and early twentieth centuries.

There were, to be sure, other important issues involved in the debate over evolution, some of which—for example, the implications for ethics—are treated in other parts of this symposium; and certain of these other issues also helped make for a lively, heated discussion. But I hope that the sampling of questions we have considered will help make clear both the revolutionary impact the theory of evolution has had on our ways of seeing and reacting to our world and some of the reasons for the general concern with the implications of evolution. Perhaps one of the most vivid ways of pointing up the differences between the world before Darwin and the new world is to note the character of some of the debates. For example, the idea that man is a part of the animal kingdom, which seemed like an outrageous proposal in the 1860's and 1870's, is now pretty much accepted as a commonplace.

Perhaps the review to which we now turn of contextualism as an expression of the changed outlook stemming from the controversies over evolution will shed further light on some of the remaining issues and afford a measure of clarification of the new evolutionary outlook.

3. CONTEXTUALISM AND THE NEW OUTLOOK

How does contextualism express the Darwinian outlook? A complete answer would require volumes, but, fortunately, the main outlines can be suggested in fairly brief compass. Central for the contextualistic philosophy in this regard are (1) change and the ways of dealing with the changing and (2) a naturalistic approach framed in terms of the biological matrix of experience; and it is difficult to discuss either of these sets of points in such fashion as to illuminate the changed outlook stemming from the controversies over evolution without assuming the other. So there will be a measure of arbitrariness as to which points we discuss under which heading as well as in the order of discussion. In the first place, then, contextualism expresses the Darwinian outlook in its wholehearted acceptance of change. For the contextualist the key fact about our world is the fact of change, and he finds it wherever he looks throughout the universe, whether in the vast galaxies of stars, the constitution of the atoms, or human affairs. Every existence is an event or history, with its own point of initiation, qualitative changes, and point of termination. Each thing in the universe comes into being, undergoes qualitative changes, and dies, making way for other individuals. A fixed world of static things is foreign to the outlook of the contextualists and post-Darwinians. Nature, on this view, is a complex of affairs, transactions, histories, marked by incessant beginnings and endings. In our changing world we may expect unpredictable novelties, qualitatively new things, incomplete things still in the making, uncertain, unstable situations as well as relatively stable ones, genuine contingencies; and to meet the challenges this kind of world presents we must try to work out ways of intelligently redirecting ongoing affairs. We need to stabilize patterns we find good and to seek ways of averting, reconstructing, or living acceptably with patterns we find bad. As Dewey argued in *The Quest for Certainty,* our concern as empirical philosophers is not the quest for the immutable or the pursuit of changelessness but rather finding patterns of change, constant or relatively invariant relations between changes.

The more we concentrate upon specific changes in the natural order, the clearer it becomes that the affairs constituting our world are many and varied; and this notion too seemed to be an expression of the Darwinian approach. Heterogeneity and diversity rather than homogeneity and sameness characterize existent things; and as William James never tired of saying, our world is not a monistic block universe but rather one with a plurality of specifically diverse, heterogeneous existences. Those views that make our world out to be more like an empire or kingdom than some sort of federation are fairly sure to be suppressing differences and underestimating the variety in it. Although, he admitted, there are many ways, mainly abstract or mechanical, in which we may refer to the universe as one, there is no way of reducing all things to a common denominator in any very significant, concrete sense; and James, like the contextualists in general, was suspicious of theories which make the world out to be all crystal clear with each part fitting into a tightly ordered logical system. The kind of world we find in experience seems rather to be one with some mud and chinks in it, with some portions shadowy and others shrouded in darkness; and although we seek to shed what light we may on these dark areas, we must not make the mistake of assuming that the significance of an item is always directly proportional to its clarity. There may, indeed, be things not dreamt of in our philosophy.

James was convinced that the pragmatic difference between pluralism and a monism of some sort hinged on the reality or unreality of novelty, and concrete novelty was for him so obvious a fact that the issue was readily resolved. It seemed clear to him that reality as perceptually experienced shows us a world of change— one shot through with novelties and risks, struggles, real losses, and genuine gains; one having a place for freedom.

Or, as James sometimes put it in a characteristically contextualistic vein, the difference between pluralism and monism or absolutism turns on the legitimacy of *some*.[11] Instead of speaking always in terms of *all* and *none*, we need to recognize that each part of the world is in some ways connected with its other parts and in others separated from them, and just how much union there is can be investigated empirically. For that matter, in discussing the range of realities in our world the contextualist feels more comfortable in speaking of *each* than of *all* in the sense of the total collection. For the former we may have a sample; for the latter our best enumeration may leave something out.

In dealing empirically with a world made up of a plurality of specifically diverse, changing affairs, there is no place for indubitables or claims to certainty. Rather a method or approach on the order of Peirce's fallibilism and probabilism seems more appropriate; and this, once more, seems to be an expression of the Darwinian outlook. Although in ordinary speech there may be justifiable ways of speaking of "practical certainty," ones which, paradoxically, leave open the possibility of our later discovering that we were in error, cognitive certainty in principle regarding matters of fact seems unattainable. Accordingly, the contextualist challenges claims to certainty, whether based on infallible authority, either of God, his representatives, a sacred book, or the established doctrine of an Aristotle or a Marx, or on self-evident principles, or on indubitable data of some sort. According to the contextualist, such claims operate as devices for shutting off or blocking further investigation, and with Peirce he wants to keep the road to inquiry open. At most the feeling of certainty is an evidential item which needs to be weighed with other evidence, and the feeling has been reported for so many highly uncertain claims that it is properly suspect. Whether in terms of current evidence or the historical record, where can we find another equally large and varied set of allegations with a poorer record of substantiation than the assertions of indubitability or certainty?

The contextualist agrees with Peirce that attempts to base our reasoning regarding matters of fact on indubitables is likely to lead to scepticism or irrationality. We neither find indubitable axioms nor need we find them. We may base our account on a number of lines of evidence each of which is only probable; but if they converge, we get a belief that is highly probable and a much better guide to action than a so-called indubitable. If, as Peirce held, our reasoning is a kind of sampling operation in which we judge concerning the nature of a whole on the basis of the proportion found in our sampling of it, our judgments are, of course, fallible; but we can by increased sampling improve on our accuracy. With appropriately selected thimbles of wheat, we can judge the quality of a shipload of wheat just as with a random selection of hookfuls of cotton we can estimate the quality of a trainload of bales of cotton, and similarly for other areas.

Mechanistic determinists have sometimes looked to scientific laws for illustrations of certainty and universality and have reasoned from them to each and every instance covered by a particular law; but according to Peirce, this procedure is largely due to the fact that the individuals who reason in this way are unfamiliar with measurement and the actual procedures involved in formulating laws. In interpreting scientific laws in the light of radical change or basic novelty, on his view, we do well to remember that these formulations are statements of uniformities, statistical averages. Such necessity as they may have is a nomic necessity, not an empirically derived one. Although they may be reasonably accurate for large classes of instances, they may be inapplicable to specific items. In this respect they may be more like the actuarial charts of the life insurance companies than the mechanistic ideal. These charts may predict fairly accurately how many people of, say, our age will die in a given year but not whether a specific person of that age will die in that year.

If we seek to avoid one set of consequences or stabilize another in a changing world, we need something better than authoritarianism, a priori formulas, simple guesswork, or merely waiting for events to run their course. Here, according to the contextualist, a frankly experimental approach is needed, and the more significant the issues at stake, the more clearly is this the case. This is something we shall comment on later in connection with an analysis or critical thinking or reflective inquiry, but here I want at least to note the importance pragmatists have always attached to experimentation as a way of dealing with change. It is a way of getting more accurate predictions and better control in a problematic situation. An experiment is a program of action to determine consequences and is thus a way of introducing intelligence into a situation. It is an intelligently guided procedure for discovering what adjustments an organism must make to its environment to ward off ills or secure goods. For that matter, of course, experimentation is relevant not merely on the individual biological level but also, as John Dewey spent a lifetime reminding us, wherever planned reconstruction of a situation may help effect desired transformations—for example, in social planning or in education. Although experimentation will not and need not provide certainty, it may help us achieve a greater measure of security in the face of a hostile environment, biological or social. For if through experimentation we can discover or introduce appropriate instruments, we may better stabilize our responses or adjust with a greater measure of success to a hostile environment.

Persistent, careful concern with specific things changing and the ways of introducing desired changes, as the contextualist sees it, introduces a basic relativism, which is also quite in keeping with Darwinism. This relativism is to be opposed to absolutism and is a way of stressing the importance of context, situation, relationships. To take things out of context is to risk distortion and loss of meaning and value.[12] Place a texture in a different context, and you change its quality. Whether the change is significant for a given purpose, however, may depend on the purpose or the context. At any rate, constancy of quality from one context to another is not something to be assumed but rather something which may have to be worked for. It may require planning and setting up appropriate control textures. Absolutes, accordingly, are ruled out on this view, and unqualified generalizations are likely to be misleading. Contextualists suggest that characterizing any affair or specifying its relations is rarely simple and usually proceeds more clearly if we can indicate for what purpose, in terms of what standards, in what respects, or under what conditions it may be so characterized. Fortunately, the context commonly makes these matters clear enough for everyday purposes.

At any rate, a response that is appropriate to one environmental situation, as Darwin's investigations repeatedly showed, may be disastrous in another. A teaching technique which works very well relative to students with one set of backgrounds and interests may prove quite inappropriate for students with different backgrounds and interests. A type of pine which flourishes under most soil conditions in Southern Illinois may fare very poorly under standard conditions in Northern Illinois. An economic policy or a plan of action is good relative to a specific situation which makes it desirable. A knife may be good for sharpening a pencil and bad for cutting a rope; but to speak of it without qualification as good or bad is quite misleading.

The contextualistic account of analysis also expresses the Darwinian mode in a number of ways: for example, in its rejection of the idea that analysis is primarily a matter of reducing a complex to permanent or fixed elements, in its acceptance of the notion that analysis is an affair of tracing patterns of change, and in its recognition of the importance for many purposes of a genetic account of events. The contextualist denies the possibility of element analysis in the sense of breaking a whole down into atomic units or irreducible constituents of some sort. Analysis for him, as Pepper notes,[13] is an affair of exhibiting the texture of an event, and this involves discrimination of its strands, but the strands derive part of their quality from the events context and have a way of leading off into it. So analysis becomes a matter of following references from one texture to another, and how far we follow them or which ones we trace depends on the problem which occasions our analytic inquiry. But we never reach ultimate elements.

We may trace the strands of a given texture into convenient control textures (schematic textures, or schemes) such as the color cone or a musical scale, and these in turn may be traced into schemes of light wave or air vibrations. These latter schemes are tied in with the system of schemes constituting physics and are interknit in various ways with the schematic structures of the other sciences. Any color may be located rather precisely as to hue, saturation, and value on the color cone, and once so located we have a basis for producing it in the future. A Chopin piano piece may be analyzed in terms of notes on a staff, thus giving us the musical score of the composition, whether we think of this score as something transcribed from hearing the piece or as Chopin's own analysis of the texture he sought to provide; and with this scheme a pianist can follow its references to the keys of a pianoforte keyboard and produce a set of tones constituting his version of the Chopin piece. Thus these schemes give us systems of references and suggested operations for producing textures of some specified character. They guide us to the objects or textures indicated.

Platonic realists and mechanistic naturalists traditionally have assumed that any whole can be analyzed completely and finally into its constituent elements. What the elements are may be debated, but that it is possible to reduce a whole to such constituents they never doubted. This possibility, however, is categorically denied by the contextualist. On his categories there is no final and complete analysis of anything; and to speak in these terms shows a lack of understanding of analysis. Analytic inquiry involves both discrimination of constituent strands of a texture and relevant contextual references, and the traditional accounts have tended to limit it to a special form of the former. When we analyze a given event into A, B, and C, this is not to say that the event is divided without remainder once and for all into constituents A, B, and C, but rather that it is possible to trace a pattern of relations connecting the event with A, B, and C. And, of course, it does not rule out the possibility of tracing another pattern of relations between the event and D, E, F, or between it and an indefinitely large number of other possibilities. Ultimate cosmic units or elements are, for the contextualist, dubious entities, and to speak of somehow reducing things into them as final constituents is to misinterpret analytic inquiry.

There is, of course, for the contextualist, no such thing as *the* analysis of anything. The texture of an event may be exhibited in any number of different ways, depending upon the purpose of the analysis or the nature of the problem generating it. The quality of an event is a fusion of the qualities of its strands, and part of the quality of any given texture accrues to it because of the quality of its environing textures, whose qualities, in turn, depend partially on the qualities of still other textures in their contexts, and so on—a process which drives the analyst beyond the event being analyzed. So depending upon which referential strand we follow and which environing texture we trace it into, we may have any number of analyses, each of which may be more or less adequate for its purpose.

If, then, analysis is an affair of exhibiting the texture of an event, a genetic account in terms of conditions and consequences is very much in order, for this is an important way of exhibiting the texture in question. Nature, for the contextualist, is a scene of incessant beginnings and endings; and initiations, consummations, blockings, with connecting references and intervening means-objects or instruments operating within a certain context, are categorial for him. These categorial distinctions, moreover, are ones in a temporal process, and an analysis which neglects this fact, contextualistically speaking, omits or slurs over something of vital importance. Although such an analysis may provide timeless logical distinctions, it is unlikely to exhibit the texture of an event with temporal spread. For that a genetic account may prove more illuminating.

If, for example, we are analyzing the death of JB, it will not do to say that whatever the cause, he is nonetheless dead, and no considerations of attendant circumstances or of anything before or after the instant of death are needed. Which considerations of this sort are relevant will depend, of course, on the purpose of the analysis. If homicide is suspected, the attendant circumstances may be crucially important for the police investigator. The act of dying, moreover, is not a matter of a knife-edge instant; it takes some time, sometimes relatively little and sometimes a fairly long time. And what happens within that time varies with the cause of death. Hence, although, to be sure, after the fact JB is equally dead whatever the cause, the police investigator's analysis of his death will be different in significant ways depending on whether the cause is asphyxiation, cyanide poisoning, food poisoning, strangulation, a stab in the heart, a blow on the head, cancer, or something else. Death from one of these causes will lead into different environing textures and will involve different control textures, observations, laboratory reports, and so on. And an adequate analysis of JB's death for the purpose cited needs to exhibit the distinctive features of the total texture stemming from whichever cause or complex of causes with whatever pertinent ramifications in more or less distant environing textures may be required.

In terms of contextualistic analysis we might well develop still other points connected with the respects in which the contextualistic philosophy expresses the Darwinian outlook in its acceptance of change and in its development of ways of dealing with the changing, for the entire philosophy centers about these topics. But the contextualist's naturalistic approach framed in terms of the biological matrix of experience is also an expression of the Darwinian perspective, and some of the key doctrines may be more conveniently discussed under this second heading than under our first or temporalistic one.

At any rate, contextualists reject supernaturalism with its doctrines of special creation, miracles, and revelation and attempt to account for natural events wholly in terms of natural principles without going outside the system of nature. They are concerned with questions growing out of affairs within nature rather than with attempts to discover some divine design for the totality of things. They reject the notion of a deity who intervenes in arbitrary fashion through special supernatural measures. One of the lessons of Darwin for philosophy, according to Dewey, is the forswearing of "inquiry after absolute origins and absolute finalities in order to explore specific values and the specific conditions that generate them,"[14] to investigate the ways in which one set of changes serves concrete purposes and another defeats them.

In spite of their forthright rejection of supernaturalism various pragmatists have used the term "God," but in each case the traditional idea has been reinterpreted in naturalistic terms, usually in terms of values, and they have sometimes noted that the ideas with which they are concerned can be formulated without using the term. Peirce, James, and Dewey all spoke of God and religion in terms of values, and even so some of his associates criticized James as too supernaturalistic. Dewey, in *A Common Faith,* spoke of God as an active relation between the ideal and the actual rather than as an eternal, completed Being. Another contextualistically oriented philosopher, Henry Wieman, speaks of God as the source of human good, the set of forces making for creativity. John Fiske thought of the deity as guiding the natural evolutionary process, but his deity works through nature.

The naturalistic approach of the contextualists assumes a basic continuity between lower, less complex forms of existence and higher, more complex ones such that the latter grow out of the former. This is not to say that higher forms and activities are reducible to lower level ones; but it is to recognize the fact that life, conscious activities, linguistic communication, and forms of culture develop in a natural order from lower forms in which these features were not present. Precisely how any one of these higher level affairs develops is a matter for reflective inquiry or scientific investigation, but contextualists adopt a genetic point of view which maintains that in principle it is possible to trace lines of development from the lower to the higher stages. Using this genetic hypothesis, moreover, it has been possible time after time to work out or follow the patterns of change involved. But the matter of levels of existence is a topic to which we shall return after noting some features of the biological matrix of experience.

Pragmatists have long recognized that one's interpretation of man and nature is intimately tied up with the psychology one adopts. If one starts with the traditional introspective psychology of Locke, Berkeley, and Hume which describes experience in terms of relatively distinct sensations or impressions and ideas somehow brought together by association, it may appear that a yawning chasm separates man and his world. If, however, one adopts a newer psychology which draws upon developments in physiological and experimental psychology and describes experience in terms of ongoing courses of purposive behavior, it becomes apparent that man is a part of nature and that the biological matrix of experience is central. This newer psychology is so focal for pragmatism or contextualism that George Herbert Mead regarded it as a primary source of this outlook. According to him, the sources of pragmatism are two : (1) a behavioristic psychology which enables one to put intelligence within conduct and to state it in terms of the activity of the organism, and (2) the research process, which comes back to the testing of hypothesis by its working.[15] And whether these be called sources or perhaps rather essential constituents of pragmatism, they are indeed basic for the view. Both, moreover, have occasioned a storm of criticism, the former sometimes on the ground that the pragmatists have confused psychology and theory of knowledge. But clearly any theory of nature or knowledge which accepts the psychological theory of its rivals places itself at a decided disadvantage.[16] Mead, moreover, himself both a social scientist and a philosopher, saw the relations between the pragmatic philosophy and its psychological base more clearly than many who have not had his combination of interests; and he early took the lead in developing a behaviorism of the sort mentioned above.

It should be noted, however, that this behavioristic psychology must be distinguished from Watsonian behaviorism, which is fundamentally individual, mechanistic, and antithetical to introspection. It is rather a purposive behaviorism which starts with behavior on a molar rather than a molecular level and finds a place equally for the usual observational material and for that which is open mainly to the observation of the acting individual. Behavior, on Mead's account, is basically social, and language as an instrument of socialization is a form of interaction within a social group rather than something to be explained simply in terms of movements within the larynx of the individual. Mind or consciousness, according to Mead, instead of being something to be ruled out of account, is something to be explained in relation to conduct; and the distinction between inner and outer, instead of being one between disparate metaphysical realms, is one made within experience.

E. C. Tolman's *Purposive Behavior in Animals and Men* offers an excellent systematic formulation of a purposive behaviourism of this general type, and the wealth of experimental material summarized in it both fits in with and uses the contextualistic categories. It needs supplementation in terms of social and cultural considerations, but the biological matrix of experience comes out clearly in this account. Tolman assumes that molar behavior-acts are correlated in significant ways with facts in physiology and physics, as might be expected from our earlier account of contextualistic analysis; but he holds that each such act has its own identifying properties. To identify it, in the first place, we describe "(a) the goal-object or objects being got to or from; (b) the specific pattern of commerces with means-objects involved in this getting to or from; and (c) the facts exhibited relative to the selective identification of routes and means-objects as involving short (easy) commerces with means-objects for thus getting to or from."[17] In the second place, identifying the act also involves specifying two additional sets of properties : (1) initiating causes, namely, (a) environmental stimuli and (b) initiating physiological states; and (2) behavior determinants, namely, (a) purposive and cognitive immanent determinants (which should include behavior adjustments, Tolman's substitute for the conscious awareness and ideas of the mentalist), and (b) purposive and cognitive capacities.

As I have indicated elsewhere,[18] the outstanding feature of such molar behavior, whether in the lower animals or in men, is its purposive character, the fact that it is motivated by some drive.

Certain states of affairs or ends are sought and achieved. Other results are to be avoided. Such behavior is carried out for the sake of satisfying various demands, to get to or from a given instance or type of environmental presence or of physiological quiescence or disturbance, to overcome problems in connection with food, water, or any of the multifarious wants and needs of a living organism. It is an attempt of the organism to maintain the organic or environmental conditions necessary, if need be, to change them to bring about the fulfillment of these demands. Upon its degree of success in fulfilling certain of them depends the very preservation of the organism or the species. Fundamental physiological drives form the core of purposive behavior, but the demands of an adult human being are far richer and more varied than this central core might suggest.

As might be expected from the above, the naturalistic approach of the contextualists places man in nature and assumes a basic continuity between him and his environment, natural, social, and cultural. As a living organism he has commerce with his environment in multitudinous ways. He enters into multiple interactions or transactions with it, some of them so integral to the events related that the distinction between organism and environment becomes a functional one which may be drawn in different ways, depending on the interaction. His environment supports some interactions and fails to support others. It enters both directly and indirectly into his lifefunctions. Thus organism and environment enter into more or less stable integrations, form uneasy equilibriums requiring frequent adjustments or modifications, with the survival of the organism depending on making some of them. Darwin and James have helped us see the focal importance of the living creature adjusting or adapting to its environment and to view intelligence as a distinctive form of behavior, one concerned with choosing appropriate means for the attainment of future ends. Whatever else mind may be, it is at least a means of controlling the environment in relation to the ends of the life process.

In terms of their acceptance of the notion of a basic continuity between organism and environment, contextualists reject such dualisms as those between the self and the world, mind and matter, and subject and object. If interaction or transactionalism be accepted, these dualisms are untenable. The relation between a self and its world, according to the contextualists, is not one of confrontation between hard and fast, sharply defined, relatively unmodifiable objects that change in little but position but rather, as we have seen, a far more intimate form of intercourse. The conception of a private inner mind set over against an external material world does not accord with the contextualists' idea of a texture and its context, with strands of reference running from one to the other in such open and aboveboard fashion that in principle there are no inaccessible textures or realms of textures. Nor is the notion of a self-contained subject which somehow knows an equally self-contained but alien object any more acceptable, contextualistically speaking. We start not with subject and object but rather with a total affair or situation within which, for purposes of practical or intellectual control, we may distinguish, say, a subject as experiencer from an object as experienced or a self from its not-self. Contextualists have similar reservations, of course, about attempting to turn a functional distinction between the organic and the inorganic into a hard and fast distinction between two separate, mutually exclusive realms of life or living things and matter or material things. And they find it not surprising that, depending on how one draws the line, viruses, say, might be in either domain.

Accordingly, when one speaks of levels of existence, from the point of view of the contextualists, this is not to assert that matter, life, and mind, say, are separate and distinct kinds of Being. Rather they might better be thought of as different modes of interconnection and operation, different ways of characterizing diverse particular fields of interacting events. In Dewey's language, they are consequences of interactions of varying complexity, scope, and degrees of intimacy.[19] And depending on the problem at hand, we might distinguish dozens of "levels" or modes of interconnection and operation, each with its own distinctive marks. In *Experience and Nature,* however, Dewey characterized the three levels mentioned above roughly as follows: (1) the physico-chemical level of mass-energy interactions on which the physical sciences seek to discover the properties and relations of things in terms of which they may serve as means or instruments; (2) the psycho-physical or organic pattern of need-demand-satisfaction activities; and (3) the level of mind or human experience in which social transactions involving language and meaning come in. These types of interactions exhibit different degrees of complexity, diverse functions, varied sets of consequences, and variegated qualitative patterns.

What we call material things, living organisms, or mind (persons, or personalities) involve in each case abstractions from concrete transactions or situations. For example, depending on the problem, we may think of material things as behavior-objects, or we may think of them as physical stimuli, as abstractions the physical scientists make from the immediacies of specific concrete behavioral situations to get generalized supports for any and all possible operations and thus provide instruments for as wide a range of interactions as possible. Physical stimuli may be described in terms of, say, air vibrations of a given frequency or in terms of mass. If we have a formula calling for a given mass, it does not matter whether that is provided by a thimble, a coin, a plant, a section of human tissue, sheets from a musical score, or pages from a manuscript on philosophy. All are equally masses. Physical schemes are constructed to provide a wide range of applications; if they do not afford such a range, we revise them to get more adequate coverage; and the possibility of locating a given strand in one of them accordingly is useful for a broad range of problems. The key thing about such a physical object or a statement involving it is whether it enables us to predict accurately and thus solve the problem for the sake of which we traced the strand into the schematic textures of physics.

Thus statements made at the physico-chemical level may be very useful for a wide range of problems, but for some problems something much more illuminating can be provided by a botanist concerning a plant or by a composer about a musical score; and the contextualists would not think of saying that something at the organic level or at the level of mind was nothing but a combination of physical or material elements or that things at these two levels should be described exclusively in terms of the laws and generalizations of the physical sciences. Nor would they wish to maintain that an organism is nothing but a material thing plus this or that additional item any more than they would wish to hold that a man is only a child plus a few additional items : such statements do not begin to do justice to the different types of interactions involved, with their diverse functions and consequences.

From the point of view of the contextualists, with their acceptance of change and fallibilism, there are new things under the sun. New qualities do emerge; and recognition of this fact, far from being simply a way of forgoing explanation, may provide added understanding. It seems to be an irreducible fact that life and mind have evolved from states of affairs in which they were absent; and *after* the fact we can show how potentialities in the former, specifiable directions of change present there, when taken in conjunction with appropriate occasions and supporting conditions, pointed toward the state which has evolved. But we make our predictions in the face of change with the knowledge that they are fallible and with the expectation that through observation and experiment we may make them better.

Another distinctive doctrine of the contextualists which clearly reflects the Darwinian biological matrix of experience at the same time that it exemplifies genetic analysis is their analysis of reflective thinking, scientific method, or critical inquiry. Although critical thinking is by no means limited to practical problems of adjustment of an organism to its environment or ones affecting either its survival or that of its species, such problems come in for special emphasis, and the characteristic pragmatic descriptions of such thinking manifestly grow out of an attempt to take seriously the contributions of biology. These accounts, moreover, express succinctly a number of ideas central for both the naturalistic and biological emphases of contextualism and provide, indeed, a kind of outline of the entire philosophy.

What, then, is critical inquiry or reflective thinking? The process has been described in somewhat different terms in various versions, but in general contextualists have thought of it as a problem-solving or doubt-resolving affair; and they are convinced that the pattern exemplified in it is applicable, with appropriate modifications, to the full range of problems facing us in this changing world. From Peirce on, however, they have maintained that these problems are specific and concrete, or can be made so; and so long as we can formulate our problems in these terms we can deal with them, work at overcoming them. If, however, we try to make a problem of everything at once or the universe at large, we have no way of solving it; and escape or despair is the response. Uneasiness about the world at large may produce a profound malaise, but it is not likely to be effectively reconstructive. Descartes' universal doubt, according to Peirce and his fellow pragmatists, is a fake: we do not doubt in general or wholesale fashion. It is only when we have some question about a specific item that doubt occurs, and this is possible because, for the moment, we accept without question a context of beliefs. This is not to say, of course, that these other beliefs are indubitable, for another time they may be questioned but once more within a context of accepted beliefs or habits of action.

Granted, then, that reflective thinking is a problem-solving or doubt-resolving process, what are the main steps or phases involved in it? The contextualistic reply finds its fullest expression in the writings of John Dewey, who has set forth his answer to this question in many different books and articles, perhaps in its simplest terms in *How We Think* (1910, 1933) and in a fuller, more sophisticated version in *Logic: The Theory of Inquiry*. For present purposes we may say that there are five main steps in a complete act of reflective thinking.

(1)   The first step is the appearance of the problem. This may be marked by a more or less vague sense of something having gone wrong, a breakdown in habitual responses or modes of action. Our usual patterns of response do not seem to be working or have run into difficulty. The situation may be one of practical friction or strain, conflicting tendencies, perplexity, ambiguity, or doubt. One of us comes thirsty to the familiar water hole or spring in the wilderness and finds no water or perhaps hears an insistent rattling as he bends to drink. (2) The second step or phase is the clarification of the problem. Through analysis and observation we gather the data needed to formulate the difficulty or define the problem. The more or less vague feeling of something amiss is replaced by a clearly formulated problem. We establish that the spring is dry or that a rattlesnake is rattling and proceed to state our problem. (3) Having it clearly stated, we are ready for the third stage, that of suggested solutions or hypotheses as to how to solve the problem. Various ideas as to how it may be solved occur to us, and these ideas suggest further observation and help point to other pertinent facts. The more adequately we have stated the problem, of course, the more expeditious our search for possible solutions. (4) The fourth step is that of deductive elaboration. We reason out the implications of the various hypotheses, noting what we may expect if we take the first, the second, the third, and so on. In some instances further information may be needed before we can deduce the implications of a given hypothesis. Then on the basis of our survey of the probable consequences of the various proposed solutions we decide which to put to the test of action. (5) The fifth and final step is that of verification or disconfirmation. Through observation or experiment we test the most promising hypothesis or hypotheses. If one of them works out, our problem is solved, the difficulty is

cleared up, the doubt or perplexity is resolved; and once more stable lines of action are resumed.

This pattern is somewhat oversimplified, and the stages do not always come one right after another in the order in which I have listed them. For example, the probable consequences of the proposed solutions may look so unpromising that we may go back to the second and third steps, rethink our problem, and see if we cannot think of other possible solutions. Nor are the stages always as distinct as may appear from this analysis; some telescoping or merging of stages may occur. But ordinarily for the solution of a difficult problem the stages need to be outlined clearly.

In spite of its simplicity, however, the above account seems to me basically accurate. This is what is involved in problem-solving activity whether it is a personal problem, a major social issue, or a significant scientific problem; and it is difficult to overestimate the survival value of this method for the human species. Indeed, if its past record of achievement should lead to its more extensive use with reference to such problems as those of war, poverty, population, and pollution, my optimism concerning our future survival would be greatly strengthened. In any event, the method of critical inquiry is a tremendous advance over the trial and error method of resolving difficulties, having the advantage of both solving a problem and providing some descriptive or explanatory statements about how it was solved.

Going hand in hand with this account of critical inquiry is a view of experience which incorporates findings and emphases of biological science and focuses upon the fact that experiencing means living in and by means of an environment, natural, social, and cultural. Experience is the entire range of our relations, interactions, and commerces with nature. It includes our enjoyments and appreciations, our consummations and sufferings, as well as our reconstructive or adjustive activities. We experience nature in manifold manners and modes, and things interacting in certain ways constitute experience.

The contextualistic interpretation of experience is very well summarized through a series of contrasts with what he called the orthodox view in Dewey's famous essay on "The Need for a Recovery of Philosophy."[20] As contrasted with the orthodox notion that experience is primarily a knowledge affair, Dewey insisted that it is a matter of intercourse between us as living beings and our physical and social environment, a process of doings and undergoings in which we employ direct environmental supports in order indirectly to help effect changes favorable to our further functioning. Far from being an invidiously subjective inner affair, separate and distinct from objective reality, it is of a piece with an objective world in which and by means of which we live, move, and have our being. Whereas the proponents of the orthodox conception have been preoccupied with what is or has been given in a bare present of some sort and have tended to look toward the past if they have passed beyond the present, the thrust of this new view of experience is primarily toward the future and a concern with what may be done to change what is given or taken to further human purposes. The new outlook stresses the contextual, situational, transactional, or field character of experience as opposed to a particularism which neglects connections, relations, and continuities and concentrates on more or less disconnected sense data, impressions, and ideas. Finally, instead of holding that experience and thought in the sense of inference are antithetical, the contextualists see experience as full of inference. For them it is loaded with references which trace directions and relations and pregnant with imaginative forecasts of future trends and movements. How could it be otherwise if we seek to eliminate environmental incidents hostile to our future life activity and to insure conditions favorable to the process? What else should we expect if we wish not merely to live but to live well, to enrich and enhance the qualities of experience and to share their enjoyments?

In keeping with this view of experience, ideas for the contextualists are not more or less miraculous psychical carbon copies of real but never given external objects. They are rather, in their distinctive role, hypotheses, plans of action, suggested solutions for problematic situations. They function as tools or instruments, anticipatory sets for guiding desired changes. With this conception goes an operational theory of truth. Truth is a matter of how ideas work out in practice, whether they lead as expected or predicted, whether they provide qualitative confirmation.

Knowledge too needs to be placed within the context of the problematic situation and reflective inquiry. But here there are some terminological differences among the contextualists. Bergson's *Introduction to Metaphysics* distinguishes between intuition or the realization of quality and analytic, relational knowledge, holding that only the former affords genuine knowledge. In *A Contextualistic*

Theory of Perception I have maintained that perception has the two limiting poles of aiming at the realization of quality or at identifying qualities and things for the furtherance of a practical drive and that it is not a question of one being more genuine than the other. Pepper declares that the intuition of quality is as informative of the nature of things as analysis and should, of course, be considered equally a form of cognition.[21] Dewey grants, in an essay on "Qualitative Thought," that Bergson is correct in contending that intuition as the realization of quality precedes conception and goes deeper but wishes to reserve the term *knowledge* for the latter, for relational knowledge;[22] and in general pragmatists have followed this usage. So long as it is clear that both of these things have important places in the contextualistic system, for present purposes I am willing to go along with this practice.

Dewey has stressed the pervasive quality which characterizes a total problematic situation and guides the course of inquiry, and it seems reasonable to hold, as Iredell Jenkins has persuasively argued, that vivid realization of this unique quality is significant not merely for aesthetics but for human survival.[23] The more novel the situation and the more rapidly changing the course of events, the more useful it is likely to be for adaptive or adjustive purposes to intuit this quality.

When all this is noted, however, it is nonetheless the case that the knowledge needed for implementing a practical drive is more than immediate awareness or the presence of sense data. Hence, as Dewey insisted, the senses are not primarily pathways to knowledge but stimuli to action; sensations are not so many pellets of knowledge but signals to redirect action, or signs of problems to be solved. Having quality before us may be a condition for knowing or the beginning of relational knowledge, but this type of knowledge is an affair of the use we make of sensory discriminations. It is inferential, and the problem is how the processes of inquiry are to be guided to trustworthy or waranted conclusions. It involves operations of controlled observation, testing, and experimentation. It is a product of inquiry —the steps in a complete act of reflective thinking.

In like fashion the contextualistic account of moral values may be treated in the context of the problematic situation and critical inquiry, but we have space for only a few brief comments. Here too, moreover, the account reflects Darwinian evolutionism in its emphasis on improving or bettering our present situation rather than upon good or bad in some absolute sense. A moral end or standard is a hypothesis as to how to overcome a moral problem. The good, if one is to speak of the good rather than the better, is what will enable us to solve the problem or difficulty, and since each problematic situation is unique, values are also unique. Problems are connected in various ways, however, and a career may be charted in terms of their sequence. Accordingly, we try to solve our current problems in ways which will not block the path to the solution of foreseeable future ones. This, I think, may be part of what Dewey had in mind in holding that if a general end is to be specified, it should be one on the order of growth, education, or continued problem-solving. In this connection we may note that our personal problems are related in various ways to the problems of others and that we have enough common problems that we may need to think in terms of our society or culture and the human situation. If shared experience is truly the greatest of human goods, the basis for this may be found in our common problems.

A naturalistic account of morals may be framed not merely in terms of adjustment, adaptation, and problem-solving, however, but also in terms of our likings, enjoyments, and desires. But contextualistically speaking, although consummations and fulfillments are categorial, there are ways and ways of achieving them; and what is good is not raw likings or satisfactions. Rather it is ones which still look good after critical scrutiny. We need to test them by checking on their conditions and consequences; and this involves running through the steps in a complete act of reflective thinking. It is not simply a question of whether as a matter of fact we do or do not enjoy or like something but whether this enjoyment or liking meets the tests of reflection. Is it worthy of being desired, demanded, or enjoyed? An affirmative answer constitutes a claim that future consequences will be such as to meet the test of reflection.

I find the context of the problematic situation and critical inquiry a highly fruitful one, and still other topics might be discussed in terms of it. For example, the contextualist also has a place for human freedom, and part of what helps bring it about is the delay between stimulus and response which makes it possible for reflective consideration of alternatives to become a factor in the decision process.

By way of summary, then, in this section I have tried to suggest some of the respects in which the contextualistic philosophy, through (1) its acceptance of change and its ways of dealing with the changing and (2) its naturalistic approach framed in terms of the biological matrix of experience, expresses the Darwinian outlook. In this connection, as we have seen, a great deal of contextualistic interpretation of evolution also has been given, thus anticipating the concluding interpretive comments of the final section of this essay.

4. CONCLUDING CONTEXTUALISTIC NOTES ON EVOLUTION

In tracing the implications of evolutionary theory for various aspects of philosophy I have already sketched the main outlines of a contextualistic interpretation of evolution, but in this section, for emphasis and further clarification, I should like to present four or five points in this interpretation in somewhat different terms and supplement briefly my earlier account of some of them. Perhaps the dominant theme running through all of them is the fact that from the point of view of the contextualists change is a central feature of the cosmos, and evolution and revolution are important patterns of change.

In a world of change novelties and contingencies are to be expected, and in terms of the contextualistic categories a number of kinds of novelty may be noted. First is novelty as uniqueness. William James insisted that every event is unique. Even if it could recur, it would be different at least in terms of its antecedents. Pepper suggests four types of novelty: intrusive, emergent-qualitative, emergent-textural, and absolute or naive novelties;[24] and there is no question about the first three. Intrusive novelty occurs when a referential strand is blocked or intercepted by a conflicting line of action. This unexpected opposition draws attention and generates enhanced perception of the situation. Emergent qualitative novelty is also to be expected. A new quality comes into being with every event as a look from this page to something over your shoulder suggests. With the constant involvement through its strands of a texture in its context and with the complex changing from moment to moment, the total texture at moment two is different from that at moment one, and accordingly, each has a different quality. One character is gone and a new one has come to be. Emergent textural novelty results from the integration or fusion of its substrands. The integration is something new. In all these cases of novelty, however, it is possible after the fact to trace strands leading up to the novel quality or texture, and in general contextualists tend to assume that continuities or connections may be traced between any two states of affairs. Nor would they hold that this casts any doubt on the genuineness or importance of the novelties.

Pepper's fourth category of novelty, however, raises a question as to whether there are not novelties of a more radical or absolute type. Are there textures or strands such that no previous event ever referred to them, one whose initiation is not an integration or fusion of other strands? And in like fashion are there absolute endings in the sense of strands with no connection with an actual present or future event, ones which disappear without a trace? There is nothing in the contextualistic categories, according to Pepper, to exclude such textures; but even if this be true and maximum allowance be made for it, as he adds, we still need more to make an affirmative case than absence of evidence to the contrary. It seems to me, moreover, that the category of context and the contextualists' operational procedure point in another direction. Loss of context in the fashion suggested by an absolute novelty or ending does not seem to me to have contextualistic ring. Insulated or isolated entities and absolutes are questionable on contextualistic grounds, and the fact (mentioned by Pepper) that many references believed to be untraceable have later been traced would seem to provide some grounds for scepticism concerning this type of novelty.

Contextualists find both continuities and discontinuities in nature, and such being the case the interpretation given either set of facts should not be such as to rule out the other. As I have tried to show earlier in this essay, contextualistic analysis is admirably fitted to do justice to both features, which contextualists find both at the level of individual textures and at that of patterns of vast complexes of events. Habits, dispositions, and regularities, say, point toward continuities, and diversities, novelties, and irregularities suggest discontinuities. A mutation marks a discontinuity, but even here there are traceable connections. A major discontinuity in some larger pattern of events we may call a revolution, but the course of evolutionary development includes revolutions as well as more gradual changes. It has a place for cataclysmic as well as for graded change. Indeed, as the history of science suggests, discontinuity and revolution are rather more common than we may have suspected.[25] Emergents signalize the combination of discontinuity and continuity of which we have been speaking. They may be novel and at least partially unpredictable relative to our best present knowledge, but after the fact it is possible to trace their coming to be, and in the light of our experience with them we work at developing better conceptual instruments in order to improve on our future predictions.

Contextualists tend to be dubious of any grand overall evolutionary scheme. If we take seriously the notion that the stars as well as the firm earth are in process, we are not likely to think that everything is all finished and done or that some pattern of cosmic development which will account for all new species is already set and that all we have to do is make note of eternal perfection or seize upon some simple formula. The cosmos in general and our world in particular are vastly more complex and diversified than some of our formulas may suggest; and we have a staggering if not almost overwhelming amount to learn about them, but through the method of reflective inquiry we are making a beginning on learning more about the world and our place in it.

Contextualists find no inevitable tendency toward progress in evolution, nor, unlike Hegel, Marx, and Fiske, do they discern any set of necessary stages through which the life of spirit, social or economic systems, or the biological order must move in the developmental process. Successful adjustment to one state of affairs may render a species incapable of responding successfully to a markedly different set of conditions, as the number of extinct species may suggest. There clearly is change, and sometimes progress, but not a necessary or inevitable progress.

In keeping with our naturalistic approach, we contextualists reject any supernatural theistic design, nor do we find acceptable an Aristotelian final cause for the universe. Purposive behavior we do observe in both animals and men, and we have no need or wish to reduce it to anything else. But we have discovered no cosmic goals or plans. The only designs or plans we know of are those of finite planners, and we are convinced that through the proper use of intelligence our human plans can be greatly improved with a consequent strengthening of our capacity to come to terms with the problems confronting us. I share the hope of John Dewey that philosophy can contribute to the betterment of our human situation through helping inform our action at all levels with vision, imagination, and reflection to bring clearly to mind future possibilities with reference to attaining the better and averting the worse.

Notes and References

1. Cf. John Dewey, "The Development of American Pragmatism," *Philosophy and Civilization,* New York, Minton, Balch and Company (1931), pp. 13–14.

2. *Collected Papers of Charles Sanders Peirce,* ed. Charles Hartshorne and Paul Weiss, Cambridge, Massachusetts, Harvard University Press (1934), **5**.414.

3. "Experience and Nature," Chicago, Open Court Publishing Company (1925), p. 71.

4. *Ibid.,* p. 28.

5. Lewis E. Hahn, *A Contextualistic Theory of Perception,* "University of California Publications in Philosophy," George P. Adams et al., eds., Vol. **22**; Berkeley and Los Angeles, University of California Press (1942), pp. 8–9.

6. For a fuller account see *ibid.,* pp. 6–19.

7. See, for example, Stephen C. Pepper's "The Conceptual Framework of Tolman's Purposive Behaviorism," *Psychological Review,* Vol. XLI (1934), 108–133, especially, 111, or his chapter on "Contextualism" in *World Hypotheses,* Berkeley and Los Angeles, University of California Press (1942).

8. Philip P. Wiener, *Evolution and the Founders of Pragmatism,* Harper Torchbooks, New York, Harper and Row (1965), originally published by Harvard University Press (1949), p. 6.

9. "The Influence of Darwinism on Philosophy," *The Influence of Darwin on Philosophy and Other Essays in Contemporary Thought,* New York, Henry Holt and Company (1910), this essay was originally published as "Darwin's Influence on Philosophy" in *Popular Science Monthly,* Vol. LXXV (1909), 90–98, pp. 1–19.

10. Cf. George Kimball Plochmann, "Darwin or Spencer?," *Science,* Vol. **130** (November 27, 1959), 1452–1456.

11. See, e.g., *A Pluralistic Universe,* London, Bombay and Calcutta, Longmans Green and Co. (1909), p. 79.

12. Cf. John Dewey, "Context and Thought," *University of California Publications in Philosophy,* Berkeley, University of California Press (1931), Vol. XII, 203–224: In this essay he maintains that "the most pervasive fallacy of philosophic thinking goes back to neglect of context," p. 206.

13. James, Dewey, and various other pragmatists have written at length on analysis, but for excellent succinct accounts of analysis for the contextualist see Stephen C. Pepper's "The Conceptual Framework of Tolman's Purposive Behaviorism," pp. 112–113, and *World Hypotheses,* pp. 248–252.

14. John Dewey, "The Influence of Darwinism on Philosophy," p. 13.

15. George Herbert Mead, *Movements of Thought in the Nineteenth Century,* ed. Merritt H. Moore; Chicago, University of Chicago Press (1936), pp. 344–359, esp. p. 351.

16. Cf. my "Psychological Data and Philosophical Theory of Perception," *Journal of Philosophy,* Vol. XXXIX (May 21, 1942), pp. 296–301.

17. Edward C. Tolman, *Purposive Behavior in Animals and Men,* New York and London, The Century Company (1932), p. 12.

18. *A Contextualistic Theory of Perception,* pp. 28–29.

19. See *Experience and Nature*, pp. 261–262.

20. John Dewey et al., *Creative Intelligence, Essays in the Pragmatic Attitude,* New York, Henry Holt and Company (1917), pp. 3–69.

21. Stephen C. Pepper, *Aesthetic Quality: A Contextualistic Theory of Beauty,* New York, Charles Scribner's Sons (1937), p. 29.

22. John Dewey, *Philosophy and Civilization,* p. 101.

23. *Art and the Human Enterprise,* Cambridge, Massachusetts, Harvard University Press (1958).

24. Stephen C. Pepper, *World Hypotheses,* pp. 255–260.

25. Cf. Thomas S. Kuhn, *The Structure of Scientific Revolutions,* "International Encyclopedia of Unified Science," Otto Neurath et al., eds. Vol. II, No. 2; Chicago, The University of Chicago Press (1962).

THE CONCEPT OF EVOLUTION AND REVOLUTION

R. PULIGANDLA

The University of Toledo

THE PURPOSE OF this paper is to examine the concept "evolution" and "revolution" in their various meanings and significance for man's understanding of himself and the world—historically, scientifically, and philosophically. The present inquiry falls into two parts, the first dealing with an analysis of these concepts and the second with their application in the fields of sciences and humanities. In particular, I shall illustrate the working of these concepts in the growth of knowledge in general and scientific knowledge in particular and in the understanding of social and political phenomena. It is hoped that the analyses and examples to be offered here will also enable one to understand the development of art, religion, and morality.

I

What is evolution? What does the term "evolution" mean? The first thing to notice is that the term has several meanings, each depending upon the particular context of discourse, although it is possible that there is some generic, overarching meaning of which the particular meanings are contextual variations. Etymologically, the word "evolution" comes from the Latin "evolutio," meaning "an unrolling." *Webster's New Collegiate Dictionary* lists the following meanings of "evolution" : (1) an unfolding; a process of opening out what is contained or implied in something; a development, especially, as leading to a definite end; as the evolution of the tragedy; (2) a movement forming one of a series of motions, as of a machine; hence, an intricate form, as if produced by such a series; as the evolution of an arabesque pattern; (3) a process of disengaging, so as to expose or free, as of gas from limestone; (4) from the biological point of view, the development of a race, species, or other group; phylogeny; broadly, the process by which, through a series of changes, any living organism or group of organisms has acquired the morphological and physiological characters which distinguish it; hence, the theory that the various types of animals and plants have their origin in other pre-existing types, the distinguishable differences being due to modifications in successive generations; (5) in mathematics, the extraction of roots—the inverse of involution; (6) in navy and military, any movement of troops or vessels designed to effect a new arrangement by passing from one arrangement to another; and (7) from a philosophical point of view, any process which exhibits a direction of change; especially, the process of the whole universe, conceived as a progression of interrelated phenomena.

A moment's reflection on these various meanings shows that, whatever the differences among them, there is a generic concept of "evolution" providing a common base for all of them. We may characterize this common base in the following propositions: 1. Initially there was something. 2. This something is not a static but a dynamic entity (or system); that is, it is capable of change. 3. It contains implicitly within itself the possibilities that will be actualized as it undergoes change. 4. Evolution is the process of actualization of these possibilities. 5. The changes the entity undergoes may be qualitative, quantitative, or both. 6. One discovers that evolution is taking place by setting up criteria for recognizing change; a property or relation (or a set of properties or a set of relations) possessed by the entity in a given state may be taken as the criteria for saying that the system is in a certain state and when these properties and relations change we say that the system has gone into a different state. 7. One may take any state of the system as the initial state (state at an arbitrarily specified time $t = 0$) and the state of the system at a later time $(t > 0)$ as the state into which the system has evolved. 8. Not every change undergone by an entity counts as evolution, for change can be in either direction, toward or away from a given state; that is, one says that a system is evolving only when changes in the system are such that the system will move toward a certain state which one has in mind. 9. For this reason, it is important to specify the character of the system at a later time in order that one may judge whether the system is undergoing evolution. Put differently, there could be no talk of evolution unless the direction of change from one state to another is somehow (on some criteria or other) regarded as unique and preferred; that is, although all cases of evolution are necessarily cases involving changes, the converse does not in general hold—not every instance of change is an instance of evolution, for the changes may be such that the process may be one of dissolution (or counter-evolution). 10. Therefore, when no preference for direction of change and the definite state to which the changes are to lead are given, one can correctly say not only that a system evolves from state A to state B but also that it evolves from state B to state A; that is, in the absence of any criteria for the uniqueness of direction of change and of the state to which the change is to lead, "evolution" and "change" are synonymous; by simply observing that a system is changing from one state to another, one cannot conclude whether it is undergoing evolution or counter-evolution. 11. If a system is changing from an initial state A to a later state B, with B having certain characteristics which neither A nor any earlier state has and if the state B is somehow regarded as preferred, we say that A has evolved from A to B.

What, then, are the characteristics in general which must be possessed by a system in a later state in order that we may say that there is an evolution toward that state? This seemingly simple and innocent question is in fact quite difficult, for an answer to it centers around the concept of order. All evolutionary discourse pre-supposes this concept. We shall see a little later what the concept of order means. But for now let us assume that we have an intuitive understanding, no matter how vague and shaky, of "order" and "degrees of order." We can then say that if an initial state A of a system is lacking in order and a later state B has order, the system has evolved from state A to state B.

To be sure, order is discerned according to criteria; and these may vary from person to person and group to group. Nevertheless, there are some generally recognized criteria of order on the basis of which one can judge whether a given state of a system is orderly or disorderly (or more or less so than some other state) and hence whether a system is undergoing evolution or dissolution. The important point, then, is that the concept of order is crucial to any discussion of evolution. I turn now to an examination of the term "revolution."

"Revolution" comes from the Latin root "revolvere" and, like the term "evolution," has several meanings. Thus according to *Webster's*

New Collegiate Dictionary, it has the following meanings: (1) a progressive motion of a body round a center or axis, such that any line of the body remains throughout parallel to its initial position, to which it returns on completing the circuit; motion of any figure about a center or axis, also called rotation; (2) motion of a celestial body going round in an orbit, or the time taken in going round in an orbit; also, apparent movement round the earth; the rotation of a celestial body on its axis; (3) a completion of a course, as of years, or of any recurring series of events; (4) a total or radical change; as, a revolution in thought, specifically, the Industrial Revolution, the change following and resulting from the introduction of power-driven machinery to replace hand labor, occurring in England after 1760; from a political point of view, a fundamental change in political organization, or in a government or constitution; the overthrow or renunciation of one government or ruler, and the substitution of another, by the governed; as the American Revolution or Revolutionary War (1775-83), the French Revolution (1789-99); the English Revolution (1688); the Russian Revolution (1917).

It is easy to see that each of these meanings is a variation of one or another of two basic and fundamentally different meanings: 1. A process which periodically repeats itself; that is, the pattern of events constituting the process recurs at specified intervals of time. The process is uniform and in recurrence everything takes place as it did earlier or will later. In particular, there are no abrupt, unexpected, and radical changes in the state of the system undergoing the process. 2. A process which takes place in such a manner that, whether or not there are any recurrences, from time to time there will be radical changes in the intellectual, social, political, religious, artistic, or other realms. Thus in a very real sense the two meanings are the antipodes of each other, for according to the one, things and events repeat themselves endlessly and according to the other there will be fundamental and radical changes from what is now. We can express these ideas in the following propositions. 1. Initially there is something. 2. This something is not a static but a dynamic entity (or system); that is, it is capable of change. 3. It undergoes changes in such a manner that either (a) over specified intervals of time the sequence of changes repeats itself and the process takes place in a smooth uniform manner; or (b) from time to time there will be fundamental and radical changes in the properties and relations of the system, no matter in what direction the changes take place.

We have said earlier that the concept of order is central to any discussion of evolution. Let us now try to unpack what this concept means. "Order" is a term used to cover a complex of concepts related to each other in a certain manner. Thus when one analyses "order," one arrives at a set of concepts and their interrelations. Herbert Spencer (1820-1903) is an outstanding student of evolution. He aimed at constructing a philosophical system centered around the concept of evolution. Regarding philosophy as completely unified knowledge, as contrasted with sciences which are only partially unified systems of knowledge, Spencer tried to show that complete unification of knowledge is to be accomplished through the concept of evolution. He considered philosophy itself as a theory of evolution taken in the broadest sense to cover the origin, development, and unification of all human knowledge. Living at a time when Darwin's epoch-making *Origin of Species* appeared (1859), Spencer felt encouraged in his philosophical endeavors and regarded Darwin's work as confirming evidence for his general theory of evolution, which he was thinking through and formulating even prior to the publication of Darwin's work. In a very real sense, Spencer's philosophy is a systematic work in that it tries to bring together knowledge in various fields such as physics, biology, psychology, sociology, and ethics in order to present a unified picture of human knowledge, the unifying concept being "evolution." Although the concept of evolution in biology has undergone considerable change (evolution!) since Darwin and Spencer, Spencer's treatment of the general concept of evolution, in my judgment, has still much to commend it for its usefulness in dealing with the concept of order. I shall therefore deal with Spencer's characterization of evolution in some detail.

> Evolution is an integration of matter and concomitant dissipation of motion; during which matter passes from a relatively indefinite, incoherent homogeneity to a relatively definite, coherent heterogeneity; and during which the retained motion undergoes a parallel transformation.

This definition reflects the characteristic thought patterns of nineteenth-century philosophers under the dominance of the physics of their time, according to which one talked of matter, mechanism, and the principles of conservation of matter on the one hand and that of energy on the other. We of the twentieth century, while retaining the term "motion," do not talk of matter in the manner of the past century; nor do we talk, thanks to Einstein, of two separate principles of conservation. On the contrary, we have the single principle of conservation of matter-energy, the term "matter" not having any of the nineteenth-century connotations. Does this mean, then, that Spencer's formulation of the concept of evolution deserves to be simply dismissed as anachronistic, irrelevant and useless? My answer is a definite "no." With certain modifications, the above characterization of evolution is still significant and helpful. The modification I suggest is that we replace "matter" by "system," whatever the area of discourse. We can then meaningfully talk about "physical system," "biological system," "social system," "political system" and their evolutions. I now state, in light of the suggested modification, Spencer's principle of evolution: Evolution is an integration of a system which takes place in such a manner that the system passes from a relatively indefinite, incoherent homogeneity to a relatively definite, coherent heterogeneity. Before proceeding further, let me explain the terms, "indefinite," "homogeneity" and "incoherent" and by contrast "definite", "heterogeneity", and "coherent." We say that something is indefinite when there are no clear and sharp boundaries by virtue of which we can demarcate it from other things. Even if there are such boundaries, we can say that the object is homogeneous in the sense that the various parts of the object are not discernible, because its properties throughout are the same. Moreover, an indefinite and homogeneous object is incoherent in the sense that, since the object does not present itself to us as made up of distinct parts, we cannot speak of how its parts are held together to form a whole. Put differently, these three characteristics describe a system which cannot be distinguished from its surroundings, whose parts cannot be distinguished from each other, and the relations among whose parts cannot be described because we do not even have the parts yet presented to us. By contrast, a relatively definite, heterogeneous, coherent system is one which can be distinguished from its environment, whose parts are perceivable as different components or subsystems of the system, and which can be thought of as a whole sustained by virtue of the relations in which its different parts stand to each other. We may emphasize here that all these three terms are to be taken in a relative sense and not in any absolute manner.

In order to be able to distinguish evolution from dissolution, Spencer elaborates upon the above characteristics by introducing three further criteria, namely, concentration (integration), differentiation, and determination. Concentration is that process by which a system which is initially diffuse, scattered, and homogeneous develops in time toward a state in which its parts are gathered together and integrated into a system which is more or less clearly separated from its environment. Thus, according to some cosmologists, galaxies, stars, and planetary systems are formed in time by progressive separation from some initial primeval homogeneous state. Similarly, in biology, the various life-forms are thought of as having originated and developed out of some primordial undifferentiated life stuff; that is, organic evolution takes place by the absorption and assimilation into the body tissue of the various elements hitherto scattered in the surrounding environment of plants and animals. Analogously, social evolution is the process in which hitherto scattered individuals, with no binding relations to each other, are integrated into family, tribe, group, society, and nation with duties, obligations, responsibilities, etc. Knowledge itself evolves by starting out with vague and incoherent sensations, perceptions, feelings, and emotions into the recognition of individuals and particulars, which are then classified under general concepts and laws. It is by these latter that we subsume under a single category or thought a variety of presentations, representations, and experiences in general.

While in the stage of concentration and integration the separation of the system from the environment takes place; simultaneously there also takes place specifiable concentrations and integrations within the separated system itself in its various parts. This is the stage of differentiation, in which the various parts of the system while closely integrated to form a whole, are further differentiated with specific functions, properties, and relations to other parts. Thus, for example, the solar system undergoes differentiation with the planets acquiring unique orbits, with their satellites each with its own properties and relations to others. Similarly, in biology, differentiation of the organism takes place into several organs with different properties, functions, and relations. This is the movement from a relatively homogeneous to a relatively heterogenous state. In social evolution, a system (be it a family, group, or whatever) is further differentiated into various classes by division of labor into providers, priests, educators, warriors, administrators, etc. Knowledge in general moves on by differentiation into various branches, such as sciences, humanities, which are further divided into mathematics, astronomy, physics, biology, psychology, sociology, literature, music, painting, poetry, etc.

Differentiation in a hitherto homogeneous system is not peculiar to evolution; it also takes place in the reverse process of dissolution. That is, differentiation is only a necessary condition for evolution but not a sufficient condition. Thus in order to be able to distinguish evolution from dissolution, we must introduce a further criterion, namely, "order," which we may define as "determinate arrangement." Thus evolution is the process by which a system proceeds from a state in which its parts are homogeneous and scattered toward a state of being a unified whole, in which its parts are not only heterogeneous but stand in reciprocal relations, thereby the system becoming a determinate arrangement. Thus the solar system, the organism, the human society are more or less ordered wholes. In the solar system, the planets and satellites are separated from each other (differentiation). Nevertheless, they all stand in definite and determinate relations to each other, eclipses occurring as a result of determinate relations, movements of planets and satellites being the result of their places in the system as a configuration. The organism is a unified whole with its several parts performing in cooperation owing to reciprocal relations in which they stand. Knowledge, too, although differentiated into various branches, is an integrated whole. Human societies are also ordered wholes, in spite of differentiation into classes, groups, each with distinct roles, functions, and relations to others, all functioning in an orderly manner toward further development of society. It should be noted that the third criterion, namely, determination, is really a conjunction of the first two, integration and differentiation. In other words, evolution is the process which transforms a system into one in which differentiation of parts does not result in disintegration of the whole; on the contrary, differentiation of the parts and integration of the whole go hand in hand. This is an important point, because in dissolution differentiation of parts does take place but results in disintegration of the whole. It may be mentioned that, like evolution, dissolution may take place in several ways.

Before considering the concept of revolution, it is important to point out that the concept of evolution cannot be meaningfully applied to the universe as a whole, although cosmologists and philosophers, ancient and modern, persistently do so, in spite of warnings by Kant to the contrary. The reason behind such warning is that evolution, as has been pointed out, presupposes (1) the notion of a whole, (2) the concept of differentiation of parts, and (3) the concept of determinate arrangement of parts. These concepts apply to limited phenomena, whether physical, biological, sociological, or religious. They do not apply to the universe as a whole not because we are somehow forced to think the universe unlimited, but because the phrase "the universe as a whole" is meaningless; given the way we are, the universe as a whole is not an object of our perception, actual or possible. Further, since evolution involves interaction between a system and its environment, what does it mean to say that the universe, being all that there is, is evolving? There is no way of saying that the universe as a whole, whatever that means, is undergoing evolution, dissolution, or just staying as it is. No wonder, some cosmologists subscribe to a static universe, others to an expanding universe, and still others to an oscillating universe. The point here is that evolution is a relative concept, and hence can only be applied to limited sectors of our experience, not to the universe as a whole which cannot be an object of our experience.

What is the relation between evolution and revolution? We have said earlier that evolution is a process which takes place in one direction only, whereas revolution may take place in two directions: the direction in which evolution takes place, and the direction of dissolution. Further, evolution is a gradual process in which differentiation of parts and integration of whole take place hand in hand, whereas revolution is usually thought of as an abrupt and radical process, in which differentiation of parts is usually accompanied by, or results in, disintegration of the whole. It should be granted, however, that a revolution may by dissolving the existing whole pave the way toward the evolution of a new whole. It is also possible that the direction of revolution may coincide with that of evolution, the sole result being simply to accelerate the process of evolution. Another possibility is that a revolution may take place in such a manner as to reverse the direction of evolution, not to a point of dissolution of the system, but to some earlier state of the system with a certain degree of differentiation and integration. Yet another possibility is that a revolution may take place in such a manner as to prevent both evolution and dissolution and maintain the system in a state of static equilibrium. (Here preventing evolution is the radical result.) The point of drawing attention to all these possibilities is to see that for something to be a revolution neither change in direction of evolution nor abruptness of happenings is a necessary or sufficient condition. We shall consider these possibilities in detail when we discuss violent and non-violent revolutions. It should, however, be pointed out that usually political revolutions are violent and abrupt, although in principle they need not be so.

II

A.   We shall discuss in this section the evolution of scientific ideas and the phenomena of scientific revolutions. We understand by the term "science" a system of propositions through which can be expressed our knowledge of man and nature, the propositions being empirically confirmable or disconfirmable. Obviously, this meaning of "science" does not concern itself with purely deductive sciences such as mathematics whose propositions are not confirmable or disconfirmable by the empirical sciences. One should, however, bear in mind that the empirical sciences make ample use of purely deductive sciences such as mathematics and logic. It is also advisable to distinguish here two senses of the term "science" : (1) science as process and (2) science as product. By the former we mean all those activities of the scientific community, such as observation, experimentation, data-collecting, hypothesis-forming, building scientific experiments, etc. Thus science as process is primarily extra-linguistic. On the other hand, by science as product we mean the body of propositions which are arrived at as a result of science as process. Thus sciences as product consists exclusively of linguistic entities such as statements, theories, etc. We shall use the word "science" in this sense.

Living in an age of science and blinded by its dazzling successes, we often forget that science itself has a career in human history. It is thus part of the total evolution of man. Science as a concept and as an activity is therefore not something which mankind has suddenly come to possess. On the contrary, it has emerged as a product of the evolution of thought over a long period of time. Human thought itself has evolved through various phases and the story of the evolution of science is part of the larger story of the evolution of thought. Let us present, then, in broad outlines the various stages in the evolution of thought leading to the emergence as well as the evolution of science.

Auguste Comte (1798-1857), the great French positivist, having studied the problem of evolution of thought in great detail, describes it in three stages. Hence Comte's law of three stages. The three stages are the theological, the metaphysical, and the positive-scientific. Comte holds that this pattern of evolution of ideas holds not only for the evolution of the general history of ideas but also for the evolution of each particular science. That is, the pattern discloses not only how the ideas of science in general have come about but also how each particular science evolves.

Let us look at these stages. In the theological stage, men tend to explain events in nature and in themselves as due to the action of some god or gods. Their minds are filled with fictional, imaginary, and fantastic beings. They feel the presence of these beings in everything—trees and dogs, seas and stars, hills and bulls, plants and planets. At a later time, they conceive of these beings as a hierarchy of divine beings; and at a still later time they come to regard the whole of existence as under the governance of some single all-powerful deity whose favors they hope to gain by prayer, worship, and submission.

In the metaphysical stage, the transition takes place from imagination to speculation. Mythological thinking and fantasy give place to abstraction, conceptual construction, and postulation of various kinds of abstract entities in terms of which to account for existence in its variety and diversity. In other words, the characteristic feature of this stage is that reason and intellectual inquiry replace myth-making, folklore, and animism. The world and its workings are no longer attributed to a god or gods or some fantastic and powerful beings but to ultimate entities and directive principles, such as the logos, entelechies, or just the nature of things. In Western Civilization, this stage is best represented by the pre-Socratics. It is important to note that although at this stage reason becomes the guiding principle, there are as yet no specific empirical inquiries but only broad metaphysical speculations.

In the third stage, the human mind itself evolves beyond myth and metaphysics. Speculations about ultimate entities, directive principles, logos, cosmic reason, and design yield place to investigation of particular processes and phenomena. That is, the search is no longer after all-encompassing concepts but after causal connections among limited areas and phenomena. The knowledge of particular data and causal uniformities leading to explanation, prediction, and control of different phenomena constitutes positive scientific knowledge. Absolutes of all kinds are renounced in preference to knowledge and understanding of specific phenomena and the power that comes therewith. It is in this manner, then, that positive-scientific thought has evolved, at least according to Comte. It should, however, be immediately granted that thought in general has not evolved in strict accordance with the Comtean pattern of the laws of three stages. There have always been ups and downs and shifts to and fro, the positive-scientific stage being reached in some fields earlier than in others. Even one and the same society has thought scientifically in some fields, speculatively and metaphysically in others, and theologically and mythologically in still others. Further, all three modes of thought may be present in one and the same person. Nevertheless, Comte's law of three stages does give, broadly speaking, a verifiable and correct picture of the evolution of human thought and hence of the human mind.

Comte classified and arranged the various sciences in the order of decreasing generality which also coincides with the historical order in which they have attained the positive-scientific stage. The arrangement is as follows: mathematics, astronomy, physics, chemistry, biology, and sociology (social physics). Psychology is absent in this list not because Comte overlooked it or did not consider it a science but because he subsumed it under sociology. For that matter, Comte includes under social physics such other sciences as political science, economics, and history. Needless to say, all these are known today as social sciences. The sequence in which the various disciplines are arranged shows a progressive transition from simplicity to complexity of observed phenomena associated with them, in the sense that the more complex the phenomena a science studies, the longer it takes for it to reach the stage of positive science. Further, the simpler the relations the science is concerned with, the more general and universal its validity, because the science which studies the complex relations presupposes and is based on the one dealing with the simpler ones. Thus while the laws and principles of mathematics (which appears first in the sequence) holds for all phenomena, those of biology and sociology hold for narrower ranges of phenomena. It is noteworthy that the sequence also reflects the methods of the various sciences. Thus the simpler the relations a science deals with and the greater its scope, the greater is the role of deduction than that of induction. Conversely, the more complex the relations and the smaller the scope of a science, the greater is the role of induction than that of deduction. Thus mathematics is deductive science par excellence, whereas sociology is the most inductive; and among inductive sciences, psychology and sociology are still in infancy compared to physics which is the model of empirical sciences.

It is worth mentioning that, according to Comte, it is impossible in the very nature of the phenomena studied by various sciences to pass from one science to the next in a smooth and continuous manner. He held that such a passage can only be affected by a leap, because of the radical differences among the subject-matters of the various sciences. Thus he was led to maintain that only a materialist could claim to reduce biology to physics. Whether or not we agree with this contention of Comte, his basic insight into the order of evolution of human thought in general and scientific thought in particular is both valuable and enlightening.

Before going further, we may make the following observations on Comte's ordering of sciences. Comte overlooked that there are more fundamental and simpler concepts than those of mathematics; for example, the logical concepts of "identity" and "difference." Comte seems to have regarded, as most nineteenth-century philosophers had, "number" and "magnitude" as the most fundamental and universal concepts. But "identity" and "difference" are more fundamental and of larger scope in that they apply not only to numbers and magnitudes but also to qualities. It thus seems appropriate that logic should precede mathematics in Comte's sequence.

Comte maintained that there is, or could be, no such thing as a purely deductive science. For this reason, he regarded mathematics not as a wholly deductive science but as the most deductive science. He believed that all sciences, both deductive and inductive, are grounded in experience and hence that all sciences are natural sciences. He also argued that if mathematics is a purely deductive science, one will be at a loss to explain how mathematics applies to reality, as is the case, for example, in physics. It is not necessary for our purposes to discuss here whether or not mathematics is a purely deductive science. It is enough to mention that the problem of the applicability of mathematics to reality is a lively controversy among philosophers of our own day.

Finally, the problem of transition from one science to another as well as that of reducing the more complex sciences to the simpler ones, for example psychology to physics via biology and chemistry, is one which has occupied the attention of both philosophers and scientists. There are as yet no generally agreed upon solutions, only schools, some holding such a reduction to be possible and others the contrary view. These schools are known as reductionist and separatist, respectively.

How do Comte's law of three stages and his sequence of sciences agree with Spencer's general conception of evolution? They agree very well indeed. As one moves from the theological to the scientific on the one hand and from left to right in the sequence of sciences on the other, human experience and knowledge become more and more differentiated while at the same time knowledge as a whole becomes integrated. Thus our experience of the world is first differentiated into the experience of the animate and the experience of the inanimate. The experience of the inanimate world is differentiated and classified as astronomy, physics, chemistry, etc. On the other hand, the experience of the animate world is differentiated into biology, psychology, and sociology. Each of these fields is further differentiated into various subdivisions (need we say this in an age of specialization?), while the entire field of knowledge gains in integration. Thus Comte's scheme concerning the order of development of knowledge in general and sciences in particular confirms Spencer's characterization of evolution as the movement from a relatively homogeneous to relatively heterogeneous state, with simultaneously increasing differentiation of parts and integration of the whole.

I come now to a consideration of scientific revolutions—revolutions in scientific thought. I shall discuss one revolution in the field of astronomy and physics and another in the field of biology. The former consists of the overthrow of Ptolemaic astronomy and Aristotelian physics by Copernican astronomy and Galilean-Newtonian physics, respectively. The latter is the triumph of Darwin's theory of evolution over the Biblical and pseudo-scientific conception concerning the origin and variety of life-forms in general and of man in particular. We may call this conception the "special-creation conception," for reasons that will become clear in the sequel.

In Ptolemaic astronomy, the earth is regarded as the center of the universe around which all else moves. Astronomical phenomena, such as planetary motion, are explained using the notions of deferents and epicycles. But every time a prediction was disconfirmed or an anomaly observed, the Ptolemaic astronomer arbitrarily altered the number of epicycles so as to bring his conceptual schema into agreement with observation. This, in main, was the way in which astronomers since Ptolemy down to Copernicus worked. Regarding such a procedure as arbitrary, unscientific, and unsatisfactory, Copernicus boldly proposed that the sun be considered the center around which the planets, including the earth, moved. This proposal rendered astronomical calculations easier and less arbitrary, although there were still some phenomena that proved recalcitrant to the new approach. Be that as it may, the important point here is that by expelling the earth from the center of the universe and making the sun the center of our planetary system, Copernicus effected a profound revolution in scientific thought. In short, Copernicus proposed a new world-view (although it is quite true that such a view was held by Aristarchus of Samos of Greek antiquity which went largely ignored owing to the unchallenged dominance of the geocentric world-system).

Copernicus' world-scheme has revolutionized scientific thought in two important ways. First, by rejecting the arbitrary and overly complicated Ptolemaic method of deferents and epicycles, Copernicus gave prominence to the principle of simplicity of nature, the belief that nature works and attains her ends by the simplest and most economical ways, without duplication, circumlocution, and by a harmonious interaction of the elements involved, thereby bringing under a single cause several effects. Hence, man's efforts to comprehend nature should not multiply causes arbitrarily and capriciously but reduce and minimize them. To be sure, the principle of the simplicity of nature is both a metaphysical and methodological principle. As a metaphysical principle, it cannot be demonstrated as such. But as a methodological principle, it can guide our description, explanation, and prediction of phenomena; and to the extent observations confirm predictions, the principle is vindicated. If a world-view based on such a principle brings order into the hitherto complex and confused experience, it cannot be absurd to believe that the earth is neither immobile nor the center of the universe.

These observations bring us to the second point, the relativity of motion. If there is motion, sense perception cannot immediately and indubitably reveal to us which is moving, the thing perceived or the percipient or both. If so, it should be possible to describe and account for astronomical phenomena by assuming the earth, the place from which we pereive celestial motion, to be in motion. At least it is worth trying whether such an assumption can give us a simpler and richer understanding of celestial phenomena. And, sure enough, it does. Such, then, is the Copernican revolution. It radically changed the direction of scientific thought by overthrowing a certain mode of thought that had governed man's efforts to comprehend the macrocosm for thousands of years before Copernicus. It is also radical in character because of the total reversal of the previous world-view it affected. Further, it has accelerated the pace of subsequent scientific evolution. Apart from being a revolution in scientific thought, Copernicus' world-view has revolutionized man's conception of himself and his place in the cosmos. The earth is no longer the majestic, unmoving center of the universe and gone is the arrogant claim of man to be the center of the cosmos. He suffered dethronement and with it came a sense of cosmic humility, which was to be the guiding theme of his future development.

Closely associated with the Copernican revolution is the Galilean-Newtonian revolution in physics. Here the Aristotelian world-view and in particular the Aristotelian conception of physical nature were expelled and replaced by a radical conception of motion, cause, description, and explanation and a new scientific method was born. According to Aristotle, the world is divided into two parts, the unchangeable and the heavenly on the one hand and the changeable and terrestrial on the other. Through his astronomical discoveries, Galileo showed that even the heavens undergo changes. Further, all changes whether heavenly or earthly, are to be comprehended by the same laws. By carefully analyzing the concept of motion, Galileo formulated the law of inertia, namely, that every body continues in its state unless acted upon by external forces, and proclaimed that changes in states of motion are to be understood through these forces. The invention and formulation of the concept of "inertia" is a great turning point in the life of physical science. The concept of inertia was totally absent in Aristotle's physics. No wonder, according to Aristotle, force is that which maintains a body in a state of constant motion. With the concept of inertia, a new foundation had been laid for the science of mechanics. And with the formulation by Newton of the famous laws of motion and the law of gravitation, both terrestrial and celestial phenomena were brought under a single system of thought. Falling bodies, ebb and tide of the ocean, the motion of planets are all to be understood through one single set of laws and principles. More importantly, explanation and prediction, not mere description, have now become the criteria for scientific validity and fruitfulness. The concept of entelechy, which played the key role in Aristotle's physics, is rejected as lacking genuine explanatory and predictive power and hence as scientifically sterile. The concern now is solely with efficient causes and their effects, which are to be related through the laws of motion. Finally, pure deduction is rejected in favor of experiment and observation, and mathematics plays a central role in establishing quantitative relations among observations and in drawing inferences. Thus was born the hypo-thetico-deductive method. Induction and deduction are both recognized as integral parts of the new scientific method triumphing over the qualitative, teleological, non-predictive modes of thought. Such, then, is the nature of the scientific revolution launched by Galileo and Newton, a revolution which has radically altered the course of science and set it on a new and exciting path.

A scientific revolution essentially consists of the replacement of an existing paradigm by a different one. By "paradigm" we mean, following Kuhn, the world-view, the basic presuppositions, the methods and models, the criteria for what is to count as a problem and what as a solution, which are accepted by the community of scientists at any given period in history as guiding their thought and work. To be sure, the replacement of one paradigm by another is always preceded by a period of doubt, anomaly, discrepancy, and dead-end, known as the crisis stage. The actual manner in which a crisis is resolved and a new paradigm replaces an old one and gains acceptance in the scientific community at large is a very complex and interesting process. We need not go into that here. But we may mention that the scientific community usually displays some reluc-tance and resistance to rejecting an existing paradigm for any number of reasons, sheer individual and institutional inertia being one. The new paradigm goes through a period of probation, a time of test and trial, before gaining general acceptance. For the new paradigm to become acceptable, it should not only successfully account for all the phenomena the old one did but also resolve the crisis to which the old one led. Needless to mention, *ad hoc* devices to resolve a crisis do not qualify for the status of paradigms.

Since the Galilean-Newtonian revolution, there have taken place in physics two revolutions, the quantum revolution of Planck and the relativistic revolution of Einstein, each in turn providing new paradigms. Both resulted in the rejection of the Newtonian paradigm and in the emergence of certain radical conceptions of nature. We need not go into these here. Suffice it to mention that the Michelson-Morley experiment on the one hand and the black-body radiation experiment and the discovery of the photo-electric effect on the other mark the crises and transition from Newtonian to Einsteinian mechanics and from classical electro-magnetic theory to quantum theory of radiation, respectively. Twentieth-century physics is physics done in the conceptual frameworks established by the revolutions of Planck and Einstein.

I turn now to the revolution in biology brought about by Darwin's theory of evolution. Although the idea of evolution of life-forms was not entirely unknown and had been in the air as vague speculation prior to Darwin, it is quite correct to say that until Darwin a certain conception of life and its variety held sway over the minds of men. This conception was that the various biological species are fixed, unchanging and immutable through the ages. Let us call this conception "the principle of fixity of species." This principle, it is obvious, is in accord with the Biblical account of life as given in the Genesis. Against this view, Darwin held that new species originate in time while some of the older ones either become extinct or cease to exist as species. The *Origin of Species* is an elaborate account, substantiated by evidence gathered from a wide variety of sources, of the conditions and circumstances of the origin and mutation of species. Let us explain the meaning of "species." By "species" Darwin means, as many other biologists and philosophers do, any group of individuals, plants or animals, resembling each other. Thus we may say that a "species" stands for a class of individuals having certain characteristics in common. That is, "species" is a classificatory term. And it is important to realize that there are many other classificatory terms, such as "genera", "phyla", "family" and "orders." In other words, a group of individuals may be classified in different ways, on different bases of classification. Thus different species may belong to the same genera, if the different species possess some characteristics in common. Similarly, different genera may belong to the same family and different families may belong to the same phyla, etc. On the other hand, a given species may be subdivided into smaller groups called varieties (races) and sub-varieties (sub-races), if in addition to possessing common characteristics the members of the species differ from each other in some respects. Finally, at the lowest level of classification, one is left with individuals differing from each other while sharing certain characteristics of the subvariety, variety, genera, phyla, etc.

The question now arises : if there are various ways of classifying organisms, the species being merely one, why does Darwin concern himself with the origin of species rather than with that of genera, phyla, etc.? This is an important question and let us answer it. It has been noticed that while the offspring of a species differ from each other and their parents in the respects by virtue of the possession of which they all belong to the same species. That is, the offspring of a given species belong to the same species, no matter how much the individuals differ from each other. The same point can be expressed by saying that members of different races of a species can breed with each other bringing forth individuals belonging to the same species, whereas members of different species cannot in general breed with each other; and if as an exception they do, the hybrid offspring will be sterile, for example, the mule. Given this fact of generation and reproduction, it would seem that it is the species which perpetuate themselves and in so doing preserve and perpetuate the larger classes of which the species are subclasses. Put differently, genera, phyla, etc. are as fixed as their species. Once the species change, the higher groups also change. Thus it is the origin of species, not that of the higher classes such as genera and phyla, that holds the key to the problem of the variety of organisms.

Further, if one accepts as fact that like breed like, then either all the species have existed from the dawn of life on earth or if new species have appeared at some stages in the history of life, they could only have been the result of special acts of creation. It should be readily granted, however, that there is no *a priori* objection to explaining the origin of new species by an appeal to special acts of creation. Nevertheless, such an explanation is unscientific in so far as it invokes some agency other than natural causes; moreover, the concept of special acts of creation raises more problems than it solves; for example, why at a given time and place species A was created rather than species B?

Darwin's aim is to provide a naturalistic, scientific explanation of how species originate. In providing such an explanation, he demonstrated that the thesis of the fixity of species is untenable. Some species, according to Darwin, have vanished from the scene of life and others exist but not as species. Without going into the technical details, we may briefly present the heart of Darwin's theory (explanation) as follows. New species arise when among the varieties of an existing species some intermediate forms become extinct and when the surviving extreme forms, owing to the circumstances, become sharply separated from each other and multiply themselves by reproduction in the course of several generations. Thus the sharply separated varieties, each breeding within itself, will become the new species. The new species in turn undergoes transformations over long periods of time resulting in yet another species and so on. At the same time, the old species become genera. In other words, a sharply separated variety is a potential species. It is important to note that with the emergence of a new species the older species from which it arose ceases to be a species and becomes a genera or simply extinct. Thus Darwin's essential insight is that as the varieties of a species become new species, the species of which they were once varieties becomes genera. The origin of species is thus the same phenomenon as the extinction of intermediate varieties of a given species. Darwin assembled an astounding amount of evidence for the extinction of intermediate forms (fossil remains of now extant species).

The mechanism by which new species arise is called "natural selection" with a two-fold function. It determines on the one hand the extinction or survival of a given species and the conditions for the multiplication of the surviving organisms and on the other the transmission of characteristics from one generation to another and the variations of offspring from each other and from their ancestors. It is in this manner that Darwin's theory explains in purely naturalistic terms the origin of species. It is a theory of evolution because, according to it, new species arise out of the old ones gradually over long periods of time. It shows how more and more complex organisms as species have evolved from previously existing forms of life. It may be mentioned in passing that Darwin's theory of evolution fulfills the criteria for evolution discussed earlier.

How is Darwin's theory of evolution a scientific revolution? For one thing, if new species arise out of the old ones by evolution and in this process some species become extinct, then clearly the thesis of the fixity of species as well as the view of man as essentially different and separate from the other forms of life is false. The theory of evolution teaches the essential unity and continuity of all life from amoeba to man, thereby affirming man's kinship with all other living forms. No longer, then, can man fancy himself as a separate and unique creation in the image of some divine creator. Moreover, the mutability and extinction of species makes it hard, if not downright impossible, to rationally hold on to the religious and pseudo-scientific belief that in view of God's infinite wisdom none of the species could vanish. It must be kept in mind that it is not God's infinite wisdom that is challenged here. What is being rejected is the appeal to special acts of creation to account for the origin of new species in the face of the stubborn evidence for the extinction of species. Like the Copernican revolution in the field of astronomy, Darwin's theory of evolution had overthrown an old paradigm and erected in its place a new one in terms of which the phenomenon of the diversity of life is rendered intelligible. Again, like the Copernican revolution, Darwin's theory had consequences which reverberated through the other spheres of human existence, such as the social, religious, and the philosophical. Thus whereas the Copernican revolution kicked the earth off from the center of the universe, Darwin's revolution shattered man's conception of himself as the unique and crowning creation of God and thus brought him back into the natural realm in unity and continuity with other living beings. Darwin's theory of evolution set a new and radical direction not only for biology but for psychology, sociology, and religion. In a word, it changed man's conception of life in general and of himself in particular from a static to a dynamic one.

B.   In this section, I shall discuss evolution-revolution in the social-political realm. Here, too, humanity has evolved over a long period of time, through many intermediary stages, into its modern forms of social and political organizations. This process is not one of smooth change from an earlier to a later stage. On the contrary, at several points revolutions broke in and have altered the course of evolution, sometimes in one direction, other times in another direction, sometimes toward progress and other times toward regress. Man has come a long way from the lonely, primitive cave-dweller through family, tribe, kingdom, city state, to nations. These units represent the various ways in which men have formed bonds of varying strength and intensity among themselves. To be sure, the emergence of such bonds and their stability would not have been possible had it not been for the culture-making capacity of man. In other words, culture arises when men form groups as distinct from other groups. In time, with the evolution of human consciousness, the various units inter-penetrate owing to several factors, thereby leading to the formation of still higher and larger units of social cohesion. Thus individuals first form families, several families a tribe, and several tribes into a still larger community. This larger community may be a small kingdom. Each tribal chief protects his tribe and promotes its welfare, with the king as the unifying agency under whom the several tribes interact with each other and live in peace. The conflicts among the tribes are resolved as per the dictates of the king, his laws, etc. Nevertheless, any form of political organization, henceforth referred to as "government", can continue to exist only in so far as the governing and the governed are earnestly committed to the discharge of their duties, obligations, and responsibilities toward each other. Also, a government is possible to the extent the governed agree to live under restrictions in the form of laws, the function of the laws being to allow each person to pursue his happiness without encroaching upon that of his fellowman. As such, there cannot be any room for unlimited freedom for a person among the governed. That is, a civil government is possible only with the conscious and deliberate abridgment of freedom on the part of the governed. To be sure, different peoples as groups will have different notions concerning the nature and extent of freedom. For example, a man living in a democracy has a conception of freedom different from that of one living under a tribal rule. Nonetheless, and this is an important point, when the ruled come to believe that the rulers, whatever the institutional form, are no longer interested in the former's happiness, welfare, and freedom and that their freedoms are being overly abridged, contrary to and beyond the contract between them and the ruler(s) and that the government has become ruthless, repressive, and tyrannical, they prepare themselves to dissolve the existing government and establish a new one in its place which they consider would serve them better. But the replacement of one government by another may take several forms. If the government resists being removed and replaced by due process of law, then the people will resort to revolution as the means to changing the government. It is in this manner, broadly speaking, that revolutions are born.

Very often when one hears the word "revolution" in the social-political context, one thinks of uprising, civil war, insurrection, rebellion, turmoil, and anarchy—violence of some form or another. No wonder the word "revolution" brings to one's mind such events as the American Revolution, the French Revolution, the Bolshevik Revolution, the Spanish Revolution, the Cuban Revolution, the

Chinese Revolution, etc. This leads us to ask to ask: 1. whether every social-political event called "revolution" necessarily involves violence and bloodshed, and; 2. whether every social-political event involving violence and bloodshed is to be called a "revolution." Put differently, these two questions are equivalent to the single question whether it is possible to speak of bloodless, non-violent revolutions. Let us try to answer it.

Take the second question first. It is obvious that there are many events involving violence which no one would want to call "revolutions." For example, a war between two nations is not a revolution, although it is certainly both bloody and violent. Another example is the pursuit by a government of a certain sector of its citizenry, such as organized crime; this surely involves violence and bloodshed but is not a revolution.

What, then, are the criteria for something to be a revolution in the political realm? I propose that we call an event "revolution" if it results either in the change of a government as a whole or in changes in some parts of a government, irrespective of the particular means employed for bringing about these changes. In other words, one can meaningfully talk of both violent and non-violent revolutions. Thus each of the above is both a necessary and sufficient condition for an event to be a political revolution. The civil disobedience movement led by Gandhi in India against the British rule is an example of a non-violent and peaceful revolution. The revolution here essentially consists of defying and disobeying the laws and refusing to cooperate with the British government, without resort to violence. The changes brought out by this revolution are the giving up by the British of their colonial rule of India and the Indian people ruling themselves.

The next question to ask is: if violence is used in a revolution, must it always be in the form of armed violence? This is a very difficult question requiring careful analysis of the concept of violence. The question really is: Is physical (armed) violence the only form of violence or are there other forms of violence? The answer, according to Aristotle, is a clear "no." He distinguishes two kinds of revolution, those by force and those by fraud. The latter kind consists of winning the people by fraudulent and deceptive means such as rhetoric and persuasion behind which lie intentions totally contrary to the ones proclaimed; the people are used to change the government and are held in bondage, unfreedom, and subjection once the task is accomplished. As pointed out, this immediately raises the question whether fraud and deception are to be regarded as forms of violence. No matter what the answer, it is clear that holding a people in subjection against their will certainly requires the use of violence. Hence a revolution accomplished through fraud and deception can maintain itself only through violence. The whole problem here, to repeat, is one of recognizing or failing to recognize various forms of violence. If by "violence" in the present context we mean forcing a group of people to lend themselves as instruments to bringing about a change of government, one of the concealed aims of such change being the subjection of these very people, then certainly fraud and deception are to be considered forms of violence. A common example of non-physical violence is verbal violence, such as abuse, name-calling, invective, innuendo, and threat of violence. Is threat of violence violence? I think it is, in so far as threats are by definition instruments to force people to acquiesce into doing or not doing something for fear of what would be inflicted on them if they did not comply with the demands of the threat.

Does the Gandhian type of revolution involve any form of violence at all? To be sure, it does not involve physical violence. Nor does it use non-physical violence by way of fraud and deception. Nevertheless, it seems to me, it does use violence in the sense that it threatens to render the government ineffective and defunct, by the methods of non-co-operation and civil disobedience. Thus even non-violent revolutions implicitly use force in so far as a government neither voluntarily changes nor is prepared to relinquish its power. But it is a matter of taste and opinion whether one wants to stretch "violence" to cover the force implicitly present in a non-violent revolution of the Gandhian variety. Thus many an Englishman looked upon Gandhi as a practitioner of fraud, deception, and the subtlest forms of violence. More recently, not a few considered the late Martin Luther King, Jr., as an extremely dangerous threat to the peace and stability of the American society. Yet there are many who regard these men and their followers as non-violent revolutionaries *par excellence*. It all depends upon which side of the fence one is.

It may be instructive now to survey the views on revolution and its varieties held by some eminent thinkers of Western civilization. Aristotle holds that when a non-violent revolution takes place, the accompanying political changes may be due to sheer accident rather than to deliberate and planned action. Thus in a given society when the gap between the haves and have-nots widens to a point where the latter overwhelmingly outnumber the former, the government has changed from democracy to plutocracy. Locke and Hobbes believe that all revolutions are necessarily violent and eventually lead to war. Thus in their view even the Gandhian kind of revolution will be violent. That is, any activity, no matter now non-violent, in so far as it divides a society and leads to schisms and factions, will lead to war of one group against another. There seems to be a basic confusion in this view. It overlooks that the dissenting group does not use violence although the government and that part of the populace which supports the government use violence to suppress the dissenters. The violence is one-sided. But Locke and Hobbes seem to regard the non-violent actions of the dissenters as violent in so far as they provoke and draw violent response from the government and its supporters. Once again, one can disagree about the meaning of "war." Some mean by it the violent activities of two nations against each other, whereas others mean the violent actions of one group of a people against another or its government. Some would call the latter "sedition" and not "war." There is the even more difficult question whether in the eyes of a conquered people the conquerors are a government at all. This question has weighty bearings upon the right of a people to revolt.

As pointed out earlier, "revolution" is usually taken to mean that activity, violent or non-violent, which radically alters the direction of evolution of a system. From this point of view, a political revolution is one which radically alters an established political order. In this context, I shall distinguish three kinds of revolution : 1. radical progressive, 2. radical retrogressive and 3. *status quo*. The first is undertaken by those who believe that the order of things has to be given a quality and direction hitherto not possessed. The second is launched by those who hold that some earlier state of the society is the most desirable and hence want to change the system from the existing state to that earlier state. The third is undertaken by those who are convinced that the present is the best of all states, past and future, and want to maintain and perpetuate that state. It is an entirely different question whether any revolution employs violence.

At this point it will be well to distinguish between political and social revolutions. Thus, for example, a political revolution need not result in economic changes, just as a revolution may be essentially economic without bringing about any political changes. The French Revolution is essentially economic in the sense that the struggle was between two propertied classes—the bourgeoisie and the feudal lords. The American Revolution is a political revolution in the sense that the struggle was between the colonists and imperial Britain, leaving the basic economic structure unchanged but resulting in the overthrow of the British rule. It must, however, be admitted that the French Revolution had certain political consequences, just as the American Revolution had some economic consequences. A purely political or social revolution is a rare phenomenon. In fact, one has mixed revolutions because of the interconnections between the political and social spheres. If the slave revolts in the United States had succeeded, they would have been both a social and political revolution—political, because of the power and privilege to participate in the governmental process that comes with newly-won freedom; economic, because as free men their economic activities are no longer tied to those of their masters. Think of what would have happened to the plantation economy if the slaves had freed themselves and walked off the plantations. The point here is that economic revolutions, besides producing changes in the conditions of production and consumption, also result in a new order of relations among the various economic classes and in a new distribution of political power. Similarly, political revolutions are likely to have some economic consequences. For Marx, a shift of political power from one group to another without change in the distribution of power, thereby leaving the political-social-economic order essentially unchanged, will be a mere continuation of the old order in new hands. This brings us to Aristotle's distinction between three kinds of revolution from the point of view of constitution : 1. those revolutions which affect the constitution, in the sense of substituting one form of government for another; for example, democracy for monarchy; 2. those that do not change the form of government but merely transfer political power from one group to another; and 3. those that result in the overthrow, establishment, or reinstatement of some particular office, for example, the impeachment of the president. It is clear from our earlier discussion that all three kinds of revolution may take place without violence, by constitutional provisions, due process of law, elections, and plebiscites.

I cannot, for want of space, go here into a discussion of the right of revolution, namely, whether a given group of people have a right to launch a revolution against their government. All I want to say is that when a people no longer believes that the existing government is in their best interests, they mobilize a revolution, violent or non-violent.

Conclusions. The concepts "evolution" and "revolution" have several meanings, each depending upon the particular context of discourse. Nevertheless, it is possible to isolate the basic meanings of which the different meanings are variations with emphasis on different aspects. The general meanings enable us to set up more or less precise criteria for understanding the phenomena of evolution and revolution in different spheres of human activity, such as science, philosophy, social and political organizations. We have illustrated in the present essay how to apply these criteria with respect to the evolution of thought in general and scientific thought in particular as well as scientific, social, and political revolutions. The criteria are equally applicable to a discussion of evolution and revolution in the religious, ethical, and artistic realms.

The notions "evolution" and "revolution" presuppose a more fundamental notion, namely that of order. Any given system, whatever it is, is said to undergo a process of evolution if it changes in such a manner that differentiation of parts is accompanied by an integration of the whole. Contrariwise, where this condition is not fulfilled, that is, where differentiation of parts is not accompanied by integration of the whole, the process is one of dissolution. Evolution and dissolution may take place in many ways, by many paths. We have shown that a revolution need not necessarily mean the movement from a state of order to a state of disorder. It is sheer confusion to equate revolution with dissolution. It is therefore necessary to distinguish between progressive revolutions, retrogressive revolutions, and *status quo* revolutions. What is common to all revolutions is the radical changes they bring about in a system, either by violent and destructive means or by non-violent and constructive means. Thus it is perfectly sensible to speak of both violent and non-violent revolutions. Finally, human evolution in various spheres is not a process of smooth and gradual change from one stage to another. On the contrary, historical inquiry, whether in the sciences or in the humanities, abundantly discloses that evolution is a process marked by periods of activity, called "revolutionary stages," resulting in radical changes. Thus evolution and revolution are not mutually exclusive processes but each contains the other as an integral part.

References

1. Aristotle, *The Basic Works of Aristotle*, ed. R. McKeon, Random House, New York (1941).

2. Aquinas, Thomas, *Basic Writings of St. Thomas Aquinas, Vols. I and II*, ed. A. C. Pegis, Random House, New York (1945).

3. Blum, H. F., *Time's Arrow and Evolution*, Princeton University Press, Princeton (1968).

4. Brinton, C., *Anatomy of Revolution*, Random House, New York (1957).

5. Buckley, William F. Jr., "The Sorry Condition of Counterrevolutionary Doctrine", *The Great Ideas Today*, 1970 (*Featuring a Symposium: The Idea of Revolution*), ed. M. J. Adler and R. M. Hutchins, Encyclopaedia Brittannica, Inc., New York (1970).

6. Calder, R., *Man and the Cosmos*, Mentor, New York (1969).

7. Comte, A., *The Positive Philosophy* (*Cours de Philosophie Positive*, 6 *vols.*), (abridged ed., 2 vols.), trans., Harriet Martineau, J. Chapman, London (1853).

8. Crombie, A. C., *Turning Points in Physics*, Harper Torchbooks, New York (1961).

9. Darwin, Charles, *Origin of Species*, Washington Square Press, New York (1968).

10. Eliade, M., *Cosmos and History*, Harper Torchbooks, New York (1959).

11. Gamow, G., *The Creation of the Universe*, Mentor, New York (1959).

12. Goodman, P., "Anarchism and Revolution", *The Great Ideas Today*, 1970 (*Featuring a Symposium: The Idea of Revolution*), ed. M. J. Adler and R. M. Hutchins, Encyclopaedia Brittannica, Inc., New York (1970).

13. Handler, P., *Biology and the Future of Man*, Oxford University Press, New York (1970).

14. Haselden, K. and Hefner, P., *The Threat and the Promise*, Anchor Books, New York (1969).

15. Hobbes, Thomas, *The Citizen*, ed. S. P. Lamprecht, Appleton-Century-Crofts, New York (1949).

16. Illich, I., "The Need for Cultural Revolution", *The Great Ideas Today*, 1970 (*Featuring a Symposium: The Idea of Revolution*), ed. M. J. Adler and R. M. Hutchins, Encyclopaedia Brittannica, Inc., New York (1970).

17. Koestler, A., *The Sleepwalkers*, Macmillan, New York (1959).

18. Koyre, A., *From the Closed World to the Infinite Universe*, Harper Torchbooks, New York (1958).

19. Kuhn, T., *Copernicus*, Harvard University Press, Cambridge (1956).

20. Kuhn, T., *The Structure of Scientific Revolutions*, University of Chicago Press, Chicago (1970).

21. Locke, John, *Treatise of Civil Government and A Letter Concerning Toleration*, ed. S. P. Lamprecht, Appleton-Century-Crofts, New York (1937).

22. Mill, J. S., *On Liberty*, ed. M. Warnock, Meridian Books, New York (1962).

23. Munitz, M., *Space, Time and Creation,* Collier Books, New York (1961).

24. Spencer, H., *First Principles,* Appleton, New York (1896).

25. Toulmin, S., *Foresight and Understanding,* Harper Torchbooks, New York (1963).

26. Toynbee, A., "Revolutionary Change", *The Great Ideas Today,* 1970 (*Featuring a Symposium: The Idea of Revolution*), ed. M. J. Adler and R. M. Hutchins, Encyclopaedia Brittannica, Inc., New York (1970).

EVOLUTION-REVOLUTION AND THE COSMOS

PAUL GUERRANT MORRISON
State University of New York College at Brockport

1. INTRODUCTION

IF, BY THE cosmos, we mean one universe or world among many
—the maximum space-time region about whose parts our scientists
can currently make testable factual claims—there is certainly ample
evidence that both evolution and revolution of various kinds go
on within it. On the other hand, one might mean, by the word
"cosmos," the unique, possibly infinite totality of all such regions.
And if this were intended, I would agree with Professor Lewis E.
Hahn's position in the thesis article of this symposium that "if . . .
we try to make a problem of everything at once or the universe at
large, we have no way of solving it . . . ." I would also concur
in his view that one may well be dubious of "any grand overall
evolutionary scheme."[1] In fact, I would champion the more extreme
position that any attempt to make theoretical sense of questions
about the nature of the cosmos as a whole must fail. And, indeed,
I shall shortly defend the view that statements of this kind are
theoretically or cognitively meaningless.

On the other hand, harking back to the general outlook of the
thesis paper, I must recognize that the *word* "cosmos," whether
or not it be lacking in theoretical reference, cannot be dismissed
so lightly. For the majority of mankind have always been moved—
despite the speculative reservations of philosophers—to entertain
various cosmic outlooks which appear to them unparadoxical, and
which, moreover, have always influenced their moral, social
and political attitudes and behavior in the most profound ways
historically. As Kant himself held, certain ideas, including that of
the cosmos, which are *theoretically* vacuous because they refer to

71

objects lying beyond possible experience, are important, nevertheless, because they can have a desirable *practical* effect on human morality.

Professor Hahn concludes his paper by observing that the contextualist school can find no inevitable tendency toward progress in evolution nor, unlike Marx and others, discern necessary stages through which social and economic systems must move. He can find no evidence, as Marxists do, of necessary or inevitable progress, nor of any cosmic design carried through by non-human forces such as the material dialectic of Marx and Engels. And while I am in close agreement with this contextualist position, I am at the same time reluctant to let matters rest there. In my own paper, therefore, after first giving serious consideration to the possibility that selected cosmic outlooks might make theoretical sense, I shall try to answer the following question : What is the mechanism by which an opaque or theoretically-vacuous cosmic outlook justifying a social evolution or revolution, by seeming to have objective or scientific substance, is able to affect profoundly, and often cruelly, the fates of those who subscribe to it?

2. DOES THE COSMOS EVOLVE?

Ever since Charles Darwin first set forth his evolutionary theory of biological development in *The Origin of Species,* imaginative and comprehensive adaptations of his theory, both scientific and speculative, have been steadily forthcoming. Prominent on the scientific side have been theories of social evolution, theories of stellar evolution and more recently a theory of the evolution of those protein molecules the most developed of which make life itself possible.[2]

Beyond these scientific adaptations, moreover, renowned philosophers of different persuasions have also been fired by the excitement of Darwin's magnificent insight to speculate about evolutionary process. Following the philosophic penchant for pushing sophisticated descriptive concepts to their limits, a pragmatist like Peirce and an idealist like Whitehead have enriched the literature of recent philosophy with speculations concerning the possibility of evolution, development or change on a cosmic scale. As one might expect, however, their respective views as to the nature of possible evolution or change of such ultimate proportions show certain dissimilarities.

Peirce, in one part of his metaphysics,[3] construes the overall process of development in nature as the operation by which purely chance occurrences eventually bring about an increasingly complex uniformity of structure. Making use of the Greek word "tyche" he dubs this ultimate generalization of evolutionary theory "tychism." And it is worth noting that the general point of view which it expresses has since proved profitable when applied to the theory of the development of an orderly economy in the human brain and in the central nervous system. In fact, W. Ross Ashby, the British brain physiologist, taking such a viewpoint has declared that order and stability in any space-time system whatever routinely result from any sufficiently large number of random changes of state in that system, regardless of the qualitative aspects of the successive states in question.[4]

From this same standpoint, Ashby has also emphasized the key evolutionary role of what he calls a breeder property to which he ascribes two essential features. (1) The first instance of a breeder property has a very low probability of occurring at all. (2) But given the occurrence of even a single instance of it, the probability that its swift replication can ever be stopped is just as low. To take an example, it was highly improbable that the basic ingredient of all living cells, DNA, would ever begin its terrestrial existence. But once even a small amount of it was formed, it became extremely unlikely that anything could keep it from multiplying relentlessly and with great rapidity.[5] The evolutionary significance of such breeder properties in any field of natural science is obvious, since it provides an almost foolproof mechanism for the diversification of the kinds of order exemplified in nature at any given time.

Also working from the general hypothesis that order develops from chance, the American neurophysiologists McCulloch and Pitts have argued that what Kant calls the categories of the human understanding are not originally present in the human psyche, but that the physiological structure required for their operation is routinely developed in each person's central nervous system as a result of a sufficiently lengthy run of random experiences.

Now may we interpret the question "Does the cosmos evolve?" as meaning "Is it plausible that order develops here and there in our world on the basis of sequences of random occurrences some of which trigger off the burgeoning exemplification of breeder properties?" The answer, I think, is that we may. Nor need the question be taken as an appeal for an *a priori* or metaphysical affirmation. For as Arthur Pap pointed out some time ago, the answers to very general questions which some philosophers take as metaphysical may often just as well be taken as broad inductive hypotheses for which some empirical evidence may be adduced, however weak the support which it may give.[6]

The cited speculation of Peirce, to which the later views of Ashby, McCulloch and Pitts are so closely akin, shows the new sophistication characteristic of pragmatism and contextualism. For these philosophic schools look upon universal qualitative and quantitative affirmations of correlation (i.e., upon "laws of nature") as generalizations of currently high probability rather than as expressions of timeless certainties about matters of fact. This is not to say that all of Peirce's hypotheses about cosmic development are susceptible to so fortunate an interpretation. For where he proposes to treat alternate hypothetical universes with incompatible structures as systems of "superlaws" which amount in effect to "superhabits," [7] it is cognitively more difficult to relate what he says, even indirectly, to what we say in our mundane informative discourse. The chief difficulty, in this case, lies in Peirce's apparent strategy of generalizing the notion of a human or animal habit to yield the concept of a cosmic uniformity. In fairness to Peirce, however, it must be admitted that in his time the distinction between a habit and a uniformity in general would have been harder to make out than it is today. But now the prevalent view in psychology is that the formation, maintenance and dissolution of habits is explicable in terms of uniformities of a more basic and pervasive sort. And the tendency to look upon even probabilistically-conceived laws of nature as cosmic habits, however warm and human its appeal, would seem to put us in the awkward posture of mistaking a special case for a general uniformity. For operant behaviorism today specifies habits as special kinds of uniformity peculiar to organisms of the same species.

In spite of these difficulties, however, Peirce's approach to the problems of cosmic order and change appear to wear better than A. N. Whitehead's. For consider Whitehead's suggestion that we are now living in a "cosmic epoch" characterized by extensive continuity which might well give way in the process of development or change, to a new cosmic epoch not so characterized. Since Whitehead holds in the same context that the ultimate metaphysical truth is atomism, and since he further declares that each atom is a system of all things,[8] it would appear that while extensive continuity is everywhere present in our cosmic epoch, it will be totally lacking in the next one. But this way of viewing matters tends to give a new, somewhat broader, meaning to the question "Does the cosmos evolve?" For Peirce's doctrine of tychism and his construing of laws as cosmic habits (i.e., superhabits) are both congenial to the scientist's feeling that while a uniform connection might be exemplified everywhere, it is far more likely to be exemplified only in limited parts of the universe. Whitehead's view of the cosmos, however, appears to call for a total discontinuity of a certain kind between each member in a serial array of cosmic epochs and its successor.

Conspicuous in the example of cosmic change just mentioned is the consideration that the thoroughgoing presence of extensive continuity in the present epoch and its total absence from the next one completely *saturate* the respective epochs to which they belong. And this raises at least two distinct problems : one about evolution, and the other about the nature of the cosmos itself.

First, let us consider the problem involved in construing evolution as a process to which the cosmos as a whole is subject, rather than as one which involves only selected time-extended strands of the universe. In order to see more clearly what this entails, it may be instructive to return to the standpoint of the biologist. Evolution as conceived in biology is always a process by which organisms of a species whose normal members are increasingly ill-fitted to survive in a changing environment become the ancestors of organisms of a new species. The genetic coding of the reproductive cells of these new organisms confers on them a structure and a behavioral potential that gives them a progressively better chance of survival in the new environment than their ancestors would have had. This is not, of course, the whole of the notion of biological evolution, but it does, I think, form an indispensable part of it.

When we try to apply this concept to the prospect of the evolution of the entire cosmos, however, at least two further questions arise. First, what is there about the evolving cosmos, or about the successive adaptation of a series of cosmic epochs, which corresponds to the avoidance of termination by death of a reproductive chain of evolving organisms? Second, presuming a satisfactory answer to the first question, what sense can we make of the progressive adaptation of the entire cosmos to a hostile environment which threatens its survival? Surely there can be no sense in speaking of any "outside" factors which "threaten" the totality of everything that there is.

It is perplexities of the latter sort[9] which bring us to raise the second main question mentioned above: What is the nature of the cosmos as a whole? How can we construe the universe—the totality of all occurrence—in such a way that its most pervasive characteristics or the manner in which these give way to one another as time passes may be definitively specified? Whitehead's way of dealing with cosmic change, unlike Peirce's tychist approach, would seem to push us into raising questions about the cosmos quite similar to those for which, as Kant argued long ago, no theoretically satisfactory answer can in principle be given. For Kant's position, briefly, was that people can talk theoretical sense only about possible objects of experience.[10] And since the cosmos or universe—the totality of possible objects of experience—is not itself a possible object of experience, Kant was able to construct logically unexceptionable demonstrations of several contradictory assertions about it. For example, he was able to demonstrate that the cosmos both is and is not limited in space and time; and that every compound substance in the world both does and does not consist of simple parts. He was also able to prove that the world or cosmos both is and is not free from the ubiquitous regulation of the laws of nature.[11] Later philosophers, including pragmatists, contextualists, and other empiricists, have found themselves in sympathy with Kant's reluctance to assign theoretical significance to affirmations or denials of characteristics of the cosmos as such. And we have already seen that Professor Lewis E. Hahn, speaking for the contextualists, holds a viewpoint not too far from that of Kant on this particular point.[1]

In a similar vein, a contemporary philosophic analyst might want to replace a Kantian question like "Is the cosmos a possible object of experience?" with the query "Can the word 'cosmos' be used as the subject of any empirically testable sentence whatever?" The answer which he would most likely give would be: "No!" Contemporary scientific theories in contrast to their counterparts in the Renaissance when theology and science were still only incompletely separated, typically refer to closed finite systems of phenomena, however immense their spatial and temporal scope. And this is because it has been found that if a theory is to be of any use in explaining, projecting, or controlling some particular aspect of whatever is currently discriminable or detectable by men, its hypotheses must at least be marginally testable. Such testability, in turn, requires that the phenomena with which the theory deals be part of a space-time region within which the discipline's current machinery and techniques—whether verbal, procedural, or mechanical in the narrow sense—provide an effective means of detection.

With the extrication of theological considerations from scientific theory, we have had to give up the notion of the total reach of any theory into all parts of the (possibly infinite) universe. In making this concession, we have also had to admit our lack, in principle, of any ability to understand theoretically the pervasive character of the totality of all things—the cosmos—in spite of the fact that we can make sense about finite systems within this incomprehensible totality. Borrowing Kant's language in this slightly altered context, the scientific empiricist would say that if a philosopher like Whitehead meant by the cosmos or by our cosmic epoch, some maximum closed finite system of occurrences accessible to the probing of the empirical science of the moment, then his assertions would have theoretical content. But if, by the cosmos or by a sequence of cosmic epochs, he meant to speak rather of some possibly unlimited totality of things or events, then he was using the expression "cosmic epoch" in a manner devoid of theoretical content. What he would be saying about cosmic epochs on the latter supposition, would be theoretically unintelligible. And the same sort of neo-Kantian distinction might be applied, *pari passu,* to Peirce's notion of the laws of nature as "cosmic habits"—depending on which way he would have us understand the word "cosmic."

To deny *theoretical* content to references to the cosmos—understood as a totality surpassing our powers of comprehension at any time—did not, for Kant (and does not for many contemporary philosophers), indicate any scorn or ridicule of other philosophers who make such references. For as Kant also emphasized, the cosmic outlook in such informatively vacuous references can express itself in important and powerful ways in the *practical* attitudes and behavior of men. It is true that he denied theoretical intelligibility to the ideas of God, of the human personality or soul as something which survives death, and of that freedom of choice which is supposedly exempt from determination by the laws of empirical psychology; and yet Kant insisted that as moral beings we can practically act on the theoretically unintelligible view that men as personalities continuing after biological death are the supreme reason for the extrascientific creation by a Supreme Being of the unintelligible totality of objects of possible experience—the cosmos.[12]

By the practical employment of reason or of rational discourse, a Kantian means a use having an effect on the will. The three notions which Kant called the ideas of reason—God, freedom and the immortal self—together with the notion of the cosmos, have figured historically in other effective uses of cosmological discourse before Kant's time. In fact, Kant's practical employment of the ideas of reason in a synthesis aimed at harmonizing Western morality and religion with modern science is, in part, an offshoot of the prescientific Augustinian use of these notions to further the Christian faith during the transition from Roman to medieval times.[13] And the cosmic outlook which Augustine developed in terms of these notions served, in turn, to suggest analogous world views which gave practical impetus to at least two other celebrated historical movements.

One cosmic outlook suggested by Augustine's, that of Calvinism, served as the basis, among other things, for practically justifying burning one's neighbors at the stake "for their own ultimate good." The other outlook, at least partially patterned on the Augustinian model, was socio-economic and political rather than religious. I refer, of course, to the world view of orthodox Marxism, proclaimed by Marx and Engels and nurtured by Lenin and Stalin. One practical effect of this world view has been to justify violence toward unbelievers and the paternalistic suppression of initiative from below in national Communist states for the sake of guarding the people's freedom. It is the rationale of this latter cosmic outlook in its practical employment which I now propose to consider.

3.   REVOLUTION AND THE COSMIC OUTLOOK

As suggested earlier, biological evolution is essentially a process by which the members of a species of organisms that are increasingly ill-fitted to survive in a changing environment generate mutations of a new kind that can cope with these changed surroundings more successfully. And it is more apparent every day that certain aspects of human societies may be construed in a similar way. It often happens that a population's mode of social organization renders it progressively unsuited to protect the majority of its members from current threats from human or other environmental sources. But sometimes it develops new forms of organization and new procedures that give the resulting structure a better chance than its predecessor of serving the population under the changed conditions. Now any particular human society is only a kind of partial ecological arrangement within a fraction of one biological species. Nevertheless, it seems appropriate to call human adaptation of the kind just mentioned *social evolution,* bearing in mind that adaptation for survival is perhaps the sole analogical basis for applying the term "evolution" to such changes in social structures or processes.

When one asks whether *revolution,* especially political or social revolution, is not just a particular kind of social evolution, the answer most usually forthcoming is: "Yes! But it is one which involves discontinuity." While this answer cannot be gainsaid, it could go further. For any change in anything, taken in context, always involves both continuity and discontinuity. To distinguish revolution from social change in general and from evolutionary social change in particular, it is perhaps appropriate to remark that revolutions involve developments that are sufficiently startling to evoke comment. And in the particular case of those political or social evolutions which are, at least in part, deliberately guided or furthered by some men in opposition to others, *ideologies* (elaborate verbal justifications of the process of change) comprise an essential part of the revolution itself. So that political or social evolution-revolution, at least, has an indispensable semantic aspect which need not be present in any random evolutionary process as such.

As shown by the historical example of those Marxist revolutions which successfully replaced feudal arrangements with industrial ones, the more effective social revolutions in our times have involved a good deal of initial violence and subsequent repression of individual freedom. Consequently, a study of the genesis and content of the revolutionary ideologies which seek to justify such extreme measures would appear to merit some attention. Now how do such measures come to be adopted by the leaders of successful revolutionary movements, and how do they seek to justify them? I shall now try to answer these two questions, starting with the genetic one.

A socio-political revolution is more than a movement aimed at bringing about a major change in the social order within one or more countries. It involves effecting such a change against the inertia and active opposition of the regime in power. A moment's reflection concerning the nature of any group of ordinary men in control of a society will show why this must be the case.

Those who enjoy the professional exercise of great power over others are typically men of "stomach." That is, they are people who readily learn to bear with equanimity a good deal of intermittent suffering among, and even permanent harm to, their subjects— hardships caused by various inadvertent or deliberate effects of the policies, tacit or explicit, which they sanction or promote. On the other hand, like all other human beings, the powerful do retain a certain limited sensitivity toward the feelings of those whom they govern. This is revealed, for example, by their natural tendency to feel more comfortable in their dominance over the multitude when it tacitly respects their enjoyment of the special prerogatives, emoluments, and satisfactions reserved to rulers and to their families.

In such circumstances, moreover, it soon becomes apparent to those in authority that they are relatively secure in the exercise of their power against occasional random attempts at encroachment by the more enterprising of their reformist subjects. They need ordinarily go to very little trouble actually to alleviate those kinds of suffering, hardship or mortal peril from which the powerful, alone, are routinely exempt. They discover that pious words and unfulfilled or ambiguous promises of social betterment will work, time and again, to quell most of the overt efforts at reform initiated by dissatisfied activist elements within the populace. It is perhaps for all these reasons that small, disciplined groups of dissidents are sometimes led in desperation to the conviction that nothing short of total commitment to a fight to the death with the establishment will lead to a tolerable social order.

Since the membership of any established regime is normally quite superior in numbers to that of a nascent revolutionary group, the latter soon realize that their only chance of success lies in a carefully planned, well-disciplined first strike at the regime. Moreover, because of the diminutive size of the hard core of the revolutionary organization, it quickly discovers that it must steel itself to the postrevolutionary prospect of using harsh measures of social repression periodically against the population which it plans to liberate. It can scarcely expect its future subjects to adopt the ways of the new order in preference to engrained centuries-old social customs without periodic prodding. In short, the revolutionary leadership, because of its small numbers, must prepare itself to employ extremist measures routinely, if it is to have even a chance of accomplishing its long-term goals.

The paradox inherent in all this, is that the required preliminary violence and the subsequent stern measures for maintaining the new order may bring even more profound long-range adversity upon the liberated masses than was visited upon them by the oppression of the former establishment. This political dilemma has much in common with a medical predicament well-known to experienced physicians. A doctor must often ask himself whether a treatment that successfully counteracts or reduces the primary deleterious effects of a serious ailment may not have side effects even more ominous than the symptoms which it momentarily combats with such spectacular success.

Moreover, just as the physician is under strain while treating patients whose medical futures are precarious, so the revolutionary who resorts to violent, harsh or repressive measures toward some of his fellow men in the name of helping larger numbers of the remainder is subject to feelings of guilt. Consequently, the members of a revolutionary cadre, otherwise no less prone to fellow-feeling toward their neighbors than are other human beings, need a means of convincing themselves and their compatriots that such unpleasant measures are justified. They must discover a foolproof way of exculpating themselves when the need arises.

Furthermore, they must recruit and maintain a following consisting of men who can bring themselves, in good conscience, to carry out further acts of violence and social repression in the achievement and maintenance of the new social order. And these recruits cannot help being as aware, as are the revolutionaries themselves, of the grave consequences of any unsuccessful challenge to the authority of the old regime. To achieve the simultaneous magic of inspiration and absolution needed to carry the revolution off, it is essential that the leaders discover some way of generating and maintaining at full pitch an invincible feeling of self-righteousness and superiority over gifted men of other political persuasions. They must cultivate the knack of routinely discounting the intelligence, talent, experience, leadership ability or personal integrity of individual members of this outgroup. Indeed, the revolutionary leaders must feel themselves to be the only true elite, justified in whatever they do to grasp and retain absolute political power. They must manage to see themselves as carrying out a program which they alone perceive to be superior to all other plans for the betterment of mankind or of some favored portion of it. If successful, they would then comprise a new ruling class, a new establishment, which would set the tone for new or revised institutional customs—political, economic or cultural— within the society.

Under these circumstances, a revolutionary leadership will need a cosmic outlook—a verbal position about things as a whole—both to strengthen its emotional solidarity and its inflexible determination to accomplish its violent program and to recruit and maintain the following necessary to help it to ascendancy. It will need a socio-political program for the postrevolutionary order, together with allegedly historically-based reasoning to show why that program, within a preferred cosmic framework, must *inevitably* be initiated and maintained under the exclusive and all-pervasive leadership of the revolutionary group which promulgates it. Reliance on a funda-mental cosmic vision is imperative if the group is to sustain itself in monopoly power over the other members of the society.

Luckily for the revolutionary, he need not start from scratch in fashioning his cosmic outlook, for precedents are already at hand. One of the more useful of these is to be found in the writings of the philosopher Augustine, who distinguished an elect group (the citizens of the "City of God") from the membership of its comple-ment, the "earthly city."[14] In the cosmic setting to which Augustine was partial, this work exemplifies a kind of philosophic outlook which can sustain the members of a determined elite in the conviction that they are justified in dealing severely with outsiders and, on occasion, in treating their own followers quite harshly, although always for their own good.

The trick in this maneuver, whether in religion or in politics, is for the small determined minority to attain, and to give verbal expression to, an alleged insight into the unique development of all or of some complete aspect of the totality of things somehow dis-cernible by, or revealed to, man. And since the historical pageant thus purportedly revealed is virtually cosmic in its spatial and temporal sweep, its uniqueness is automatically guaranteed. For the sequence of events on a cosmic scale cannot be just one of a number of concurrent sequences, one going on in one contingent way in one region, while another simultaneously takes place in another fortuitous way somewhere else.

The success of this particular ploy in convincing oneself and one's public depends to a large extent, of course, on the ignorance of all concerned of the theoretical vacuity of the cosmic outlook from which the requisite practical results are to be expected.[15] For a failure to grasp clearly the *theoretical* paradox and emptiness of the cosmic outlook of the elect is essential to the maintenance of the zeal required in the initiated who are to be inspired by its implications to carry out the difficult *practical* mission of the group.

Let us suppose the members of a small group somehow to discover themselves to be the only people who grasp a unique overall cosmic pattern within which all human development—presumably leading toward freedom—must take place. How would this discovery be likely to affect their individual and collective attitudes? Because they alone had been vouchsafed such an ennobling vision, they would perceive themselves to be "the elect." Moreover, because of its total temporal coverage, the cosmic pattern unfolded to them would extend into the future as well as the past, simultaneously giving them a superior stature and the exclusive foresight needed to lead mankind into the fulfilment of its destiny. Furthermore, all this will seem to the elite group to justify in advance whatever measures it may take to quell the opposition of dissidents. It will also seem to underwrite whatever official means they employ to subdue the ineradicable human impulses of followers whose own vision of the hallowed and inevitable march toward human freedom occasionally tends to fade.

The advantage of discovering a unique course of events, cosmic in their sweep, is that there would appear to be no point in considering (always unrealized) alternatives to them. Seemingly, they would be *necessary* events. Seen from another standpoint, of course, any event is a chance occurrence. Friedrich Engels, one of the chief spokesmen of orthodox Marxism, tried to bring the two aspects together by remarking that "necessity determines itself as chance, and, . . . on the other hand, this chance is rather absolute necessity."[16] If the members of the elect can look upon themselves as always carrying out a necessary or inevitable part of the cosmic plan, they may count themselves absolved from guilt when committing insensitive or brutal incursions upon the lives, health or personal stability of other men. Only what is avoidable can be a matter of moral concern. And no one, presumably, would blame himself or others for what is, in the nature of the cosmic scheme, inevitable.

When, in spite of this, recruits or outsiders still feel anxiety or revulsion at the prospect of inevitable brutality or insensitivity, the elitist theoretician has a time-honored answer. He will observe that almost any really beautiful or worthwhile thing is apt to have structural parts which strike one, in isolation, as ugly or disagreeable— but which, if lacking, would mar the beauty or perfection of the whole.[17] And, of course, the awesome totality of human history is compatible with, and perhaps may require, occasional or even routine ugly incidents as parts. Thus those who are excluded from the elite are purportedly "mistaken" in feeling indignation, revulsion or loathing for various officially tolerated or instigated acts or conditions fostering violence, social repression or human misery as part of the attainment or maintenance of the new "freedom." Unlike the elect, the uninitiated recipients of the benefits of the new order presumably do not appreciate how local deformity in the static and dynamic aspects of their reformed and liberated environments is indispensable to the splendor of the new social order as a whole. In short, the members of the elite need not worry about condemnation for the violence or social repression which they perpetrate or tolerate, whether it be self-condemnation which they feel in weak moments or condemnation by their uninitiated subjects. The underlings of the new order see only a small, unedifying part of the big picture— only a part which, in isolation, gives its spectators a paltry outlook, in contrast to the magnificent overview of those whose intelligence and cosmic insight make them members of the elect. And lest it be thought that the orthodox Marxist does not have a cosmic outlook, it might be worth observing that the material dialectic, to which the inevitable course of events is due, is presumably operative everywhere —as seems implicit in the very title of Engel's work, *Dialectics of Nature*.

4.   INEVITABILITY AND DETERMINISM

The official world view of a contemporary revolutionary party which seriously aspires to seize power by violence will typically construe the revolutionary process whose justification it helps to provide as a discernible part of the regular and indeed inevitable development of man's world. The aspect of grandeur and inevitability which this outlook gives to the violent program of political, economic and social change advocated by the revolutionary group is introduced by a process of transfer from whole to part.

The cosmic outlook shifts the feeling of awe which the believer already has for events of unimaginably-large proportions to certain of their small localized parts. From this standpoint, as remarked earlier, even the most sordid personal episodes of a violent local revolution and the most callous arrangements deliberately or inadvertently brought about by the revolutionary leadership take on the mien of hallowed, if modest, aspects of a majestic evolutionary development within the cosmos.

More than this, the representation of such a limited spatiotemporal process as part of an overall cosmic scheme has the added advantage of transferring the sense of mystery about the cosmos to the revolutionary process itself. One of the chief values of a feeling of awe is that it is typically generated by a thing that is so complex and so pervasive as to make us feel that we can understand it only very imperfectly. This is part of what makes an actual or proposed revolution, even when viewed against so minuscule a backdrop as that of world history, seem simultaneously both impressive and imperfectly intelligible. From this vantage point, it seems more plausible that whatever understanding we can gain concerning the revolution must be gathered under the tutelage of the initiated, that is, of the revolutionary leaders themselves. It is presumably they who, by the magnificent breadth and scope of their vision, first perceive the imminence of the revolution as part of the grand scheme of nature. And this would naturally seem to qualify them better than anyone else to inform us concerning the character and importance of the revolutionary effort. In fact, it is almost as though their original discovery entitles them, by right of prescription, to be the sole *bona fide* dispensers and interpreters of the truths thus discovered.

In bygone days, it was the function of wise men, poets, prophets and medicine men to interpret the awesome to their fellowmen. In recent times, revolutionaries and other political visionaries have made good use of the same procedure. They have done so by identifying proposed evolutionary social changes as parts of the pattern of world history, by endowing those projected developments with an aura of mystery, and investing themselves as the official interpreters of the otherwise inadequately intelligible cosmic marvels by part-whole transfer of which they speak. In the case of orthodox Marxist accounts, the predicted utopian era—in which the state, and even democracy, are to wither away as human oppression vanishes forever—serves in good Augustinian fashion to enhance the merely transitory aspect of violence and paternalistic repression in the unfolding cosmic development.[18]

Now even though an occasional reader might accept everything claimed thus far for what I have called the Augustinian approach to justifying a program of revolutionary violence, he would be quite correct, should he insist, that a crucial ingredient is missing from it. However convincing any individual claim may appear, the overall suggestion of the foregoing is that the tacit psychological appeal of this modern revolutionary ideology is basically a medieval one—and hence, incapable of attracting as broad a spectrum of contemporary intellectuals as, in fact, the orthodox Marxist ideology has done. I now suggest that the missing ingredient is the claim that the justificatory account given in this ideology is a *scientific* one. Indeed, it has been claimed by at least one influential insider, Milovan Djilas, that the best-developed contemporary revolutionary ideology, that of orthodox Marxism, is scientific—is social science, or at least, is a social science.[19]

Indeed, if a social science, or any empirical science, could deal—as the orthodox Marxist ideology appears to do—biographically, with the development of the whole of things perceived by, imagined by, or revealed to men, then, in dealing with a unique course of events, it might appear to deal with the inevitable. And in claiming to know the inevitable—to know what cannot fail to happen—the Marxist seems to be claiming certainty in advance about matters of fact, where the rest of us, allegedly from an unprivileged vantage point, think that we perceive only contingent occurrences. But to say that what happens is inevitable is only to say that what happens does happen. It is only to utter a substitution instance in the mathematically-certain, and hence factually-empty, assertion : "For any x, if x is a member of F, then x is a member of F." And this, in turn, can be proved on the basis of the even more general theorem, "If p, then p," together with the rules of quantification.

As observed earlier, Engels held "that necessity determines itself as chance, and, on the other hand, this chance is rather absolute necessity."[16] His observation was prompted, in turn, by Hegel's remark that ". . . The Contingent . . . has no Ground because it is contingent; and, equally, because it is contingent, it has a Ground."[20] In another place in the *Dialectics of Nature,* Engels speaks of certain events as occurring "more or less by chance, but with the necessity that is also inherent in chance."[21] When one compares Engels' remarks with the passages from Hegel which suggested them, it seems fair to conclude that what he calls chance may just as well be called contingency, and that what he calls necessity is what a later orthodox Marxist like Lenin will call inevitability.[22] But to hold as both Hegel and Engels apparently did that actualized contingencies are inevitable occurrences, and *vice versa*, is to confuse compound assertions of the form "If p, then p," which *cannot* be false, with categorical assertions of the simpler form "p," many of which *are* false. More than this, however, if an inevitable occurrence is one referred to by an assertion which is necessarily true, while a chance occurrence is one referred to by an assertion whose truth can never be fully guaranteed, there will be no inevitable occurrences at all—but only chance ones. Assertions which cannot be false are true regardless of what occurs in the cosmos, and hence, not true on the basis of the occurrence or non-occurrence of any chance event. But then the cosmos, considered as a totality of events, is a totality of purely chance events. So that any talk of inevitable occurrences that purports to be justified by a logic like that of Hegel or Engels in the quotations cited turns out to be completely vacuous.

Again, a revolutionary ideology conceived on such a grand scale typically purports to justify a proposed program of violent revolution and its monolithic aftermath by means of a *historical* account. It makes the credentials of the proposed upheaval depend on a preferred account of a sequence of events in the history of the society or of mankind in general leading up to the time at which revolution is proposed. This historical account will be so drawn up as to suggest that the only satisfactory way to alleviate or remove the persistent and intolerable disorders or inequities which beset the society is to implement the program of violent change proposed by the revolutionaries.

By the party theoretician's account, this violent effort will succeed in bringing about the required social changes only if his group is put totally in charge. To establish that the leaders of that group, alone, are fitted to eradicate the main disorders in the social mechanism, the Marxist account distinguishes the orthodox Communists from the rest of mankind as having discovered "laws of history." These are allegedly laws which *prescribe* the manner in which dominant patterns of human relationships have changed and will change as successive events occur. Milovan Djilas recently expressed doubts as to whether the classless society predicted by the orthodox

Communists is possible, and as to whether they could have an "exhaustive knowledge of the laws of society and history and drive the living social reality along in accordance with these laws. . . ."[23]

He suggested in the same passage that the revolutionary Communists consider themselves to be the *elect*, to which, alone, the "laws of history" are revealed. But since these are always laws of society, they turn out to be not just historical, but scientific as well.[19] Thus, Djilas, raised as a Marxist himself, who once served in the quadrumvirate which ruled the Yugoslav People's Republic, clearly understands the crucial nature of the orthodox Communist belief that scientific laws actually indicate the inevitable way in which mankind, through violent struggle, will achieve an ungoverned utopian society.

Let us consider another view of what this notion of historical inevitability might entail. As indicated earlier, the attempt to identify necessary or inevitable occurrences on the basis of a logic like Hegel's as interpreted by Engels, fails because of its conflation of factually-empty mathematical assertions with contingent claims of matter of fact. But this is not the only stratagem by which the orthodox Marxist has tried to give a clear meaning to the notion of inevitable sequences of events perceptible to those who have discovered "laws of history." For as Djilas suggests, a great part of the ideological appeal of Marx's account of human history leading up to the era of revolutionary uprisings by the proletariat lies in the aura of inevitability emanating from the depicted procession of momentous historical events. He observes that in reading Marx's account, one is reminded of the classical tragedies of Sophocles[24]—a clear indication that Marx was conveying a *fatalistic*, rather than a merely logical, sense of the inevitable in dealing with the main course of events in human history.

Moreover, in another part of his appraisal, Djilas declares that Marxism is a science—". . . not a science in the sense of . . . the exact sciences," however, "but primarily a social science and a course of action based on its own findings. . ."[25] Further reading reveals that Djilas, despite his break with much of the fundamental outlook of orthodox Marxism, tends himself to interpret historical inevitability as a species of determinism in social science conceived along Marxist lines.[26]

But if Marxist determinism in social science is Sophoclean, determinism in the other empirical sciences, including the empirical social sciences, most definitely is not. Indeed, the determinism of the contemporary empirical scientists stands in the sharpest kind of contrast to the fatalism of classical Greek tragedy, as Hans Reichenbach observed some years ago.[27] Nor was Reichenbach alone in seeing this. The great American pragmatist, C. I. Lewis, also made it quite explicit that the empirical sciences in no way dictate what *must* happen in the careers of the various occurrences to which they apply.

As Reichenbach put it, if in a given situation the antecedent of a universal conditional law is fulfilled, the consequent will also be fulfilled. Thus cases of determined concurrence or sequence are always *hypothetical* in character. According to classical fatalism, on the other hand, a specific event is *categorically* destined to occur on any hypothesis or supposition whatever, and hence, regardless of what else occurs. If the occurrence of a fated event is prevented from happening in accordance with one or any of several laws of nature, it will always manage, on a fatalistic account, to occur anyhow in accordance with some other natural law.

C. I. Lewis made the contrast between determinism and fatalism in a more personal way. He expressed it by saying that "the practical value of foreseeing what inevitably will happen, is in order to make sure that it does not happen to *us*; or that it does; according as the happening means a grievous or a gratifying experience. The use of making *categorical* prediction of *objective fact,* is in order to translate this fact into *hypothetical* predictions of *experience,* the hypothesis in question being one concerning some possible way of acting." [28] But surely there is nothing fatalistic about determinism of this hypothetical sort.

There is no sense, in the passage just cited, of the unconditional sort of inevitability which Marxist social science seems to require, but only of an experimental or optional sort of "conditional inevitability." For the hypothetical determination of one event by another is certified, in empirical science, only by a law-like generalization which never has, in any case, more than a given *probability* of being correct. Thus, in empirical determinism we are always left with only a *pro tempore* or provisional "if-then" connection between events of two different kinds. To the student of empirical science, then, as against the devotee of orthodox Marxist social science, it makes little sense to speak of laws which unconditionally prescribe that one particular sequence of major social changes rather than any alternative sequence must occur. An orthodox Marxist might allege that the course of history is destined to be controlled by a superior breed of men whose perception of historical laws determining the inevitable development of social patterns entitles them, alone, to be in charge. A person of a more empirical bent is apt to side, instead, with the sentiments of John Stuart Mill, who once remarked that "submissiveness to the prescriptions of men as necessities of nature is the lesson inculcated by all governments upon those who are wholly without participation in them." [29]

In fine, whether necessity or inevitability be construed in terms of logic or of fatalism, the resulting interpretation of the view that there are laws which govern the inevitable course of history makes it, either way, a view with no empirical reference to anything that takes place. The main difficulty about this Marxist thesis, perhaps, is that it attempts like much of the metaphysics of the medieval and early modern periods to enunciate certainties about matters of fact. Only if there is historically a unique overall pattern within which social evolution must take place can we speak meaningfully of any social action as correct or incorrect because it contributes or fails to contribute to the next step in the unfolding of this unique pattern. But then, how could *incorrect* social actions occur at all? It would be impossible for what *must* take place—what *must* help to fill out the unique—to interfere with it instead. The problem in this historical setting is analogous to the problem of evil in natural religion. Just as the devout ignore the problem of evil and continue to speak of sinful human actions which contravene the irresistible will of the Almighty, so orthodox Marxists continue to speak of incorrect personal or group actions which contravene the cosmically efficacious control of the material dialectic.

If we ignore the inconsistency and give a somewhat strained sense to the word "objective," then only if a single unique overall social development is realizable can moral norms be objectively derived from the pattern of social change. And only those, then, to whom this unique pattern of social evolution is perceptible would be the genuine social scientists. But then, they would also be the high priests of the unfolding new order of social events. To question the morality of the expedients, whether gentle or harsh, which they advocated to help the millenium to its realization, would be simultaneously a theoretical mistake and an act of sacrilege.

On the other hand, if, instead, we deny sense to the notion of an

elite insight into a cosmically-derived unique and inevitable course
of social evolution-revolution, then the political acumen of revolutionary leaders does not, of itself, make them either cognitively or
morally superior to other able men. Nor does it exempt either
their world view or the program of action to which it purportedly
leads from the same kind of criticism and testing through experience
to which theories and programs of comparable sophistication are
subject.

5.   IDEOLOGY AND EMPIRICAL SOCIAL SCIENCE

While, as we have seen, a sophisticated ideology customarily includes
a sustained analysis of social processes, it is by no means unique in
this respect. The empiricist student of social phenomena, albeit from
a different standpoint and for somewhat less partisan reasons, is also
seriously concerned with the nature of social process. The distinction
which I propose to draw between ideological and empirical
(scientific) accounts of social occurrences is made harder to establish
by the fact that the most voluminous and well-read of the ideologies,
the orthodox Marxist one, claims a scientific status for its social
analyses. Indeed, any contemporary revolutionary ideology which
is to accomplish its practical mission of recruitment and of maintaining its adherents "steadfact in the faith" must hallow its social
descriptions with at least one of the labels "objective" or "scientific."
Nevertheless, there are several crucial differences between ideological
sociology and empirical sociology which merit close attention :

(1) An ideological account, unlike an empirical account of social
phenomena, can not rest content to *discuss* value statements and to
estimate their probable effect on other social behavior. It must
champion certain of them. In other words, in addition to mentioning
value statements, an ideologist must also use some of them.

(2) Unlike the declarative assertions of the empirical social
scientist, those which form an ideological account need not all be
testable by way of value-neutral test sentences.

(3) Unlike an empirical treatise in the social sciences, an ideological treatise is incomplete without a *program* recommending
social arrangements of one kind in preference to all other competitive
kinds of social arrangements.

(4) Both ideological and empirical analyses of social phenomena
profess to be deterministic. But while the determinism of a scientific

theory of social process is always a probabilistic and causal one, hypothetical in character, the determinism of an ideology is rather of the logical or fatalistic variety, and hence, unequivocally categorical. For the ideologist, the determinism of social events tends to have a cosmic point of departure. Sequences of momentous social occurrences are flatly prophesied by him, rather than predicted as merely probable.

From the standpoint of rhetoric, of course, it is quite understandable that the categorical, fatalistic determinism underlying the allegedly inevitable march of events should be more appealing and convincing to the unreflective majority of mankind than the more modest hypothetical, causal determinism of the empirical sociologist. For the empiricist's concern is to take pains to establish probable, testable correlations between social phenomena of different kinds. Furthermore, the categorical account of the revolutionary ideologist will naturally seem to those who are only casually concerned with matters of the intellect more sincere and forthright than what appear to him as halfhearted generalizations by the empirical sociologist. This unfortunate appearance results from the layman's misunderstanding of the merely hypothetical or tentative character of all respectable generalizations in the empirical sciences.

(5) There is nothing objectionable, of course, in combining an empirical theory of social phenomena with one of a variety of technical or technological recipes for its application. In such cases, however, the scientific theorist does not represent the application as part of the scientific theory itself, but only as an adjunct to it. For a clear separation between basic science and its possible practical applications is fundamental to the thinking of empirical scientists and technologists in all fields in which connections between space-time phenomena are investigated.

The revolutionary ideologist, however, cannot afford the luxury of making this particular distinction. For him, an adequate theory in social science cannot even be meaningfully discussed in abstraction from its applications among which presumably he distinguishes one correct or objective application in distinction from its complementary set of incorrect ones. Indeed, many of the revolutionary ideologist's most fundamental "scientific" pronouncements are value statements.

(6) In any case, however, the goal of an empirical science of social phenomena is alien to the goal of an ideology. The purpose of a social science, like that of any other empirical science, is to set

forth a system of regular connections between *kinds* of events—even regularities whose embodiment in actual events, although observable in principle, may, in certain cases, *never* be observed in fact. Unlike the connections of events of which the revolutionary ideologist speaks, nothing requires law-like empirical connections of social phenomena *always* to be embodied in the actual course of history. The relentless, fateful, inevitable or *necessary* (dialectical) succession of events prophesied by the revolutionary theoretician—a succession which must always become actual, rather than remain merely possible—has no place at all in an empirical science— whether or not it deals primarily with social phenomena as such. In dealing with hypothetical regularities of connection, a genuine empirical science treats always of what *would* happen *if* something of another kind *should* happen—even in those cases in which nothing of the first and second kinds ever does, in fact, take place.

6.   CONCLUSION

Immanuel Kant held that we can make a reflective though theoretically-empty judgment that man is the ultimate reason for the Divine creation of the cosmos.[12] He also believed that this teleological judgment could be applied constructively to affect human morality—leading men to increasing charity toward, and consideration for, their neighbors. I have indicated my agreement with Kant concerning the theoretical vacuity of any absolute notion of the cosmos. I further agree with his position that uninformative affirmations about the cosmos can play an important role in guiding the attitudes and behavior of men. In this essay, however, I have tried to show that such practical employment of the idea of the cosmos need not always be constructive. In fact, the cosmic outlook which has guided the successful Marxist revolutions, in my opinion, even though it has led constructively to the industrialization of several backward nations, is currently serving to obstruct, rather than to promote, greater human freedom.

In conclusion, I should like to consider whether there may not be a way to avoid the kind of dogmatic inflexibility which too literal a reliance on a cosmic outlook may otherwise impose on an evolutionary program of social action which it allegedly justifies.

If the foregoing analysis is correct, those interested in the justification of social change will fall into two groups. On the one hand, there will be *absolutists,* holding that a scientific, historical account, objectively binding on all reasonable men, will dictate which of a number of rival programs is the sole correct and morally appropriate one to follow. On the other hand, there will be *consensualists,* holding that while an acceptable justification of any particular program of social change will be partially grounded in objective fact, it will also contain an irreducible minimum of prescriptive principles. The consensualists will hold, moreover, that the basic prescriptive component of a justification will vary from one group of people to another. They will maintain that no particular set of basic prescriptive principles can be established once for all over its rivals on the basis of extranormative considerations alone and, especially, not on the basis of scientific ones.

Now a cardinal requirement in any consensualist approach to the manner in which adaptive social change is to be effected is that it make a clear cognitive separation between statements held to be objective or scientific, and statements held to be normative. The primary vice in this context is to blur the distinction between what is prescribed and what is merely described. If he does not make this separation, the consensualist is in danger, himself, of turning absolutist—of believing himself to have discovered scientific reasons for adopting one social program over all alternative ones *at any cost.* From there on, it is an easy jump to a view which justifies routine violence and systematic widespread repression as rational, scientific techniques for first enlarging, and then maintaining, human freedom.

Notes and References

1. Hahn, Lewis E., "Contextualism and Cosmic Evolution-Revolution," in this volume.

2. Cf. Fox, S. W., "The Evolution of Protein Molecules and Thermal Synthesis of Biochemical Substances," *American Scientist,* **44**: 347–359 (1956).

3. Cf. Hartshorne, C. and Weiss, P. eds., *Collected Papers of Charles Sanders Peirce,* Cambridge, Massachusetts; Belknap Press of Harvard University (1960). Vol. **6**, sections 47–65, and section 102, on tychism.

4. Ashby, W. Ross, *Introduction to Cybernetics,* New York; John Wiley and Sons (1957), pp. 167–169.

5. *ibid.,* pp. 70–71.

6. Pap, Arthur, *Elements of Analytic Philosophy,* New York; Macmillan (1949), p. 439.

7. Cf. Hartshorne, C. and Weiss, P. eds., *Collected Papers of Charles*

Sanders Peirce, Cambridge, Massachusetts; Belknap Press of Harvard University (1960) Vol. **6**, section 490.

8. Whitehead, A. N., *Process and Reality,* New York, Macmillan (1929). Reprinted as a Harper Torchbook (1960). Cf. p. 53 of the Torchbook edition.

9. Philosophic perplexities of this kind leading men to ask the cosmological question did not, of course, have to await the birth of Darwinianism. Any attempt to apply predicates of limited scope to the whole of things will do just as well. The predicate which raises the question need not be the word "evolution." In fact, I must hasten to add that I am not aware that Whitehead, himself, explicity invoked the notion of an *evolutionary* sequence of cosmic epochs. Nevertheless, he did say (*ibid.,* pp. 139–140) that "the laws (in the current epoch) are not perfectly obeyed . . . (so that) . . . there is . . . a gradual transition to new types of order" which supervenes on the dominance of the present natural laws. And this would appear to be the process by which one cosmic epoch gives way to another.

10. Kant, Immanuel, *Critique of Pure Reason* (1781), tr. F. Max Mueller, reprinted in New York; Doubleday and Co., Anchor edition (1966), p. 303 : "If we apply our reason, not only to objects or experience . . . but venture to extend it beyond the limit of experience, there arise . . . propositions, which can neither hope for confirmation nor need fear refutation from experience. Every one of them . . . can point to conditions of its necessity in the nature of reason itself . . . unfortunately, its opposite can produce equally valid necessary grounds for its support."

11. *ibid.,* p. 306 ff.

12. Kant, Immanuel, *The Critique of Judgment,* tr. James Meredith, Oxford; Clarendon Press, 1952; Second Part, pp. 75–163, esp. p. 92.

13. Cf. Augustine, *The City of God,* tr. Marcus Dods, New York; Random House (1950.)

14. *ibid.,* Books XII ff.

15. Cf. the discussion of Kant's position on the idea of the cosmos in Section 2. above.

16. Engels, Friedrich, *Dialectics of Nature,* tr. Clemens Dutt, New York; International Publishers (1940), p. 233.

17. Cf. Augustine, *op. cit.,* Book XI, Section 22, p. 365, in which the point is made that a human eyebrow, presumably ugly in itself, if shaved off, would detract greatly from the attractiveness of an otherwise beautiful face.

18. Cf. Lenin, V. I., *State and Revolution,* New York; International Publishers (1932), pp. 15–20.

19. Cf. Djilas, Milovan, *The Unperfect Society,* tr. Dorian Cooke, New York; Harcourt, Brace and World (1969), pp. 8, 44, and 45.

20. Cited in Engels, *op. cit.,* p. 358, as coming from Hegel's *Logik,* A., I/2, pp. 205–206, B., **II,** p. 177.

21. Engels, Friedrich, *Dialectics of Nature,* tr. Clemens Dutt, New York; International Publishers (1940), p. 22.

22. Cf. Lenin, V. I., *"Left-Wing" Communism, An Infantile Disorder,* Moscow; Foreign Languages Publishing House (1920). On p.7, Lenin remarks, in effect, that ". . . understanding international significance to mean international validity or the *historical inevitability* of a repetition on an inter-

national scale of what has taken place in . . . (Russia) . . . ," one can understand what it means to speak of the international significance of the Russian Revolution.

23. Djilas, *op. cit.,* p. 131.

24. *ibid.,* p. 46. " . . . in . . . (Marx) . . . visions of the future, one can sense, as in the frenzied acting out of destiny in a play by Sophocles, the forceful changing of whole societies for the sake of a new prospect, because of a fateful destiny . . ."

25. *ibid.,* p. 124.

26. *ibid.,* p. 12. At this point, Djilas speaks of those who "fail to understand the determinacy and inevitability of Communism's triumph in certain countries . . ." And on p. 63, Djilas refers to Marx' "proofs of the inevitability of the new society." Again, emphasizing his contention that Marx was a social scientist who spoke of what was historically necessary, Djilas remarks (*ibid.,* p. 61) that "Marx . . . took his ideas of the ideal society from the laws of its development, i.e., from *historical necessity,* . . . (thus expressing) . . . a knowledge of these social laws."

27. Reichenbach, Hans, *Atom and Cosmos,* tr. from the German, New York; Macmillan (1933), Chapter 18.

28. Lewis, Clarence Irving, *Analysis of Knowledge and Valuation,* La Salle, Illinois; Open Court Publishing Co. (1946), p. 206.

29. Mill, John Stuart, *Considerations on Representative Government,* New York; Henry Holt and Co., p. 76.

PART TWO

THESIS :   Alastair M. Taylor

COUNTERTHESES : Lee Thayer

                            Edmund F. Byrne

EVOLUTION–REVOLUTION, GENERAL SYSTEMS THEORY, AND SOCIETY

ALASTAIR M. TAYLOR
Queen's University

> The theory of relativity has justly excited a great amount of public attention. But, for all its importance, it has not been the topic which has chiefly absorbed the recent interest of physicists. Without question that position is held by the quantum theory. The point of interest in this theory is that, according to it, some effects which appear essentially capable of gradual increase or gradual diminution are in reality to be increased or decreased only by certain definite jumps. It is as though you could walk at three miles per hour or at four miles per hour, but not at three and a half miles per hour.
>
> *Alfred North Whitehead*[1]

TERMINOLOGICALLY derived from the Latin *quantus* (meaning "how much"), Planck's theory recognized quantization in the exchange of radiation by electromagnetic oscillators. In contrast to the traditional concept of a continuous mode of exchange of energy by waves, it argued for a discontinuous mode in which energy left one oscillator in a fixed amount or quantum and was absorbed by another oscillator, again as a quantum. Subsequently, Einstein applied Planck's hypothesis to suggest that radiation travels through space in minute bundles of energy (photons), while Rutherford and Bohr in turn extended the quantum theory to explain the discontinuous nature of atomic spectra, i.e., the appearance of spectral bands in addition to continuous bands.

Inasmuch as the phenomenal world is ultimately composed of energy, it can be claimed that the latter's behavioral characteristics are likely to be of consequence to an understanding of structure and process at levels above the sub-atomic as well. Moreover, the phenomenal world can also be viewed as functioning as a number of interacting systems,[2] or, if we prefer, as a series of subordinate systems organized and integrated within a superordinate system of universal proportions. Thus, particles of energy are organized as systems at the atomic level, atoms are organized into molecular systems, and so on *seriatim,* including the faunal level (where we encounter man). As a biological system, Homo remains a basically unspecialized primate whose cognitive and tool-making capabilities, however, enable him to construct his own specialized systems of environmental and societal equilibration.

Employing Whitehead's phraseology above, we shall be concerned in this essay with attempting to answer the question: does the quantum phenomenon apply to these man-made systems so that "some effects which appear essentially capable of gradual increase or gradual diminution are in reality to be increased or decreased only by certain definite jumps"? In short, we shall be examining continuity *vis-à-vis* discontinuity in human affairs. In so doing, we shall also have to concern ourselves with such related matters as indeterminacy (probability) and causality.

I.   "EVOLUTION": GENESIS AND DEVELOPMENT OF THE CONCEPT

All events occur simultaneously in space and time. But while we can speak of spatial symmetry, we have also to recognize the asymmetrical nature of experienced time,[3] so that "time's arrow" is irreversible: its flight is into the future, and hence the human landscape displays monuments to the past only. Time is inextricably linked in turn with change, whose existence in nature and human affairs alike has seldom been denied, except by Parmenides and his school. Rather, it is the nature of change, rather than its existence, which calls for explanation, whether one is dealing with the physical or social sciences. Of concern at this juncture are two long-standing concepts: "one, linked with the notion of chance, implies discontinuity in the causal fabric; the other, connected with the idea of evolution, implies on the contrary a global and oriented movement."[4] We might begin by tracing briefly the genesis and development of this second concept.

As its Latin root suggests, "evolution" represents a process of unfolding. According to the dictionary, it may be one of "continuous change from a lower, simpler, or worse to a higher, more complex, or better state," or, again, of "gradual and relatively peaceful social, political, and economic advance." However, as Morgan points out, the evolutionary process can also be regarded in two senses: as "the unfolding of that which is enfolded; the rendering explicit of that which is hitherto implicit" or, instead, as "the outspringing of something that has hitherto not been in being."[5] Both senses—what we might distinguish as "development" and "emergence"—require examination.

The observation in classical times of hillside deposits of marine fossils has been taken as implying early recognition of the evolution of terrestrial life. Lucretius reasoned that the earth "first put forth grass and bushes, and next gave birth to the races of mortal creatures springing up many in number in many ways after divers fashions. For no living creatures can have dropped from heaven . . . It follows that with good reason the earth has gotten the name of mother, since all things have been produced out of the earth." Inasmuch as "time changes the nature of the whole world and all things must pass on from one condition to another," while "many races of living things must . . . have died out and been unable to beget and continue their breed," still others were "protected and preserved" because of natural qualities.[6] Meanwhile, Aristotle had written that "Nature proceeds little by little from things lifeless to animal life"; that "there is observed in plants a continuous scale of ascent towards the animal"; and that "throughout the entire animal scale there is a graduated differentiation in amount of vitality and in capacity for motion."[7] For his part, Aquinas agreed that the world of living organisms comprised a graduated scale ascending from less to more perfect forms of life. But whereas he saw this graduated scale as a hierarchy involving essential differences,[8] or what we might term quantum changes, Locke perceived a virtually perfect continuity marked by differences only in degree. Thus "in all the visible corporeal world, we see no chasms or gaps. All quite down from us the descent is by easy steps . . . There are some brutes that seem to have as much knowledge and reason as some that are called men: and the animal and vegetable kingdoms are so nearly joined, that, if you will take the lowest of one and the highest of the other, there will scarcely be perceived any great difference between them: and so on, till we come to the lowest and the most inorganical parts of matter, we shall find everywhere that the several species are linked together, and differ but in almost insensible degrees. And when we consider the infinite power and wisdom of the Maker, we have reason to think that it is suitable to the magnificent harmony of the universe, and the great design and infinite goodness of the Architect, that the species of creatures should also, by gentle degrees, ascend upward from us toward his infinite perfection, as we see they gradually descend from us downwards . . ."[9]

Whether the "graduated scale" of terrestrial phenomena comprised differences of kind, as Aquinas suggested, or only of degree (Locke), a hierarchy in nature had been observed. It remained for Kant to call for the discovery therein of "a system following a genetic principle . . . When we consider the agreement of so many genera of animals in a certain common schema, which apparently underlies not only the structure of their bones, but also the disposition of their remaining parts . . . there gleams upon the mind a ray of hope . . . that the principle of the mechanism of nature, apart from which there can be no natural science at all, may yet enable us to arrive at some explanation in the case of organic life. This analogy of forms, which in all their differences seem to be produced in accordance with a common type, strengthens the suspicion that they may have an actual kinship due to descent from a common parent. This we might trace in the gradual approximation of one animal species to another, from that in which the principle of ends seems best authenticated, namely from man, back to the polyp, and from this back even to mosses and lichens, and finally to the lowest perceivable stage of nature. Here we come to crude matter; and from this, and the forces which it exerts in accordance with mechanical laws (laws resembling those by which it acts in the formation of crystals) seems to be developed the whole technic of nature . . ."[10]

Kant's "ray of hope" attained full illumination in Darwin. Rejecting special creation as a scientific theory and Aristotle's static taxonomy, he erected a general theory of evolution based on his explanation of adaptation. Darwin's theory, as finally developed, involved four major factors which were, in order of his opinion as to their decreasing importance: natural selection; inherited effects of use and disuse (as suggested by Lamarck); inherited direct action on the organism by external conditions; and "variations which seem to us in our ignorance to arise spontaneously."[11] As the title of his most famous book indicates, Darwin was primarily concerned with the origin of species by means of natural selection. But so powerful was its impact upon contemporary thought that the Darwinian theory influenced the psychology of William James, the writings of Freud, the philosophies of Bergson and Dewey, the plays of Shaw, and various theories of economics, political science, and progress.

Because of its implications for our essay, we have singled out for special mention the phenomenon known as Social Darwinism. The biologist's theory was employed to legitimize a variety of doctrines. Thus the struggle for survival was seized upon by the advocates of *laissez-faire* to sanction "intra-species" competition, i.e., capitalistic rivalry within a given societal structure, as well as "inter-species" competition, namely, the ideologies of nationalism, imperialism, and militarism. Darwin himself professed to find these deductions ludicrous—yet shortly before his death he wrote in a letter : "I could show fight on natural selection having done and doing more for the progress of civilization than you seem inclined to admit . . . The more civilized so-called Caucasian races have beaten the Turkish hollow in the struggle for existence. Looking to the world at no very distant date, what an endless number of the lower races will have been eliminated by the higher civilized races throughout the world."[12]

Meanwhile, Marx had also found an ally in Darwinism, and proposed to dedicate *Das Kapital* to its progenitor who, however, declined the honor (writing to a German scientist that he considered any connection between evolution and socialism a "foolish idea."[13]). "But Marx saw nothing foolish in the idea. Indeed he looked upon the *Origin* as a basis for his own views, the struggle of species in nature being paralleled by the struggle of classes in history, with nature and history evolving in the same natural, inevitable fashion."[14] Eugenicists, such as Darwin's cousin Francis Galton, in turn sought to "further the ends of evolution more rapidly" by discovering and expediting the changes "necessary to adapt circumstance to race and race to circumstance,"[15] while geographers were influenced by Darwinism to advance organismic theories. Thus Ritter regarded continents as the primary organs of the living globe, while Ratzel considered the State a living entity which, in its struggle to survive, is involved in an endless competition for space, or "living room" (*Lebensraum*). In this century, the doctrine of *Geopolitik* linked the concept of the organic State to military objectives and strategic techniques, as "justified" by appeal to Darwin's natural selection. Under Haushofer and his contemporaries, Geopolitics became a significant weapon in Hitler's arsenal to further the cult of Aryan racism and extend German territorial suzerainty.

That a scientific theory can be employed to legitimize doctrines that are either contradictory—as in the case of *laissez-faire* and Marxism—or intellectually questionable (to say nothing of being morally repugnant to many who accept the functioning of natural selection within Darwin's own frame of reference) raises important questions. First, does a fallacy inhere in a simple transposition to human societies of concepts that were originally formulated to explain phenomena restricted to physiological and/or biological organization? If so, wherein does the fallacy lie? Upon the outcome of these questions may depend whether it is possible, or desirable, to structure a "science" of human society. *Inter alia*, must any such attempt require the reduction of human actions and societal phenomena to mechanistic determinism, i.e., to the "causal" and hence exclusion of the "causal" and the indeterminate, to say nothing of those variables which must be classified as "unknowns" on the basis of our current understanding of man and his world? We shall return to these central issues later in the essay.

While recognizing in the evolutionary process the presence of variations which seem "in our ignorance to arise spontaneously," Darwin shared with Locke—and various contemporary "naturalists" —the view that these variations are so imperceptible as to justify the traditional canon of *Natura non facit saltum* ("Nature does nothing by jumps") : "I have been astonished how rarely an organ can be named, towards which no transitional grade is known to lead . . . Nature is prodigal in variety, but niggard in innovation . . . Why should not Nature take a sudden leap from structure to structure? On the theory of natural selection, we can clearly understand why she should not; for natural selection acts only by taking advantage of slight successive variations; she can never take a great and sudden leap, but must advance by short and sure, though slow steps."[16] However, this insistence upon evolutionary continuity and gradualism was all but shattered by the researches of Weismann, Mandel, and de Vries. True, the germ cell transmits through reproduction a continuous stream of protoplasm from one generation to the next, and the hereditary characteristics carried by the chromosomes are unaffected by changes in the somatic cells (which die with the individual). But sudden and unpredictable mutations within the chromosomes can be transmitted by heredity to produce new species. In effect, within the over-all evolutionary continuum occur abrupt discontinuities. Thus, the quantum phenomenon found at the sub-atomic level manifests itself no less at the biological levels. What happens in sociocultural systems?

II. "REVOLUTION": DEFINITION AND UBIQUITY

Revolvere signifies to "roll back" or "cause to return"—and in this sense "revolution" refers to rotation or recurrence. But we are concerned rather with its second meaning: whereas evolution is associated with gradual change, revolution signifies "radical transformation"—a concept which applies in our thesis to any system, be it physical, biological, or sociocultural. Specifically, in this context evolution signifies change to an extent and at a tempo that permits retention of the system's given parameters, whereas revolution implies fundamental transformation of its internal and/or environmental relationships accompanied by—or, again, because of—an accelerated tempo of activity (i.e., expenditure of energy).

"Revolution" as a concept crops up in sociological, political, historical, and economic theory alike. Chalmers Johnson points out that in its political sense, "revolution" was not employed until the late Renaissance. However, in Chinese—the oldest continuously extant language—*ke-ming* complements *t'ien-ming*. The latter signifies "mandate of heaven" and refers to the Emperor's right to rule or authority, while *ke-ming*—which can be translated as revolution—means "to withdraw the mandate." This pairing of terms is important because "revolutions must be studied in the context of the social systems in which they occur. The analysis of revolution intermeshes with the analysis of viable, functioning societies, and any attempt to separate the two concepts impairs the usefulness of both."[17] societies should precede that of revolution. "This is true because social organization itself is intended to restrict or minimize violence among the people united in a society, both purposefully in terms Hence it follows logically that the study of the sociology of functional of the conscious policies pursued by a society's members and functionally in terms of one of the unintended consequences of the value-coordinated division of labor."[18] Hobbes had made basically the same point: in their natural condition men live in a state of war, "where every man is enemy to every man"; "without a common power to keep them all in awe," there can be no industry, knowledge, arts, letters, or society—only "continual fear, danger of violent death, and the life of man solitary, poor, nasty, brutish, and short."[19]

Leaving aside Hobbes' or any other theory of "social contract," we can agree with social theorists that the existence of a society

involves mutually acceptable forms of behavior and mutual expectations among its members—as demonstrated, for example, by the
concept of a division of labor. Force, i.e., the expenditure of energy,
is employed in a sociocultural system for the purpose of its maintenance—even as it can also disorient social behavior or, in its massive
application, even destroy the system entirely.[20] When employed to
bring about such radical societal transformation, force is often described as "violence," and the resulting phenomenon as "revolution."
What is important to recognize is that force is related to societal
mechanisms that make for stability and instability alike—which
Parsons argues are "related configurations in any human grouping."
In other words, sociocultural systems are involved continuously in
processes of equilibration—both intra- and inter-systemic—a subject
to which we shall later return.

Our definition of revolution—involving as it does the fundamental
transformation of a system's existing parameters, accompanied
generally by acceleration in the rate of systemic change—underscores
the element of discontinuity, whether within the system itself or vis-
à-vis its external environment. Discontinuities in turn call for
boundaries of some kind, spatial, temporal, technological, institutional, conceptual, or ideological. The recognition or, again,
imposition of boundaries often presents problems, especially to the
analyst of sociocultural systems. Such problems are familiar to the
geographer and historian as "regionalization" and "periodization"
respectively. Thus, historians have argued among themselves whether
there ever was a "Renaissance," with Michelet and Burckhardt, for
example, responding affirmatively, and the medievalist Thorndike
taking the other side.[21] Numerous archeologists and cultural anthropologists have introduced a three-age classification of society, an
approach first advanced by Thomsen in 1836 when he published a
guidebook to northern antiquities.[22] His division, based on materials
employed in tool industries, comprised the Stone Age, the Bronze
Age, and the Iron Age. This tripartite classification has been
extensively modified since Thomsen's day,[23] but we still make use of
his "ages"—though some scholars have suggested that these temporal
divisions should be replaced by the term "stages" to reflect technological or "functional-economic" shifts instead. For example, Gordon
Childe has stated that "the archaeologist's ages correspond to
economic stages. Each new 'age' is ushered in by an economic revolution"[24]—such as the "Neolithic Revolution" and the "Urban

Revolution."

For its part, historiography is also replete with tripartite forms of classification. Thus Hegel's *Philosophy of History,* in which he sought to record the unfolding of the self-consciousness of freedom in the human spirit, sets forth three great epochs or stages: the Oriental, the Graeco-Roman, and the Modern or Germanic. To Comte, the intellectual progress of mankind experienced three distinct stages: the theological, metaphysical, and scientific. For his part, Spencer discerned three major stages of social development: (1) that of tribal society, emerging out of small and sporadic groupings; (2) the military period, during which small tribal groups were welded together into states by war; and (3) the industrial period. Meanwhile, "Marx fitted his philosophy of history into a mould suggested by Hegelian dialectic, but in fact there was only one triad that concerned him: feudalism, represented by the landowner; capitalism, represented by the industrial employer; and socialism, represented by the wage-earner. Hegel thought of nations as the vehicles of the dialectic movement; Marx substituted classes."[25] In summary, these various tripartite theories share a view of quantum shifts in societal organization and process, but little else.

The political sphere—where "revolution" abounds in theory and history alike—recognizes the primacy of "power" or force. In Weber's words, "A state is a human community that (successfully) claims the *monopoly of the legitimate use of physical force* within a given territory."[26] Thus, within that state, "power" comprises the capability to act, while "sovereignty" asserts the legitimacy of this capability—*sans* accountability to any other authority. Power embraces a spectrum of actions, ranging from influence and consensus-building to coercion of various kinds and total physical destruction. Force is the most extreme form of power, and because it is the *ultima ratio,* its use is a highly prized responsibility of state authorities. "When force is used by authorities in a manner understood or expected by those sharing the system of values—i.e., in a way to which all value-sharers are committed—it is said to be legitimate. Any other use of force, including that by persons who exercise power but not authority, is condemned as violence and is itself subject to a negative obligation (exercise of power) and to suppression by legitimate force (exercise of authority)."[27] Conversely, loss of authority can be symptomatic of a revolutionary situation.[28] This loss may derive from various causes, such as failure of the social system to adjust to changing environmental conditions, failure of political and economic institutions to keep pace with technological innovation; conflicts of material interest among competing groups or, again, over social goals—in short, the creation of societal disequilibria, which in turn generate not only tensions between the governors and the governed but also the formulation of alternative social values and strategies of action that run counter to the interests of the authorities. "As action and reaction can be equal and opposite in physical motion, so in social change revolution and counter-revolution can aim in opposite directions. In either case, whether revolution reverses the direction of change or precipitates a radical transformation toward which things are moving too slowly, revolution seems to involve *overthrowing* the established order rather than *developing* its latent tendencies."[29]

In revolutionary situations each side attempts to maximize its power potential, and to legitimize its position.[30] In other words, *t'ien-ming* versus *ke-ming*—yet the ultimate prize remains the "mandate", i.e., the re-equilibration of the societal system whether in its traditional or in a new mould. This is one of the uniformities discernible in political upheavals designated as revolutions.[31] Others have been discovered by Crane Brinton in his well-known comparative study of four major revolutions. Three of the four— the English, French, and Russian—"have courses in general surprisingly similar. All have a social or class rather than a territorial or nationalistic basis, though Oxford and Lancashire, the Vendée, and the Ukraine, suggest that one cannot wholly neglect these latter factors. All are begun in hope and moderation, all reach a crisis in a reign of terror, and all end in something like dictatorship—Cromwell, Bonaparte, Stalin."[32] The American Revolution does not quite follow this pattern : it was predominantly a territorial and nationalistic revolution; it never had a reign of terror "though it had many terroristic aspects, usually soft-pedalled in school and popular histories"; and it was an incomplete social revolution—though "it was also in part a social and class movement, and as time went on its social character came out more and more strongly."[33] In short, after all due allowances have been made, there remains in Brinton's judg-ment "a residue of uniformities, some of which . . . can have some application to the world of the 1960's"[34] (when he added an epilogue to his original work). In other words, "revolution" at the socio-cultural level has its own "anatomy" or recognizable pattern—like the quantum phenomenon at the physical and biological levels, societal discontinuities can be analyzed in probabilistic terms.

Yet not to the same degree of precision or predictability. It is germane to our thesis that Brinton considers it virtually impossible to obtain an "exact definition" of "revolution." He likens the difference between a revolution and other kinds of societal change as logically nearer to that between a mountain and a hill than to that, say, between the freezing and boiling points of a given substance. "The physicist can measure boiling points exactly: the social scientist cannot measure change by any such exact thermometer, and say exactly when ordinary change boils over into revolutionary change."[35] But why is this the case? To dismiss the difference between the respective approaches of the physicist and the social scientist as commonplace or obvious is to skirt a difficult but central question for which some kind of answer deserves to be attempted.

Moreover, while it is obviously hazardous to seek to measure quantitatively the nature of any sociocultural systemic quantum, actually a political "revolution" may be easier to recognize, and to demarcate temporally and spatially, than other forms of massive societal change. While few will deny that the four revolutions analyzed by Brinton all had an incubating period extending over many years or decades, certain specific events can be singled out for their catalytic, or "triggering" role. Thus, we can recall the events of April 19, 1775, at Lexington and Concord where, in Emerson's words, "embattled farmers stood, and fired the shot heard round the world"; the storming of the Bastille on July 14, 1789; and Petrograd's February Revolution of 1917. Again, the spatial dimensions of those revolutions were coterminous with the boundaries of the rebelling colonies and the European countries in question (though the revolutions set off shock waves which subsequently crossed territorial boundaries and oceanic expanses). Contrast these relatively sudden and precisely located upheavals with Childe's Neolithic and Urban "Revolutions," with Adam Smith's "great revolution" resulting from the shift from an agrarian to a manufacturing economy,[36] or, again, with that major phenomenon of modern times known as the "Industrial Revolution."

What we might describe (with apologies to Brinton) as an anatomy of economic revolution has been attempted by Rostow[37] who distinguishes five successive stages of development. In the "traditional society" (comprising the "whole pre-Newtonian world"), limited means of production place a ceiling on the level of attainable output per head. The second stage applies to societies when certain preconditions for subsequent development occur—such as in the late seventeenth and early eighteenth centuries with overseas expansion, construction of effective centralized national states, and the translation of "the insights of modern science . . . into new production functions in both agriculture and industry." Next comes ' the great watershed in the life of modern societies . . . the take-off" when traditional blocks to steady growth are finally overcome and this growth becomes the "normal condition." However, the "take-off"—Rostow's graphically-chosen synonym for "quantum"—is dependent upon certain specific related conditions.[38] When it occurs, there follows a long interval of sustained if fluctuating progress as the now regularly growing economy drives to extend modern technology over the whole front of activity; 10-20% of the national income is steadily invested—enabling output to outstrip population increase—and the economy finds itself in the international economy. "Some sixty years after take-off (say, forty years after the end of take-off) what may be called maturity is generally attained." At this stage the economy possesses the technological and entrepreneurial skills to produce what it chooses. Finally, the economy reaches the stage of high mass-consumption, where the leading sectors shift towards durable consumers' goods and services, and increased resources are also allocated to social welfare and security. Beyond this fifth stage, Rostow believes it is impossible to predict, "except perhaps to observe that Americans, at least, have behaved in the past decade as if diminishing relative marginal utility sets in, after a point, for durable consumers' goods . . ."[39]

More recently, however, the thesis that economic growth is technologically assured and open-ended or, again, can be employed as a universal yardstick of the "good life," has been questioned by Galbraith and others.[40] While sharing their doubts—given the boundaried condition of the phenomenal world which sets limits on growth of any kind, to say nothing of the qualitative judgments raised by such mounting problems as the population explosion, diminution of non-renewable resources, and environmental pollution—we shall content ourselves with questioning the over-all historical perspective of Rostow's five-stage developmental thesis. We can agree that some kind of technological-economic-societal "take-off" has occurred in various Western countries and Japan during the past two centuries, and that these take-off periods are of relatively short duration (even though we may not share the reasons given by Rostow for these sudden shifts or the dates which he assigns [41]). But when Rostow contends that the Industrial Revolution "may be said to have begun" as from the launching of Britain's canal and cotton-textile boom of the 1780's,[42] he raises the familiar problem of periodization.

Economic historians have disagreed among themselves whether any such "revolution" occurred to overturn the existing economic and social order so as to create "overnight" or "suddenly" a modern ultra-industrial Britain. There is little doubt that the massive use of steam and iron supplanted existing production techniques and created a large-scale factory system. Yet a machine technology and incipient factory system had emerged as far back as the Middle Ages (or centuries before the termination of Rostow's "pre-Newtonian world")—a technological-economic development which Lewis Mumford has termed the "Eotechnic phase."[43] Moreover, inventions in the past hundred years have created what is sometimes called the "Second Industrial Revolution." Surely all this attests to the relativistic nature of terms such as "revolution," "discontinuity," and "quantum" inasmuch as what may be described as "sudden" or a "radical transformation" in any given time-span appears to become an integral part of an ongoing evolutionary process when viewed from a broader historical perspective. *Plus ça change, plus c'est la même chose.* Have we been conceptualizing "revolution" and "evolution" in fallaciously—albeit traditionally—dichotomous terms instead of recognizing their complementarity?

III.   ORDERING CONCEPTS UNDERLYING EVOLUTION-REVOLUTION

That dualities abound in the phenomenal world can be easily demonstrated. In the force fields we find electro-magnetic "positive" and "negative" charges and north and south magnetic poles, while protons and electrons relate also to each other as polarities, and energy's behavior can be viewed in terms of "wave" and "particle" phenomena. All systems exist within the dualistic parameters of "negentropy" and "entropy"; again, in biological systems the binary principle occurs as sexuality. In short, we find duality everywhere : heat and cold, the two directions of electron spin, equilibrium and disequilibrium, symmetry and asymmetry, and left- and right-handedness[44]—as well as in the 0–1 mechanism of a computer and the synonyms-antonyms of a thesaurus.

Even as organisms equilibrate within their environment in terms of both negative and positive feedback, so sociocultural systems in turn display stability (negative feedback dominant) and instability (positive feedback dominant). In Whitehead's words : "Throughout the Universe there reigns the union of opposites which is the ground of dualism."[45] This binary principle is present not only in "being" but equally in "becoming"—because the phenomenal world exists in a space-time manifold, in which the multi-directional symmetries of space ($+$) are fused with the uni-directional asymmetry of "time's arrow ($-$) so as to create an overall asymmetry ($+ \times --- $).
Process or change is therefore ubiquitous, an invariant, as Heracleitus pointed out, which manifests itself no less in dualistic terms, either as "evolution" (continuum) or "revolution" (quantum).

Continuity-discontinuity requires, as we have seen, the presence of boundaries of some kind. All systems, which have been defined as "bounded regions in space-time, involving energy interchange among their parts," possess boundaries that may be simple and clearcut (as in the case of a tree) or, again, non-material.[46] In recent years, ecological studies have shown that the behavior of many animal communities is closely associated with a sense of territoriality and activities localized in terms of boundaried areas. In human societies, in turn, boundaries are employed both to delimit and to formalize their spatio-temporal organization and activities. But whereas with sub-human species the delimitation of territorial boundaries is part of a Darwinian genetically-imprinted process of adaptation to the extra-dermal environment, in human societies boundaries alter in consonance with the dynamics of percept concept transaction and man's technological capability to implement the results of that continuing process of transitional equilibration, i.e., progressive adaptation of the extra-dermal environment to *his* needs and goals. The term "goals" is appropriate here, because alone of all the species, man is continuously—and self-consciously—involved in making choices and judgments as a concomitant of his construction of sociocultural systems, which requires the imposition of boundaries. Hence, because in the equilibrating process man has to take account of his total environment, including his fellow man, his choices necessarily—and uniquely—involve him in moral or ethical considerations, and commit him to constructing value systems.

In recognizing the presence of boundaries throughout the phenomenal world, we have suggested a greater fixity in non-human systems. For example, in employing the quantum theory to explain the discontinuous nature of atomic spectra, Bohr reasoned that in circling its nucleus, an electron had a specific energy so long as it was restricted to a certain orbit; by making a single leap it could jump to another orbit, but in the process gave off or absorbed energy. Subsequently, Pauli showed that only a limited number of electrons can occupy each of the atom's concentric "shells," and that chemical properties derive from the number of electrons in the outer shell. In addition to revealing the presence and operation of indeterminacy and probability at the most fundamental levels of organization in the phenomenal world, quantum mechanics demonstrates that the chemical properties of the elements comprising matter are based upon a mathematical ordering principle that involves orbital boundaries. Again, we can precisely determine the presence of boundaries in the structure and behavior of solids, liquids, and gases. Experiments by Bernal show that whereas regular three-dimensional solids have symmetries limited to multiplicities of 2, 3, 4, and 6, liquids possess essentially asymmetrical assemblages of molecules that are reflected in polyhedra in which pentagonal faces predominate. This morphological difference expresses the discontinuity between liquid and crystalline phases. (It is relevant to note that when water crystallizes, it acquires an hexagonal structure, as seen in the snowflake.) According to Bernal, "We now have ample evidence that the essential nature of the liquid state is the existence of statistical molecular configurations with varieties of coordination patterns geometrically necessarily different in kind from any that can occur in a regular solid."[47] He also characterizes the three states of matter in terms of molecular or atomic structure as follows: "crystalline solids have regular and coherent structure; liquids irregular and coherent structure; gases irregular and incoherent structure."[48]

At this point it is pertinent to ask : can sociocultural structures also be regarded from the standpoint of "coherence-incoherence" or, again, "regularity-irregularity"? Specifically, do human societies equilibrate within a given environmental "shell" which sets boundaries upon the activities of its members, so that to attempt unprecedented types of activity may require a quantum "jump" to yet another environmental framework? If so, can these societal discontinuities be historically delineated? In other words, while recalling our earlier remarks about apperceptional relativism when imposing spatial or temporal boundaries upon sociocultural phenomena ("regionalization" and "periodization"), and recognizing the presence of many more variables in sociocultural as compared with physico-chemical systems, can we demonstrate that mankind's societal continuum has been marked periodically by "revolutions" (as we have earlier defined that term)?

To all these questions we reply affirmatively. We shall begin by relating quantization to yet two more ordering concepts, with each of which it is logically compatible. Thus far, we have been shifting back and forth between atomic or, again, biological phenomena on the one hand, and sociocultural phenomena on the other. Unless we are content to reason only analogically, we must (1) find shared characteristics among all these phenomena, and (2) subsequently account for their different degrees of organizational complexity. In our first endeavor we are assisted by the principle of isomorphism; in the second by application of the principle of successive levels of organization and integration.

In the interwar years, important advances were made in understanding the relations of the sensory, cortical, and the motor neurons in the nervous system. This knowledge enabled scientists to learn much about information theory, and to make spectacular postwar advances in automation and the science of cybernetics. Important applications flow from these developments. On the one hand, interwar neurological research explained how organisms, in addition to transforming energy as mechanisms, also function as systems capable of processing, storing, and retrieving information, and of making decisions. On the other hand, these attributes also characterize *non-living* systems. Consequently, it becomes possible to generalize "the concept of 'organism' to the concept of 'organized system.' Organized systems include organisms."[49] Thus the principle of isomorphism enables us to group, and compare, one-to-one correspondences between living and non-living entities on the basis of *characteristics or attributes shared as organized systems*—while conversely, we can also employ the principle so as to reveal and analyze those attributes which make living and non-living entities dissimilar. As Ashby points out, at the core of "organization" is the concept of "conditionality," namely, once the relation between two entities becomes conditional upon the value or state of a third entity, a necessary component of "organization" is present.[50] Consequently, the study of organization has inevitably to concern itself with holistic concepts that include relationality, interaction, equilibrium by means of feedback (i.e., circular causal trains), decision-making, and purposeful or goal-seeking behavior. Isomorphism (and hence General Systems Theory) provides us with a valid epistemological tool by which to correlate sociocultural phenomena with both inanimate and biological phenomena.

Isomorphism helps explain how systems share similarities of structure and function, but we have also to account for degrees or stages of complexity among organized phenomena. As we have seen, Pauli's exclusion principle shows that since no more than two electrons of opposite spin may occupy the same shell, the electrons excluded from the filled inner shell have to occupy shells further removed from the nucleus. In this manner, simple or "light" atomic structures build up into more complex forms of symmetry, based upon a mathematical formula that is basically simple in itself yet enables matter to be structured in terms of progressive complexification. That the phenomenal world can be arranged as a hierarchy of successive levels of organized complexity (as Aristotle, Aquinas, Locke and Kant had surmised) involves the interaction of a given entity or "individual" of some kind with its environment—be it an electron, atom, molecule, crystal, cell, plant, animal, man, family, tribe, state, or world government. Gerard has employed the term "org" to describe these entities or systems which are individuals at a given level while in turn serving as subordinate units—or "lower-level orgs"—in superordinate entities, or "higher-level orgs." The attainment of each new level of organization is accompanied by an "explosive increase in richness of pattern . . . an emergence of unpredictable novelty"[51] (the factor central to Morgan's *Emergent Evolution*). Importantly, "the addition of new levels of integration has not involved the abandonment of integration at lower levels."[52] Thus, the appearance of sociocultural systems is marked by the continuance of "lower-level orgs" serving as subordinate units, while the former are characterized for their part by attributes and functions unique to their own level.

We might now list some of the uniformities found among integrative levels:

1. Each level organizes the level below it plus one or more emergent qualities (or "unpredictable novelties"). Consequently, the

integrative levels are cumulative upwards, and the emergence of qualities marks the degree of complexity of the conditions prevailing at a given level, as well as giving to that level its relative autonomy.

2. The *mechanism* of an organization is found at the level below, its purpose at the level above. Therefore, explanation of phenomena is continuous from below, discontinuous from above. Knowledge of the lower level implies an understanding of matters on the higher level; however, qualities emerging on the higher plane have no direct reference to the lower-level organization.

3. The higher the level, the greater its variety of characteristics, but the smaller its population. This is accounted for by the increase in the number of emergent qualities, and the fact that a given unit is composed of subordinate units. From the standpoint of population, the integrative levels form a pyramid.

4. The higher level cannot be reduced to the lower. Since each level has its own characteristic structure and emergent properties, any such attempt results in the fallacy of reductionism.[53]

5. Every organization, at whatever level, has some sensitivity and responds in kind. Examples of characteristic behavior include: action-reaction at the physical level, combination-rearrangement at the chemical level, sensitivity-reactivity at the biological level, stimulus-response at the psychological level, and contact-adaption at the cultural level.

The structure of the levels below the sociocultural stage can be diagrammed sequentially as shown in Figure 1.[54]

For some two thousand million years the environment has acted to adapt the organism, so that the resulting symbiosis has been due to the Darwinian process of "natural selection"—which we shall call "adaptive equilibration." But with the advent of, say, *Homo habilis* who, according to Leakey, was a tool-maker, we find a progressive shift from adaptation to manipulation. This stage of "manipulative equilibration" is made possible by "unpredictable novelty" emerging at the human level of biological organization, namely, man's unique capability to engage in high-order abstractions, to communicate symbolically, to possess self-identity and self-direction, and by means of tools to adapt and order the external environment progressively to his specialized purposes. This distinction between sub-human and human levels has been described as a "cerebral/conceptual Rubicon"

FIGURE 1

LEVELS OF ORGANIZATION (Adaptive Equilibration)

LEVEL	SYSTEM	PROPERTIES/EMERGENT QUALITIES		EXPLICATION: CONT.-DISCONTINUITY/INUITY
		(Sociocultural-Technological Levels) "Cerebral/Conceptual Rubicon"		
L₆ ANIMATE	Open Flora & Fauna (multi-organic)	 BELOW	+	Integration of internal and external environments; biotic equilibration
L₅ ANIMATE	Open Organisms, Metazoa (multi-cellular)	 BELOW	+	Division of internal functions; neurological codes
L₄ ANIMATE	Open Cells, Protozoa (multi-molecular)	 BELOW	+	Negative feedback (homeostasis); biochemical codes
L₃ INANIMATE	Closed macro-molecules, Crystals (multi-atomic-molecular)	 BELOW	+	Internal molecular forces; crystal structure; replication
L₂ INANIMATE	Closed Molecules (multi-atomic)	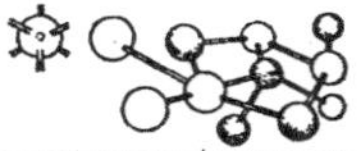 BELOW	+	Chemical bonding
L₁ INANIMATE	Closed Atoms (Multi-particle)	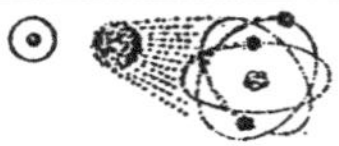 BELOW	+	Electrical attractions; Pauli's exclusion principle
L₀ INANIMATE	Closed Particles	 Positions, Velocities Forces		

—and its place in planetary evolution is represented schematically in Figure 2.

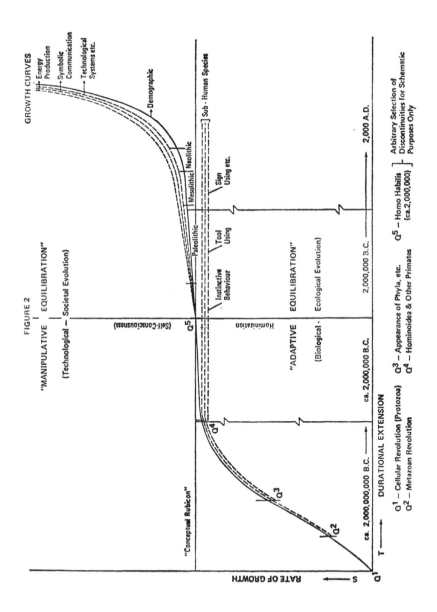

FIGURE 2

FIGURE 3   LEVELS OF ORGANIZATION (Manipulative Equilibration)

LEVEL	SYSTEM OF ENVIRONMENTAL CONTROL		PROPERTIES	EMERGENT QUALITIES					Explication: Cont-inuity / Discont-inuity
	EXPLOITED SPACE	IMPLETED SPACE		TECHNOLOGY	SCIENCE	COMMUNICATION	GOVERNMENT	CONCEPTUAL/SYMBOLIC STAGE	
L_{11} TS^3	Three-dimensional (extra-terrestrial)	Megalopolis ("Ecumenopolis")	BELOW +	Automation Electrical-nuclear energy Cybernetics	Einsteinian Relativity	Electronic transmission (Simultaneity throughout exploited space)	"Ecumeno-cracy" (Supra-national political systems)	"Kosmos": Holism, systems theory, feedback contracts, multi-relationality Differentiated universals Interdependence	
L_{10} TS^2	Two-dimensional (oceans, continents)	City	BELOW +	Machine technology Transformation of energy (steam)	Scientific Method (Newtonian world-view)	Mechanical transmission (printing) Alphabet	National State system Democracy	"Logos": Anthropomorphism ("Man is the measure") Salvation by God-made-man Two-valued logic Independence	
L_9 TS^1	One-dimensional (riverine societies)	Town	BELOW +	Metal tools Non-biological (wind, water) prime movers	Proto-Science	Writing	Theocracy	"Theos": God-King ("Mandate of Heaven") Solar cultus Male principle Rectilinear Gestalten Human dependence	
L_8 $TS^0(n)$	Particulated Universal (sedentary)	Village	BELOW +	Domestication of plants, animals Animal energy		Ideograms	Biological-territorial nexus	"Mythos": Earth-Mother ("Magna Mater") Telluric cultus (womb/tomb) Female principle	
L_7 $TS^0(p)$	Undifferentiated Universal (Nomadic)	Cave/tent (intraterrestrial)	BELOW +	Stone tools Human energy		Pictograms	Biological nexus (family, clan, tribe)	Curvilinear Gestalten (Spatial, Aesthetic, conceptual -including temporal)	

"Cerebral/Conceptual Rubicon"

The ubiquity of the presence of the principle of integrative levels in sociocultural constructs can be demonstrated by examples from major institutions : the administration of the early Christian Church, the three tiers of government in a federal system, and the pyramidal structure present in every military establishment, governmental department, and private corporation. Nowhere can this organizational principle be more clearly seen than in the educational system. The learning process from kindergarten through the post-graduate school is carefully structured in terms of precise levels, i.e., "grades," with each assigned its specific functions ("curriculum"). Note that while each grade makes use of the information provided by the grades below, its own curriculum displays unique concepts and goals, in turn serving as a sub-level for "higher-order" grades above to which it furnishes its information and methodological tools.

As in the case of Figure 1, the question of delineating the number and categories of organizational levels for the stage of manipulative equilibration must depend largely upon an individual's particular conceptual approach. We have approached this question from the basic standpoint of the man-environment nexus, i.e., since Homo is a biological and therefore an "open" system, it follows that his sociocultural constructs will exhibit feedback stabilization with their overall environment. However, at the manipulative levels, this relationship assumes the form of environmental *control* systems. Within this conceptual framework, we have structured what we consider to be fundamental organizational levels, as found in Figure 3. In the remainder of the essay, we shall be concentrating upon data depicted in Figure 3, and in particular assessing relationships as viewed from both the horizontal and vertical dimensions. *Inter alia,* it is central to our overall thesis that the right-hand column in both Figures 1 and 3 displays the paired configuration of continuity-discontinuity (in keeping with the second uniformity listed under the principle of integrative levels)—extending from L_0 "upwards" to L_{11}, and presumably thence to L_n.

IV.   SOME APPLICATIONS OF GENERAL SYSTEMS THEORY

We might begin by recalling Morgan's point that the evolutionary process can be regarded in two senses : (1) the "unfolding of that which is enfolded," i.e., making explicit what has previously been implicit; and (2) the "outspringing of something that has hitherto not been in being." Relating these two concepts to Figure 3, the "unfolding" process has affinity with the horizontal dimension inasmuch as sociocultural development calls for actualizing the potential of the transacting components—societal, technological, economic, political, institutional, aesthetic, religious, etc.—as enclosed by the boundaries or parameters of a given level (L_7 . . . 9 . . . 11 . . . n). Conversely, the "outspringing" process is "vertical" in that this type of change occurs across the boundaries of a given level so as to quantize a new level possessing its own parameter. Once quantization has taken place, the "unfolding" process again proceeds to actualize the potential of what is now the highest level. In terms of our thesis, the "unfolding" process, i.e., developments within a given level of organization, can be construed as "evolution," while the "outspringing" process becomes synonymous with "revolution." However, even as Mendel's "mutations" can be viewed as an integral part of an overall Darwinian evolutionary continuum, so socio-cultural quantization can also be subsumed within an ongoing, over-arching societal process that at L_{11} has become not only planetary but even extra-terrestrial in its environmental control capability.

We would make two points at this juncture. First, while planetary history attests to an overall thrust from simpler to more complex levels of organization and integration, the process is neither pre-determined nor necessarily uni-directional. True, in Figure 1, biological information is genetically encoded at L_{4-6} and transmitted with almost complete certitude in "impleted" space, i.e., in space that is intra-dermal, so that the Darwinian-Mendelian process attests to progressive complexification. Above L_6, however, cerebrally derived information is transmitted across "expleted" or extra-dermal space—and history is replete with examples of this information being distorted or completely lost. Hence it becomes possible for socio-cultural systems to quantize from a more complex to a simpler level of organization. Secondly, at all levels (L_{0-11}) a system may be able to equilibrate indefinitely within its given environment. Just as the coelacanth was supposed to have become extinct in the Cretaceous period (135-63 million years ago), but has continued to exist at an optimal depth and temperature in the Indian Ocean, so various sociocultural systems have demonstrated their capacity to remain viable over long periods of time after having become stabilized within their respective physical environments. Such, for example, are the Eskimos, surviving as L_7 systems, i.e., societies forced to remain at the food-gathering level because their habitat is north of the tree-line, or, again—the L_8 shifting cultivators in the interior of New Guinea. (The penetration of L_{10-11} technology and other alien sociocultural technics can of course remove them from their respective traditional levels of organization; in consonance with the dynamics of the principle of integrative levels, the "higher" tends to dominate the "lower.")

Horizontal stabilization vis-à-vis *vertical* quantization among systems calls in turn for explanation of the binary nature of the equilibrating process. It involves feedback which can be described as either "negative" or "positive." In its first form, it serves to correct deviation, namely, to close the circuit, whereas positive feedback amplifies deviation. For example, while the overall global evolutionary process is mutagenic and open-ended, and hence exhibits positive feedback, negative feedback dominates in organisms at the stage of adaptive equilibration. Conversely, organisms with sensory-cognitive circuits are at the stage of manipulative equilibration precisely because they possess deviation-amplification capabilities. Applying the principle of integrative levels, we can understand why adaptive, i.e., instinctive homeostatic, equilibration can be retained at the lower levels of an organism such as Homo who, at the same time, functions consciously at the highest, or cognitive level in order to adapt the external environment to fit his own constructs. In fact, the concomitant presence of negative and positive forms of feedback free him from having to devote all his conscious efforts to taking care of basic physiological needs (as his autonomic nervous system attests).

This process applies also to the structuring and behavior of societal systems. At the lowest level (L_7), that of food-gathering, mankind has to devote virtually all its time and energies to physical subsistence. In effect, it remains for hundreds of thousands of years at a stage primarily of adaptive equilibration, in which negative feedback dominates. However, as he develops technics of one kind or another to obtain greater environmental control, he moves to a more advanced level of societal organization—and as a consequence, the lower level then serves as "mechanism" (it shifts from positive to negative feedback stabilization) while the higher level assumes responsibility for purpose and direction. For example, the food-gathering stage becomes subordinate to that of food-production, while societies at this latter (Neolithic) level subsequently become part of the mechanism of still more advanced societal systems in turn.

Although both forms of feedback are present no less in the manipulative than in the adaptive stages, and therefore interact at every level (L_7 onwards), we would suggest that negative feedback has demonstrably greater potency than its positive counterpart on the horizontal dimension, i.e., within the parameters of any given level, while on the vertical dimension the converse holds true. Naturally, change (deviation amplification) will occur within any societal system, often in the direction of actualizing the potential of its various components; however, deviation-correcting mechanisms will also operate so as to ensure that form and function retain their correlational viability within the parameters of that system. Quantization occurs when deviation is amplified to the point where no deviation-correcting mechanism can prevent the rupturing of the basic systemic framework, i.e., when the latter can no longer contain and canalize the energies and thrust which have been generated.

All systems exist simultaneously in a fused time-space continuum. In progressively organizing his environment, man has therefore to involve himself with the continuous interplay of temporal and spatial phenomena. When his functions relate primarily to spatial factors, he conceptualizes models and fabricates tools essentially in order to *attain* environmental control. These can be designated as *material technics*, and may be as rudimentary as an Acheulian hand-axe or as complex as the telemetry that keeps a space-capsule on its flight-path to Mars. Thus the function of material technics has a special relevance to the organization of space. Once some degree of environmental control is attained, how is it maintained? Implicitly, this question recognizes not only the relevance of the temporal factor but also the need to ensure continuity beyond the life-span of any one individual. Therefore, from the outset men have devised *societal technics,* i.e., institutional structures, value systems, and methods of persuasion and coercion by which a community seeks to organize and retain environmental equilibrium. These technics include religious and philosophical concepts, law codes, governmental administrations, educational and economic systems, and the like.

Let us now correlate material and societal technics with stabilization (negative feedback dominant) and quantization (positive feedback dominant). While it is true that the thermostat on the wall is designed to correct temperature deviations and hence is a negative feedback mechanism, historical evidence attests to the environment-expanding role of technology since lithic times; moreover, as we shall see, it has served as perhaps the major single factor in quantization. Conversely, while societal technics may sometimes act as catalysts to re-structure a system, their primary role has been to serve as negative feedback mechanisms.

Lithic (L_{7-8}) societies provide examples of the presence of negative and positive feedback alike, with the former dominant so as to give them what, by our modern standards, are remarkable stability and longevity. At L_7, spanning some two million years, advances certainly occur in technology, i.e., in the direction of progressive specialization and miniaturization of tools (microliths). However, these changes occur so slowly as to result in alterations within the societal system itself, rather than transforming it. Meanwhile, as Figure 3 indicates, L_7 societal constructs are based primarily upon a biological nexus—family, extended family, clan, and tribe—and negative feedback predominates in biological systems. Logically, therefore, in lower-level societies the blood-ties present in the family and larger groupings act as potent social pressures to ensure conformity and, where deviant action has occurred, to enforce again the traditional order. For example, stealing and other offences call for restitution so as to restore, or approximate as closely as possible, the *status quo ante*. Still another form of potent collective pressure is the use of *tabus,* which are prohibitions that custom (the socio-temporal continuum) has placed on various actions or words so as not to break the "cake of custom." The appropriate conceptual *Gestalt* for lithic societies is the circle, whether one thinks of a negative feedback (closed) circuit, the utilization of space for obtaining food, the conceptualization of time as cyclical, or of architectural and sculptural forms. It is a telluric world-view, symbolized by the Magna Mater and fertility cults—and with all aspects of the lithic culture pattern logically compatible and mutually reinforcing, empathizing in aesthetic and religious (animistic) expressions alike with lower-level physical and biological systems.

Had we the space, we could examine "horizontally" the other levels in turn so as to demonstrate the inner cohesion and transactions of all the major components comprising each system—and how each has conceptualized its own unique world-view. Instead, we must content ourselves with making two points. First, Figure 3 should be viewed in terms of two correlated time-and-space factors: acceleration and agglomeration. The time intervals become shorter as we ascend the diagram while, at the same time, spatial control rapidly increases. In others words, whereas the lower level socio-cultural systems may be described as "long on time but short on (controlled) space," the converse holds true at the upper levels. We have already noted a logical compatibility between durational continuity and the dominance of societal technics which are negative-feedback oriented. Understandably, therefore, lithic societies are stable and conservative almost to the point of rigidity, yet their impressive longevity permits a gradual "evolution" of their simple but sturdily equilibrating systems. In contradistinction, L_{10} and L_{11} systems demonstrate progressively their respective capabilities to accelerate physical and sociocultural change alike because a different configuration—of positive-feedback material technics and spatial agglomeration—has now become dominant. Hence these levels are marked to an extent unprecedented in the past by societal innovation, instability, and "revolutionary" (quantum) change.

Secondly, Figure 3 presents an idealized schematic presentation of sociocultural levels, as if all the components on each level were neatly compartmentalized, thereby enabling any given societal system to function smoothly and holistically, and so eventually actualize its potential. In point of fact, given the durational sweep of L_7 especially, and to a much lesser but still extensive degree, of L_8, the relatively few variables comprising these systems are able to equilibrate harmoniously—and virtually indefinitely. But as new levels of societal organization emerge, and the numbers of variables increase, there is a tendency for an uneven tempo of change to take place. This is especially marked between the area of positive-feedback oriented material technics, on the one hand, and that of negative-feedback oriented societal technics, on the other. Precisely because the first is innovational and radical, while the second tends to be traditional and conservative, there develop asymmetrical relationships, sometimes across the boundaries of our levels—a technological quantum may often precede a societal quantum, so that we encounter the phenomenon of "culture lag." Such sociocultural asymmetries are accompanied by, or in turn create, areas of pressure within the cultural pattern, disequilibria which often take the form of conceptual, institutional, economic, or political "revolutions." It is on the dynamics of the quantizing process, i.e., the examination of Figure 3 in terms of discontinuities on the vertical dimension, that we might now focus attention.

As might be expected from a conceptual approach that takes as its universal invariant the man-environment nexus, an intimate relationship exists between the role of tools in manipulative equilibration and fundamental shifts in the organization of the environment. Pertinently, analysis of lithic technological development shows not only progressive specialization and miniaturization but increased efficiency of cutting edge *vis-à-vis* size and weight of flint—so that Upper Paleolithic and Mesolithic tool industries exhibit impressive sophistication in form-and-function. However, there comes a point where the limitations inherent in flint prevent any further significant development, i.e., when the law of diminishing returns applies. In effect, man has now encountered boundaries which preclude continued linear technological "evolution"; to proceed further, *Homo fabricans* must innovate by making a fundamental shift in concept and materials alike. This quantum jump takes place—after some two million years of lithic development—with the advent of metals, followed in turn by progressive knowledge and exploitation of a potential far greater than that which inheres in stone.

Metal tools represent an L_9 advance over L_{7-8} lithic industries—but meanwhile another technological quantum has occurred so as to revolutionize stone-age societal organization itself. In L_7, man is a nomadic food-gatherer who has also to rely on his own muscles as prime movers. However, in the Old and New Worlds alike—independently and in far-removed millennia, yet exhibiting parallel conceptual invention—men domesticate wild grasses and animals. In addition to being a food-gathering hunter or fisherman, man has now also become a food-producing farmer or herdsman—and his increased technological capability enables him to "stay put," i.e., to *localize* his environmental control. This results in an increase of population numbers and densities, new settlement patterns (in particular the village node), as well as concomitant changes in societal attitudes and institutionalization, in which the man-environment nexus assumes a progressively less "biological" in favor of greater "territorial" importance. In effect, the parameters of the existing sociocultural system have been altered so fundamentally as to justify Gordon Childe's term "Neolithic Revolution." The sociocultural equivalent of a biological mutation has taken place in the man-environment nexus.[55]

Building upon these fundamental innovations, men subsequently develop a technology capable of exploiting the rich bottom lands of various river valleys—such as the Nile, Tigris-Euphrates, Indus, and Huang-ho. This new riverine technology provides harvests sufficiently large both to sustain unprecedented population densities and to permit sizeable numbers of people to engage in specialized non-agricultural pursuits within an urban environment. Again we can accept Childe's term "Urban Revolution" inasmuch as the social, economic, and political centers of gravity shift to the cities which henceforth control the "lower order" countryside. From a geographical standpoint, the technological capabilities of this environmental control stage are still so limited as only to provide control over the river valley and its immediate hinterland (where irrigated by lateral canals). Hence we find a one-dimensional (TS^1) environment, organized as L_9. These fluvial societies, sometimes referred to as "hydraulic civilizations," are noteworthy for their longevity and conservatism alike, for though they are marked by important technological and conceptual advances—including the advent of writing and what has been described as "proto-science"—negative-feedback societal technics long dominate in their purpose to perpetuate "Theos"; a theocratic world-state with its solar cultus and pyramidal societal and decision-making structure of organization.

As from the second millennium B.C., especially, we find environmental control systems acquiring a major two-dimensional (TS^2) capability so as to obtain dominion over seas, then oceans, and finally entire continents, culminating in the "conquest" of the two Poles by the end of the nineteenth century. *Inter alia,* this historic process of exploring, mapping, and controlling the earth's surface is intimately connected with the Greeks' discovery of the scientific method of higher-order abstraction, so that Euclidean geometry could be employed by Eratosthenes to compute the earth's circumference (to a remarkable approximation) and by him and Ptolemy to apply a grid system to the plotting of maps, thereby placing cartography on a scientific basis. Once again the sociocultural parameters have been transformed, together with conceptual goals, epistemological tools, and aesthetic and symbolic *Gestalten.* In L_{10} constructs we find that "man is the measure" so that he employs *logos* to develop institutions and a *Weltanschaung* which culminate conceptually in Newton's model of celestial mechanics.

The last hundred years in turn have been marked by vast conceptual, scientific, and technological innovations. Information is transmitted by electronic forms of communication around the globe and through extra-terrestrial space at the speed of light. Man's environmental control capability has been extended vertically so as to create a three-dimensional system (TS^3). Conceptually, Newtonian-Euclidean absolutes have been replaced by Riemannian geometry, Planck's quantum and indeterminacy, and by Einstein's theory of relativity and a new model of celestial mechanics. In short, we have made another quantum shift, conceptually and technologically, to a new level of organization—L_{11}—which is global-*cum*-extra-terrestrial in its scope and requiring an appropriate shift from our traditional view of reality.

However circumscribed the preceding description, we have sought to demonstrate why in our view these levels result from shifts so seismic or fundamental as to affect all sociocultural parameters: spatial dimensions and control, temporal constructs (including progressive quantification), and the roles of material and societal technics alike, so that each emergent *Weltanschaung* creates its own apperceptional modalities and symbolic imagery. If space permitted, within each of these revolutionary shifts, or sociocultural quanta, we might undertake to distinguish subordinate, or "evolutionary" developments. For example, we could trace factors of technological specialization and societal complexification in the so-called Lower, Middle, and Upper Paleolithic periods, relating these changes also to climatic and other environmental factors. Or turning to, say, L_{10}, we might apply the principle of integrative levels to analyze stages of socio-political organization, beginning with the city-state concept of the Greeks, proceeding next to the Roman world-state, followed by its "fall" in the West—so that the socio-political system quantized "down", i.e., from equilibrating as part of an *imperium* extending from Britannia to the Tigris, Western Europe had to re-equilibrate at the grass-roots level by means of such institutions as feudalism and manorialism. But following this temporary "Dark Ages regression"—in large measure explicable in terms of failure of negative-feedback technics and loss of information alike within the existing system—we can perceive the renewal of the processes of progressive spatio-political agglomeration and complexification—culminating in the apogee of L_{10} juridico-political conceptualization, the national state system.[56]

As viewed, too, from our particular vantage point, Brinton's four "Revolutions" can be regarded as major societal transformations within the parameters of the national state system, itself a manifestation of an L_{10}-level of organization. Thus, in each case, these four political upheavals seek to express the sovereignty of the people and their primacy in government (as interpreted by Locke, Jefferson, the philosophes, and Lenin), while keeping the national state concept intact. In short, all four Revolutions were lineal descendants of Protagoras and Pericles who for their part had also inveighed against L_9 rule by "divine right." Again, our conceptual approach can account for such other previously-raised questions as : was there a "Renaissance" or, again, an "Industrial Revolution"? The former represented a major cultural and intellectual transformation within a conceptual construct that begins with the Greeks (in the Western world), while the latter is also to be seen within the same L_{10}-stage of organization, in the technological and socio-economic spheres. It may be relevant to note that steam—inextricably associated with the "Industrial Revolution"—was first employed as a prime mover by Hero of Alexandria, while the factory system and a machine technology antedate, as we have already noted, the Industrial Revolution by many centuries. The latter might also be more accurately described as a transformation—this time socio-economic—rather than as a conceptual revolution.

Our thesis contends that we are now witnessing, and participating in, one of history's revolutionary stages, or conceptual quanta. One has to remember the distortions resulting from propinquity so that contemporary events always loom preternaturally large and significant. Nevertheless we feel justified in speaking about an L_{11} stage, three-dimensional and extra-terrestrial in its spatial expletion, and planetary in its technological organization and potential societal integration. Yet we must also recognize that if the conceptual thrust on the vertical dimension of Figure 3 is toward such a sociocultural level, examination of the horizontal dimension reveals serious discrepancies. In short, a major "culture lag" exists. Our societal technics have lagged dangerously behind our material technics; while the latter have propelled us into a "global village" of interdependence and multi-relational orientations, our major institutions —government, law, management and labor, and even education— are still largely two-valued (either/or) in their orientations and logic, placing a higher value upon L_{10} absolutes and "independence" than upon L_{11} relativity and probability—in consonance with the findings of Einstein and Planck—and the acceptance of "interdependence" as an ecological and societal necessity alike. Given nuclear energy as an L_{11} prime mover, the critical factor would appear to be our capacity to update—or rather "up-stage"—our socio-political constructs so as to create a supra-national authority capable of controlling conflict before it escalates to a thermonuclear flash-point and transforms our planet into a radioactive fireball.[57] If, as we believe, we can live through these dangerously uncertain, i.e., disequilibrating and dysfunctional, days of transition, twenty-first-century global society can well become firmly established "across the boundary" in L_{11}.[58]

V. CONCLUDING OBSERVATIONS

We are now better equipped to attempt to answer questions posed earlier by our discussion of Social Darwinism. Specifically, we can suggest where the fallacy lies in any simple transposition to human societies of concepts originally formulated to explain biological phenomena. It lies in reductionism, as demonstrated by application of the principle of integrative principles. Even as the structure and organization of organisms are more complex than inorganic systems, so in turn human societies make use of the inorganic and organic levels of organization as a physico-biological foundation while, in addition, functioning at a new and more complex negentropic level. Consequently, to describe a human society as an "organism" is to reduce its stage of organization to a lower level and, concomitantly, to lose those very qualities of structure, self-regulation, and self-direction unique to its own plane.

The unique properties inhering in self-direction require men to make choices—because even the shape to be given their tools involved our lithic ancestors in goal-seeking and problem-solving—and these choices represent judgments and values. Therefore only at the levels of manipulative equilibration will there emerge ethical and moral concepts and value systems. Yet it is precisely in this area that Social Darwinism is deficient—dangerously so. Thomas Huxley had contended that evolution as a "cosmic process" was antithetical to ethics, since its brute force and cunning provided a guide only to immorality. Yet his grandson, Julian Huxley, was to argue a contrary thesis, namely, that evolution provided an objective basis for determining human values and social progress (and that man is now in a state of transition from the biological into the psychosocial area of evolution).[59] That these two eminent relatives could arrive at seemingly opposite conclusions underscores our objection to yet another simplistic and fallacious transposition, namely, employing anthropomorphic concepts and language to explain structural and behavioral patterns at pre- or sub-human levels of organization. Whether we adopt T. H. Huxley's vocabulary, or Sir Julian's "love," "beauty," and "selfless morality," we risk applying to sub-human evolutionary processes value judgments and a teleology characteristic only of the human level of organization and self-direction. If conscious teleology characterizes the stage of manipulative equilibration, i.e., if it is to be applied to the behavior of sociocultural systems, we cannot extrapolate backwards and expect to have it apply also to purely ecological-biological levels of equilibration. Conversely, the fallacy of reductionism should equally caution us against equating human society with biological "organisms" so as thereby to divorce the former from ethical and moral factors which are exclusively its possession. Yet this line of action is exactly what many Social Darwinists have advocated in their attempts to justify the rule of force, war, and territorial expansion in the name of "natural selection" and "survival of the fittest."

We suggest, too, that the principle of integrative levels can also assist social scientists and historians in clarifying their concepts and terminology regarding "revolution." The second section of this essay sought to demonstrate that social anthropologists, historians, and economists—whether they engaged in tripartite divisions of planetary societal experience or, again, wrote of political "revolutions" or economic "take-offs"—did not share any common criteria upon which to base their various theses to explain apparently sudden or unusual change—indeed they did not even share a common definition of the term "revolution" itself. We will recall that Brinton had also made the point that it was not possible for the social scientist to measure change with the exactitude of a physicist measuring boiling points—but he offered no explanation. Were Brinton's metaphors employed only as a literary device, or did they not also at least imply that the traditional approach toward social phenomena, i.e., the non- (or pre-) general systems approach, is incapable of proceeding with any assurance beyond analogical reasoning because it has not been able to provide a conceptual framework and epistemological tools by which to discover valid, isomorphic relationships between inorganic and organic phenomena? Our chosen approach enables us to account for the problem encountered by Brinton: simply put, the number of variables increases exponentially as we ascend Figures 1 and 3.[60] Despite the admittedly crude state in which general systems theory still finds itself, it can nevertheless help us to define "revolution" and "evolution" in some consistent manner, i.e., within a dynamic, ongoing relationship of process that shares a common denominator, namely, a "discontinuity-continuity," with Mendelian-Darwinian, mutational-gradualistic natural selection, or, again, with particle-wave behavior at the inanimate L_1 level of organization associated with Planck and Pauli.

We have suggested that "evolution" represents growth within the parameters of an existing system so as to actualize (ideally) that system's capabilities, whereas "revolution" represents a transformation so fundamental as to rupture the systemic framework and quantize to a new level of equilibration. In either expression of change, function and form are paired; in other words, we encounter the presence of boundaries. These are related, as we have seen, to two types of equilibration, negative and positive feedback. If space permitted, general systems theory could be employed to analyze in depth the relationship of these types of feedback not only to the roles of material and societal technics (as we have briefly adumbrated) but, specifically, to societal attitudes and forms of behavior. At this juncture, we shall simply hypothesize that a logical affinity exists between negative-feedback-oriented technics and traditional forms of religious belief, conservative politics (the "right"), emphasis upon "law and order," and retention of the *status quo* in national and international political behavior—while converse attitudinal and behavioral patterns can in turn be correlated with positive-feedback-oriented technics. This kind of analytical approach can in our view be profitably employed to examine the basic roles of our societal institutions in the political, economic, and juridical spheres.[61]

"Evolution" and "revolution" are in the last analysis but two complementary components of change, i.e., process in time-space. When we regard process from the standpoint of the principle of integrative levels, two related questions may arise: is this a planetary process that is pre-determined (or pre-programmed) and therefore "inevitable?" And does it also represent "progress"?—a question of continuing interest to modern man, despite the presence of many apparently regressive societal phenomena (world wars, gas chambers and genocide, etc.). We think the first question has already been answered by Mendel's mutations, Planck's indeterminacy, and Morgan's and Gerard's "emergence of unpredictable novelty"—as well as by our own emphasis upon the uncertainty factor present in the transmission of information across expleted space in levels above the Conceptual Rubicon.

All too often, "progress" is confused with progression, or mere movement. To board a plane in Miami and land in Havana is hardly likely to be considered progress by those passengers who had bought tickets for New Orleans. In other words, progress would seem to imply a goal—but cultural anthropology and sociology abound with different goals, value systems, and world views. Actually, much of our concern with the concept of progress and our desire to believe in its inevitability reflect Western ethnocentricity. Again, the principle of integrative levels can be of assistance. It warns against any linear approach toward evaluating organized phenomena in terms of some specific or preconceived values system. Each level has its *raison d'être* and qualities—and to suggest that the culture of twentieth century New York is superior to that of Periclean Athens, or Neolithic Jericho, is about as meaningful as to argue that a rose is "better" than slime-mould. What we can say in each case is that there has been an advance in complexity of organization—and here our conceptual principle provides us with at least a partial resolution of the question of "progress." Speaking of "progress in science," Collingwood suggests that this consists "in the supersession of one theory by another which served both to explain all that the first theory explained, and also to explain types or classes of events or 'phenomena' which the first ought to have explained but could not." Thus Darwin propounded a theory to account for phenomena which the theory of fixed species could not explain. Again, we have "the new more familiar relation between Newton's law of gravitation and that of Einstein, or that between the special and general theories of relativity."[62]

Here we have examples of the evolution of the conceptualizing process as demonstrated in the movement from lower- to higher-order abstractions capable of controlling increasingly complex amounts of phenomena and organizing them into systems possessing their own inner logic and cohesion—and, as a result, of adding to man's capacity to engage in manipulative equilibration. Put into different words, the move from lower- to higher-order levels of abstraction represents progression toward universalization of organization and control. And this is precisely what is happening in regard to sociocultural levels of organization as well. Hence the historic quantum shifts from L_7 to L_{10} and, in our century especially, the transition in turn towards L_{11}, namely, in the direction of planetary equilibration at once ecological and societal.

L_{10} is pre-eminently the level of two-valued (either/or) orientations and Aristotelian logic with its principle of the excluded middle. It was superb for the initial stages of the scientific method, such as in the classification and inductive study of empirical data. Whether one starts in L_{10} with the Judaic-Christian concept of Genesis, in which God gave man dominion over nature, or with the Graeco-Roman concept that "man is the measure," Homo was conceived as basically an entity apart from his extra-dermal environment. In such a relationship the binary principle was logically interpreted and applied in terms of a dichotomous duality—a dialectic of thesis *versus* antithesis, a confrontation and struggle of inimical forces that lie at the core of the world-views of Hobbes and Marx alike, of struggles between kings and commoners, between "independent" nation-states, and exploiting versus exploited economic classes—and it also accounts for the rationale of Social Darwinism. Here we find a central component in an L_{10} paradigm or model that emphasized the *independence* and isolation of man in his relationships with nature and his fellow man alike.

If our thesis has any validity, L_{11} will possess a very different paradigm: one whose basic physical parameters have already been delineated by Planck, Einstein, de Broglie and Pauli. In terms of our subject matter, the binary principle as applied to process must in turn be reconceptualized as a duality that is not dichotomous and antagonistic, but complementary and synergistic. Instead of the either/or of Adam's search for knowledge at the expense of the beasts, fishes, and fowls (and culminating today in the danger of self-expulsion from a polluted Paradise), we might well explore the multi-relational orientations and logic implicit in Einsteinian relativity and made explicit in various non-Western systems of thought—such as Jainism's "Sevenfold Division" of predication (*saptabhangi*), or again by Taoism and the concept of Yin-Yang whereby complementary forces transact at every level of the phenomenal world, the Tao itself representing universal equilibrium.

The paradigm for L_{11} recognizes the continuous, and ongoing, interdependence of all systems from the microcosmic atom to the macrocosmic galaxy, with *Homo concipiens et fabricans* positioned appropriately in the middle of this vast equilibrating process.

We began with a quotation from Whitehead; we might close with another :

> The Universe is dual because, in the fullest sense, it is both transient and eternal. The Universe is dual because each final actuality is both physical and mental. The Universe is dual because each actuality requires abstract character. The Universe is dual because each occasion unites its formal immediacy with objective otherness. The Universe is *many* because it is wholly and completely to be analysed into many final actualities—or in Cartesian language, into many *res verae*. The Universe is *one*, because of the universal immanence. There is thus a dualism in this contrast between the unity and multiplicity. Throughout the Universe there reigns the union of opposities which is the ground of dualism.[63]

Notes and References

1. *Science and the Modern World* (New York: New American Library, 1948), p. 130.

2. We define a "system" as: a whole functioning as such by virtue of the relationships between its components (or parts) and their attributes; by "relationships" we mean those states or conditions which tie the system together in terms of the interdependence of its component parts.

3. As Margenau points out: "A body can be in the same place at different times, but it cannot be at the same time in different places. Here we find the essence of the one-way character of time, and in this respect time *is* different from space. Only by supposing that objects do occupy different places at the same time can this difference between time and space be removed." *The Nature of Physical Reality, A Philosophy of Modern Physics* (New York: McGraw-Hill, 1950), p. 160.

4. Raymond Aron, *Introduction to the Philosophy of History*, trans. by George J. Irwin (London: Weidenfeld and Nicolson, 1961), p. 39.

5. C. Lloyd Morgan, *Emergent Evolution* (New York: Henry Holt, 1923), pp. 111–112.

6. On the *Nature of Things,* Book V, 783–855.

7. *History of Animals (Historia Animalium),* Book VIII, 1.

8. Mortimer J. Adler, "Evolution," *A Syntopicon of the Great Books of the Western World* (Chicago: Encyclopaedia Britannica, 1952), Vol. I, p. 452.

9. *Concerning Human Understanding,* Book III, Chapter VI, 12.

10. *The Critique of Judgement.* Second Part, Second Division ("Dialectic of Teleological Judgement"), Appendix ("Theory of the Method of Applying the Teleological Judgement"), 80.

11. George Gaylord Simpson, *The Meaning of Evolution* (New York: New American Library, 1955), p. 127.

12. *Life and Letters of Charles Darwin,* (ed. by Francis Darwin, Appleton, 1887), Vol. **I**, p. 316 (July 3, 1881).

13. *Ibid.,* Vol. **III**, p. 237 (December 26, 1879).

14. Gertrude Himmelfarb, "Varieties of Social Darwinism," Historical Critique in T. W. Wallbank, Alastair M. Taylor, Mark Mancall, *Civilization: Past and Present* (Chicago: Scott, Foresman, 1969), Vol. **II**, p. 397.

15. Francis Galton, *Inquiries into Human Faculty and Its Development* (J. M. Dent, Everyman, n.d.), Vol. **I**, p. 218.

16. *The Origin of Species,* Chapter VI ("Difficulties of the Theory").

17. Chalmers Johnson, *Revolutionary Change* (Boston: Little, Brown, 1966), p. 3.

18. *Loc. cit.*

19. *Leviathan,* Part I, Chapter 13.

20. On the subject of the relationship between the integration and disintegration of social systems, Johnson quotes Talcott Parsons: "The maintenance of any existing system, insofar as it is maintained at all, is clearly a relatively contingent matter. The obverse of the analysis of the mechanisms by which it is maintained is the analysis of the forces which tend to alter it. *It is impossible to study one without the other.* A fundamental potentiality of instability, an endemic possibility of change, is inherent in this approach to the analysis of social systems. Empirically, of course, the degree of instability, and hence the likelihood of actual change, will vary both with the character of the social system and of the situation in which it is placed. But in principle, propositions about the factors making for maintenance of the system are at the same time propositions about those making for change. The difference is only one of concrete descriptive analysis. There is no difference on the analytical level." Talcott Parsons and Edward A. Shils, *Toward a General Theory of Action: Theoretical Foundations for the Social Sciences* (New York: Harper Torchbooks, paperbound edition, 1962), p. 231. Emphasis in the original.

21. See Lynn Thorndike, "Renaissance or Prenaissance?" *Journal of the History of Ideas,* IV (1943), pp. 65–66, 74; also Lewis W. Spitz, "The Renaissance: A Historical Controversy," Wallbank, Taylor, and Mancall, *op. cit.,* Vol. **I**, pp. 477–479.

22. Christian Jurgensen Thomsen, *Ledetraad til Nordisk Oldkyndighed* (Copenhagen, 1836); Thomsen further developed his ideas in his *Skandinaviska Nordens Urinvânare* (1838–43).

23. For an account of these developments, see Glyn E. Daniel, *The Three Ages, An Essay on Archaeological Method* (Cambridge: University Press, 1943).

24. Quoted in Daniel, *op. cit.,* p. 45; for a fuller exposition of Childe's views, see his *Social Evolution* (London: Fontana Library, 1963), Chapter II, "The Classification of Societies in Archaeology."

25. Bertrand Russell, *History of Western Philosophy* (London: George Allen and Unwin, 1947), pp. 815–816.

26. *From Max Weber: Essays in Sociology,* H. H. Gerth and C. Wright Mills, trans. (New York: Oxford Galaxy Book, 1958), p. 78. Emphasis in the original.

27. Johnson, *op. cit.,* pp. 30–31.

28. Hannah Arendt contends that "No revolution ever succeeded, [and] few rebellions ever started, so long as the authority of the body politic was truly intact." *On Revolution* (New York: Viking Press, 1963), p. 153.

29. Mortimer Adler, "Revolution," *A Syntopicon of the Great Books of the Western World* (Chicago: Encyclopaedia Britannica, 1952), Vol. **II**, pp. 627–628.

30. Even as in colonial rebellions, the metropolitan authority employs *force majeure* in the form of a "police action", as sanctioned by the "principle of domestic jurisdiction," to restore the *status quo ante,* whereas the insurgents take up arms in a "military action" for the avowed purpose of implementing the "principle of self-determination," and thereby create a new sovereign state.

31. When *force majeure* is employed within existing systemic boundaries the upheaval can be designated "revolution"; when it is used *across* boundaries it becomes an inter-systemic conflict to which we apply the term "war." In either situation, re-equilibration is being attempted.

32. Crane Brinton, *The Anatomy of Revolution* (New York: Vintage Books, revised and expanded edition, 1965), p. 24.

33. *Loc. cit.*

34. *Ibid.,* p. 269.

35. *Ibid.,* p. 25.

36. *The Wealth of Nations,* Book III, chapter 4.

37. *The Stages of Economic Growth* (Cambridge: University Press, 1960), chapter 2.

38. "(1) a rise in the rate of productive investment from, say, 5% or less to over 10% of national income (or net national product);

(2) the development of one or more substantial manufacturing sectors, with a high rate of growth;

(3) the existence of quick emergence of a political, social and institutional framework which exploits the impulses to expansion in the modern sector and the potential external economy effects of the take-off and gives to growth an on-going character." *Ibid.,* p. 39.

39. *Ibid.,* p. 11.

40. For a detailed analysis, for example, of the dangers inherent in an uncritical application of Western economic theorics, models, and value judgments to the problems of poverty and development in South and Southeast Asia, see Gunnar Myrdal, *Asian Drama: An Inquiry into the Poverty of Nations.* 3 volumes (Penguin Books: 1968).

41. See Table I ("Some tentative approximate take-off dates"), *op. cit.,* p. 38.

42. *Ibid.,* p 61; this period is equated with Britain's "take-off period" to which Rostow assigns the dates 1783-1802.

43. "What is usually treated as the technological backwardness of the six

centuries before the so-called Industrial Revolution represents in fact a curious backwardness in historical scholarship. Significantly, the great technical advances of the eighteenth century took place in the earliest neolithic or chalcolithic industries, textiles, pottery, metallurgy, canal building." Lewis Mumford, *The Myth of the Machine, Technics and Human Development* (New York: Harcourt, Brace and World, 1967), caption under Plate 26.

For a short but useful historiographical treatment of this general subject, see Eric Lampard, *Industrial Revolution: Interpretations and Perspectives* (Washington: A.H.A. Service Center for Teachers of History, Pubn. No. 4, 1957).

44. For an interesting introduction to the implications of left-right symmetry and asymmetry for the physical and biological sciences, see Martin Gardner, *The Ambidextrous Universe: Left, Right, and the Fall of Parity* (New York. Basic Books, 1964).

45. Alfred North Whitehead, *Adventures of Ideas* (London: Penguin, 1948), p. 222.

46. For example, a squadron of fighter planes flying in formation over enemy territory functions as a system organized by communication signals; here the system's boundaries are non-material, complex, and in continuous flux; a breakdown in the communications system, accompanied by a loss or dispersal of the component aircraft, and the system ceases to exist.

47. J. D. Bernal, "A Geometrical Approach to the Structure of Liquids," *Nature,* No. 4655, 17 January, 1959, vol. **183**, p. 146.

48. Bernal, "Structure of Liquids," *Scientific American,* August 1960, p. 124.

49. Anatol Rapaport, "Foreword," *Modern Systems Research for the Behavioral Scientist* (Walter Buckley, editor, Chicago: Aldine Publishing Co., 1968), p. xix.

50. W. Ross Ashby, "Principles of Self-Organizing System," *Principles of Self-Organization* (H. Von Foerster and G. W. Zopf, Jr., editors, New York: Macmillan, 1962) p. 256.

51. R. W. Gerard, "Units and Concepts of Biology," *Science,* 125 (1957), pp. 429–433.

52. Robert Redfield, "Introduction," *Levels of Integration in Biological and Social Systems* (New York: Jacques Cattell Press, 1942), pp. 5-26.

53. In this connection see Arthur Koestler and J. R. Smythies (editors), *Beyond Reductionism: New Perspectives in the Life Sciences* (The Alpbach Symposium, New York: Macmillan, 1968).

54. The diagrams employed in this article are taken from the author's contribution to *Integrative Principles in Modern Thought* (Henry Margenau, editor, New York: Gordon and Breach, 1971).

55. That we speak of "equivalent" and not "identical" is in consonance with our discussion of "Social Darwinism" later in the essay.

56. For a more detailed discussion of the national state system from the standpoint of the application of the principle of integrative levels, see the author's "Toward a Field Theory of International Relations," *General Semantics Bulletin,* No. 35, 1968, pp. 9–43.

57. The previously cited article (see footnote 56) deals at length with the

respective capabilities of the United Nations system and emerging regional economic "communities" to advance from an L_{10} to an L_{11} stage of organization and control in the political, juridical, and military spheres.

58. Conversely, of course, a thermonuclear holocaust, if it did not wipe out mankind, might well quantize human society from L_{10-11} back to L_{7-8}.

59. See "The Evolutionary Vision," *Evolution after Darwin* (Sol Tax and Charles Callender, editors, Chicago, 1960), Vol. **III**, p. 251.

60. Which helps explain in turn how it was possible for Newton and, subsequently Einstein, to provide their mathematically elegant models of celestial mechanics—concentrating on the most fundamental levels of phenomenal organization—prior to the construction in 1954 by Crick and Watson of their working model of the DNA molecule which is of course organized at a more complex level. Elementary extrapolation indicates that the social scientists will require the computer and many additional quantifying tools before they can begin to emulate the physicist's or biologist's control and predictive capabilities in the decades—or possibly centuries—ahead.

61. To take one example: what should be the appropriate role for the United Nations once we accept the invariant presence of change in the phenomenal world? Specifically, to what extent do the Charter and activities of that Organization call for the application of deviation-reducing, as opposed to deviation-amplifying, technics when the United Nations addresses itself to national state system—or even the system itself. Conversely, negative-feedback can render the Organization open to the charge of being "revolutionary" because it disturbs a *status quo* favorable to one or more members of the national state system—or even the system itself. Conversely, negative feedback action can be employed by Member states to use the United Nations to reinforce the *status quo*—and this process in turn can raise the charge of "counter-revolutionary." In our view, both forms of feedback are required if the United Nations is to be empowered both to reduce inter-national conflict (at L_{10}) while at the same time raising global standards of living and dignity (in consonance with the requirements of L_{11}). General systems theory provides both a fresh conceptual framework and new analytical tools for the study of societal attitudes and behavioral patterns in the international political environment.

62. R. G. Collingwood, *The Idea of History* (Oxford: Clarenden Press, 1949), p. 332.

63. Alfred North Whitehead, *Adventures of Ideas* (London: Penguin, 1948), p 222.

COMMUNICATION
LEE THAYER
The University of Iowa

ONE OF THE occupational diseases of the intellectual is a sharply heightened belief that he is going to win at the "explanation game." The symptoms are a faltering sense of humor, the assumption of "objectivity," and a sense of destiny.

This is but a special case of the more general occupational disease of everyday life: that of knowing-why.[1] The essential difference between Marx and the man-on-the-street, who talks himself into a plausible theory of why his favorite football team lost last week's game, is but one of reach. The non-intellectual limits himself, typically, to current events of personal interest. The intellectual, if his fever is running high, is seized to explain not only those events of his personal interest, but all of history for the sake of all mankind.[2] If his fever is running high, he plays the game as if he wanted to "win."

I stand directly with Abraham Heschel on this. He writes, "... the truth of a theory about man is either creative or irrelevant, but never merely descriptive." This is based on the deceptively simple proposition that the nature of man and of men depends upon the images of themselves they adopt. "We become what we think of ourselves."[3]

An explanation or a theory about past events may add to or subtract from the way we think of ourselves. It may even enrich our thinking about the images others have fashioned themselves by. Above all, it may aid us in fixing those arbitrary distinctions between those who are our friends and those who are our enemies, between those who think right (like us) and those who don't. But it never merely describes what happened.

What is clear is that events do not explain themselves. What is equally clear is that the explanation of events is a creative human act. What is not so clear is how, not the past, but our creative

141

explanations of the past affect, not the future, but our creative conceptions of the future. The nexus is, of course, the present. And the present, of course, is constructed by those who have some image of themselves as being "in" it.

The problem, therefore, is not how to account for the past or for the future, but for the present. Further, the problem is not that of predicting the future, but of accounting for those conditions by which men are free, or are not free, to fashion themselves in the present and hence to invent their futures.

What we need to concern ourselves with initially, therefore, is perhaps neither evolution nor revolution, but devolution. But then words are catchy, and one always risks winning some minor prize in the "explanation game" by catching words. Nonetheless, I will try to show how this sidestep may be useful.

All I can claim for this essay is that it is the only attempt I know of to talk about the conditions of man's present devolution solely from the point of view of communication. I do so not opportunistically, but because it is the only frame of reference I feel competent to bring to bear. I claim no competence with respect to the many controversies that seem to persist in the philosophy of history.[4] Nor can I claim to be able to bring to bear all or, in all likelihood, even a respectable proportion of the relevant arguments which exist in the literatures of the various-ologies : sociology, anthropology, ethnology, ecology, politicology, psychology, and so on.

I propose, rather, to look at these all-too-familiar issues from yet another point of view. Frankly, I see no logical or empirical way of avoiding this point of view when we want to observe human or social behavior. Yet I cannot be sure how high my own "explanatory" fever runs. I feel somewhat reassured by the fact that I undertake this task with considerable humility, and with the full confidence of my apparently immunized critics that the future of man will neither devolve nor revolve on what we say here.

"In the beginning was . . . the word . . ."

The fact that the concept of communication was not central to explanations of human nature and human societies in earlier times can be accounted for, in part, by the prior fact that it was not a concept which was popular with the intellectuals of earlier times. That communication is not presently central to explanations of human behavior and social evolution can be accounted for, in part, by the fact that it has recently been too popular with the non-intellectuals.[5]

I do not intend to be, nor do I intend to seem to be, *unduly* facetious about this condition. I will readily admit to the grotesque trivializations of the concept to which it has been subjected by intellectuals and non-intellectuals alike. I will also readily admit to the pretty esotericizations of the concept by many writers (e.g., words get translated into "symbols," conventional ways of understanding things into "symbol systems," their study into "semiotic," and everything that people have ever said to each other into a "noosphere"). I am not oblivious to the fact that the history of our abuse of the term has left it well-nigh useless for intellectual purposes.

Nonetheless, there are still vital phenomena of human and social behavior to be accounted for and, with all of its residual and rich connotations, I can find no better term in my lexicon for pointing to and indexing those phenomena.

So let not the word separate us in mind or in spirit.

COMMUNICATION AND THE CONSTRUCTION OF REALITY

I propose to begin by setting forth some basic propositions and, in brief, their rationale.[6]

1. In its most rudimentary form, that is, of some living system taking something-into-account to some end, communication is one of the two basic life processes. All living systems exhibit some capacity for being-communicated-with.

2. Throughout the phylogenetic scale, communication subserves the function of genetic actualization *vis-à-vis* the ecosystem and, in the case of those creatures that must capture their food, the function of threat and opportunity vigilance (i.e., of "survival").

3. Intercommunication—i.e., mutual regulation and control through language—is commonplace in aggregates of mobile and nomadic creatures, from plankton to bees to apes. What is unique about human communication is not that people talk to each other, but that they talk to themselves. Man's unique characteristic, from a communication point of view, is this biological capacity for self-reflexiveness, from which, exercised, he derives his "consciousness," and through which he constructs his reality, not only of himself and his environment, but of himself in his environment.

4. At the level of individual man, therefore, communication

subserves the additional function of enabling the creation, alteration, confirmation, or exploitation of some one or set of reality constructs.[7] Men have selective abilities and susceptibilities for being-communicated-with not only by the event-data of their environments through their particular reality constructs, but by themselves.

5. All ecosystems may ultimately be "closed" energically, but only man's ecosystem is "open" informationally, or communicatively. A lily must open to the sun as long as it lives. Not so, man. For men have construed the sun to have a wide and widening range of significance from benevolent god to malevolent devil to the indifferent fulcrum of all life.

As I hope to show, this difference is a significant difference. And it is one which renders questionable, if not irrelevant, the application of conventional "scientific" methods to the study of human and social behavior.

COMMUNICATION, INTERCOMMUNICATION, AND CONTROL

The relationship thus implied between communication and control can perhaps best be explored by looking more closely at the conditions and consequences of the "adaptive" function subserved by *communication* (i.e., the individual process of selectivity taking-something-into-account to some end), and then at the functions subserved by *intercommunication*.

1. There is a *necessary* relationship between the "take-into-account-abilities" (the capacities for being-communicated-with) of all "adapted" living systems and the conditions of the environments within which they "adapt" or "survive," or are genetically actualized. To the extent that a species had evolved perfect capacities for being-communicated-with by its environment, or to the extent that the species and its environment had evolved in perfect integration (i.e., to the extent there is a perfectly integrated ecosystem)—to that extent we would be justified in concluding that that species was also perfectly controlled by its environment.[8] To the extent that these capacities were imperfect or lacking, one would expect to observe in the one case incomplete genetic actualization and, in the other, either extinction or mutation. This is, indeed, what can be observed.

2. The same conditions and consequences would hold for individual man, were he not maintainable in "artificial" environments.

To the extent that a man were capable of perfectly construing the realities of the environments in which he must or would operate, we would be justified in concluding that he was being perfectly controlled by his environments. But the nature of the relationship for man is not a necessary one,* given that his ecosystem is informationally or communicatively "open." It is either a *sufficient* relationship, or it is not.

3. The more man becomes conscious of himself, the more his functional constructions of himself *vis-à-vis* his environments take on primacy in his articulation of himself into those environments. His adventure begins there, where he ventures forth, guided by his own ideas of the world, into the unforeseeable.

Man's loss of innocence was not, therefore, a matter of his groin, but of his consciousness. Once deprived of *necessary* communicative relationships with his environment, he was free to construe himself and his world in any way that seemed—at least expediently— sufficient. His consciousness gave him freedom from *necessary* control, freedom from *necessary* constraint. But it gave man also his dual nature : his capacity is his incapacity; his advantage, his disadvantage. For, to the extent that he has found his constructions of the things and events of his environment insufficient, he has always had two choices—to modify his constructions of reality, or to modify the environment to fit his existing constructions. Where his wisdom has failed him, he has reshaped the world in the image of his own *accidie*. Faced with the task of articulating himself into his environment via his own consciousness, the species *Homo* continues to survive through some delicate balance between controlling and being controlled in a precarious ecosystem which is energically "closed" but communicatively (informationally) "open." He must define himself as he goes for, as Condorcet noted, there is no other instance in nature where two self-conscious creatures confront one another. The issues involved take on their full significance when we consider the functions of intercommunication in human sociation.

4. The primary functions subserved by intercommunication in human sociation are those of normative regulation and control, and of the creation, alteration, maintenance, or exploitation of human

* That is, whatever relationship exists between the "realities" of man's environments and his ways of taking-them-into-account is necessarily *not* a necessary one.

institutions—which are, in turn, formal or informal "epistemic communities"[9] organized around one or a set of "communicational realities," or guiding metaphors.[10]

5. When intercommunication "works," i.e., when, in the very conventional sense, one person achieves his intent or purpose with respect to one or more others, we can posit as the "machine" which does this "work" an operating system in which the minimum components are an individual capable of producing a message, and one or more individuals capable of "taking-into-account" that message in a way which is sufficient to the goal or *raison d'être* of the system. The components are therefore interdependent and mutually defining, as they are in all systems. The components must be, in fact, for most social transactions, essentially *interchangeable*. There is no magic in "messages." Any utterance* which makes a difference requires at least two interdependent components (e.g., a "persuader" *and* a "persuadee"), and these components must be designed or equipped to be, if the system is to do the work for which it has been struck, functionally complementary with respect to that system.[11]

6. If people are to articulate one another into social organizations (institutions) of two or more persons, their enabling intercommunicative transactions must be minimally facilitative of the creation, maintenance, or exploitation of those institutions. That is, whatever else an intercommunicative encounter is "about," it must *at least* create or confirm or exploit the metaphor (belief, "reality," etc.) which makes that encounter possible. Therein lies the source of all social control, and hence of social devolution, evolution, and revolution.

SOCIAL CONTROL AND SOCIAL DEVOLUTION

To function within a social institution, whether that be a "meaningful" conversation, a marriage, a community, an economy, or a nation, one must be sufficiently in-formed. One must be made competent both to produce those cultural artifacts (utterances) and to take-into-account in some "acceptable" fashion those social artifacts (utterances) which are, first, vital to the maintenance of the established institutions in which he must, or would, participate. Second, one must be made competent to produce and to "consume"

---

* Which is, itself, a particular sort of cultural artifact.

those cultural artifacts (utterances) which might be useful to the exploitation, enhancement, or proliferation of those institutions.

One must also be sufficiently uninformed. He must be made relatively uncompetent either to produce or to take-into-account those cultural artifacts (utterances) which might, first, make possible the

One must also be sufficiently un-in-formed. One must be made relatively uncompetent to produce or consume those cultural artifacts which might enable the creation of conflicting institutions.

1. It is upon these conditions that the persistence, the integrity, and the viability of social institutions rests.

2. The in-forming of humans is, initially, a product of homogeneous communication channels. It is perhaps not coincidental that 80% of the child's basic structure in all cultures is set during that period when his in-forming communicative experiences are likely to be the most redundant and impoverished—in the period between birth and about 4–5 years of age.[12]

3. To the extent that some conflicting inhomogeneity in the child's in-forming communication channels occurs, there is the *power* with which traditional institutions (e.g., motherhood, the family, formal education, etc.) have been vested for legislating the consequences of venturing "too far" from the guiding metaphors of that society—i.e., of legislating the consequences for behaving or not behaving as if "properly" in-formed.

4. There are, third, the sanctions and dissanctions of the "properly" in-formed upholders of established institutions. Whether subtle or vicious, the reactionary behavior of others to one's improper conduct is rarely, if ever, carried out in the spirit of saving the jeopardized institution. It is rather a matter of personal jeopardy when one person's construction of reality is obliquely threatened by behavior which reveals (to him) the existence of a contradictory construction of reality. If the disconfirming behavior is blatant enough, and if the reality jeopardized is vital enough, it becomes a matter of personal continuity, of one's very identity.

5. For the sum total of one's "reality" constructions, of the "communicational realities" and metaphors by which he lives his meanest moments as well as dreams of his most glorious lifetimes— this is itself a living thing. It persists in its whole or in its parts only so long as it is adequately nourished through sufficient confirmation. This requires maintaining throughout one's life a pattern of communication channels with others which can be depended upon to be

at least mutually confirming of one's guiding metaphors, and to be sufficiently reliable. They must be depended upon to require neither an over-sufficient rate-of-change nor, at the outset, an over-sufficient redundancy. One lives, in the sum of his "reality" constructions, a precarious existence between the Charybdis of too little opportunity for growth and the Scylla of too much to handle. Too much uncertainty, too little confirmation, may jeopardize the health and the actualization of one's conscious life no less than does too much confirmation or redundancy, too little uncertainty.

6. It is trivializing to refer to these dynamic conditions of man's conscious life as mere "attitudes" or "beliefs" or "values." The more integral he is, the more these are his conscious lifeblood. They are reality processes. At the social level, reality processes are as vital to the life of society as a man's metabolism is to his physiologic existence. At the individual level, they are the nexus of conscious life. A man's capacities for, and his susceptibilities to, being-communicated-with must be structured in *some* way. The ways in which they do get structured establish, except in rare instances, the limits within which they can and must be confirmed, altered, or exploited, and within which new ways of knowing or seeing or feeling might be adopted or created. The thrust of conscious life for continuity within those precarious limits in the company of other men endows those institutions which men mutually create and perpetuate with an inertial force which necessarily transcends, once in motion, the deflective reach of individual man.[13]

7. Thus society devolves. The overlapping of generations necessitates the sufficient in-forming of one's successors in order to perpetuate those "communicational realities," those metaphors, by which one organizes a continuous conscious life. Social institutions, once infused into existence, must be perpetuated in order to maintain the continuous existence of their constituents. If these reality processes which make possible the continuous conscious existence of individual humans were over-sufficiently fragile or ephemeral, conscious life would quickly cease to exist. If they were over-sufficiently rigid or eternal, conscious life would gradually disappear. Man lives in the precarious margin between the two.

To treat either "too slow" or "too rapid" change in human institutions as merely a "problem" to be solved or an obstacle to be overcome greatly oversimplifies the rather more profound issues at stake.

THE "OPEN" SYSTEM AND DIVERSITY

Unlike his energic ecosystem, man's informational or communication ecosystem is "open." It is the relative closedness of the energic system which permits of ever more continuous refinement of its control. It awaits but further incisiveness of definition.

Contrarily, the more incisive are a man's definitions of himself and his multitudinous artifacts, the more his consciousness expands, increases, grows, enriches. This is in the nature of the difference between an "open" and a "closed" system. We have yet fully to recognize and, finally, to learn to cope with, that difference.

The tools by which man understands the nature of the energic system within which he lives permit him—indeed, impel him—to control it. The tools by which man comes to understand the nature of the communicational, the conscious world which he creates—those tools, potentially at least, recreate or complexify that world in infinite ways.

Were it possible to in-form one another perfectly so that all might be perfectly informed, then a single homogeneous society could be perfectly devolved. But perfect social control would not only be self-defeating; it is unlikely. It is unlikely because the very tools which enable man's in-forming and informing of himself and of others equally enable the diversity of imperfection.[14] Men are consciously different, communicatively diverse, not only because there are imperfections in their in-formation of one another, but because the conditions which enable man to control and be controlled socially are the very conditions that set *him* free. To talk to and to understand others, he must as well be able to talk to himself. His social world may be, temporarily, finite; but his communicative world, the product of individual consciousnesses, is, once enabled, infinite.

Man moves in three worlds simultaneously: the "closed" world of his energic ecosystem, the finite world of the temporary institutions through which he substantiates and exercises the products of his consciousness, and the infinite world of his consciousness. Rarely does man move boldly. And only then out of fantastic conviction that what he knows or believes is the ultimate end of knowing and believing. He moves boldly only when he is convinced that his third world is, after all, "closed," and that it was he who closed it, or that it was he who had finally discovered its outer limits. He has always been wrong.

But, boldly or timidly, he moves. Rarely does he refuse to move at all, for to do so he must "drop out" altogether. He cannot move certainly, but only with faith. And faiths rest on the most precarious artifacts of all man's inventions—his metaphors, his "communicational realities". Always imperfectly in-formed, men create and recreate their metaphors in isolated consciousnesses; and every metaphor exists as ten or a hundred or ten thousand or 200 million variants of those consciousnesses, each differing from the other in some ingenious or accidental way. And diversity provokes diversity.

Diversity is given in the nature of the conditions by which man comes to conceive of himself and of himself-in-social-relation-to-others. What reveals the stuff of society is not the presence but the absence of diversity, not heterogeneity but homogeneity, not many-mindedness but one-mindedness.[15] Were individual man totally free to construct his realities independently, there could be no society. What is revealed in the discrepancy between what diversity exists and what diversity *might* exist is the efficacy of social (or communicative) control.

"COMMUNICATIONAL REALITIES" AND THE ARTICULATION OF HUMAN SOCIETIES

Civilization rests upon the process of two or more men relating themselves to each other via one or a set of "communicational realities." A "communicational reality" is anything which has conscious existence for two or more humans because it can be and is talked about. The nature of its existence, its reality, inheres in the way or the form or the manner in which it can be and is talked about. The "I" who addresses you as well as the "you" I address: these are the basic communicational realities upon which all others are built. Those who cannot handle consciously the metaphors of an "I" or a "you" have no admission into human society. If one says to the other, "I love you," exactly what or who is the "I" or the "you"? They exist only in the conscious life of the humans involved and nowhere else. All other so-called "referents" for the terms by which we index our metaphors are but the creative metaphors of a third consciousness.

Good and bad, right and wrong, correct and incorrect, ultimately even *is* and *isn't*: these are the dimensions along which man's grandest communicational realities have been woven. Yet, no metaphor, no communicational reality by which a man conceives of himself and relates himself to his fellows, no matter how petty or how odd, can be looked upon as "mere" metaphor.

One man alone might articulate himself with his environment by the implacable "realities" of that environment. Short of a totally artificial environment, a farmer who wants wheat in the spring must plant it in the fall.

But two or more men must articulate themselves with each other and mutually into their institutions in another way. There is no "ultimate reality" behind their consciousnesses. The essential and vital realities of man's social environment are the constructions of his diverse consciousnesses. These are what I prefer to call *communicational realities* in order to keep clearly in mind that this is the only existence which guiding social realities have.

Yet they are no less "real." They simply serve a different function. The gestation period of winter wheat would be of no vital concern to a man who had no wish to produce it or otherwise to profit by it. A "reality" is significant only in the function which it subserves. Communicational realities subserve the functions of man's sociation. Can we much longer depreciate those realities which invent and energize human institutions while idealizing those which invent and energize but one of man's institutions—that of "science"? That the one reality seems less "real" to us than the other reveals, more than anything else, the irrationality of some of our most central, current metaphors.

Two or more men who articulate their behavior *vis-à-vis* one another on the basis of complementary prescriptions and proscriptions, these deduced from a metaphor in which they at least tacitly profess faith, comprise an institution. Institutions emerge as metasystems on the basis of which individual men may devise "appropriate" utterances and "appropriate" take-into-account-abilities, metasystems on the basis of which individual men may define the operational systems by which they have contact with one another and by which they may define their freedom of input and out-take. But always imperfectly. Imperfections which do not deny other guiding metaphors are met with irritation, impatience, disapproval, or avoidance. Those imperfections which cannot be made to articulate in men's minds are dealt with by disenfranchisement or by excommunication.[16] But if two or more individuals relate themselves on the basis of the "imperfection," a new institution is born, whose lifespan and whose scope can never be predicted at the outset. Thus institutions complexify and diversify, and new variants are born out of voluntary or involuntary excommunication. Thus society evolves at the level of men and institutions.

What is invariant is some complementarity of definition. Any communicational contact between or among humans which is ordered, i.e., which "works," is one in which the participants have brought to it or have invented in it complementary definitions of each other (i.e., complementary metaphors). These definitions are invariably related to the mutual definitions they bring to or give to the system itself. An institution is therefore a tacit conversation which persists.

The anticipated threat to the tolerable range of uncertainty-redundancy in communication channels may be taken to account for the fact that such complementary definitions—particularly with respect to the most important evidences of our underlying humanness, such as love and death—sooner or later become bureaucratized. What is sufficient becomes ritualized; what is prescribed becomes etiquette; what is proscribed becomes taboo or "sin;" and so on.

Social institutions persist so long as they serve to provide a sufficient relationship between their constituents and their total communicational environments and so long as they are articulable one with the other. As components of a system in a hierarchy of systems,[17] changes in one occasion an adjustment somewhere in the whole. The imperfections of consciousness and of in-formation ultimately account for the evolution of the whole (the comprehendable society of man) as well. Mutually noncontradictory institutions must be made to accommodate each other, for they must be articulated in the minds of those men who must or would cross their boundaries. A man who conceives of two institutions as mutually contradictory does not willingly cross their boundary. If by some third metaphor he feels compelled to do so, he does so violently, for one reality must be "wrong" if the other is "right."

And thus societies evolve, through the imperfections of men of the same mind, and through the contamination of men of other minds. But the hurt and the travail and the agonies (and even sometimes the violence) of change—these are not to be borne by man's institutions or his societies, for they are but senseless creations of his own consciousness. These he must bear himself as long as he would hold to the freedom of conscious existence, of life in the infinite third world.

For what constitutes at once both the precariousness and the tenacity of social institutions is that they provide man with the only confirmation he can have of the "reality" of his third world. Thus they must be not only mutually defined but, for these men who would realize themselves through two or more social institutions, they must be articulated (in the consciousnesses of those men) as defining each other in some complementary way. That a man cannot bring himself to be a practising Jew and a practising Muslim at the same time reveals very little about the differences between these two extended metaphors. What it reveals is the temporary (or perhaps permanent) incapacity of that man to adopt or to invent or to invoke a third metaphor that would articulate the other two.

It need not be answered why man lives by and for—or even dies for—his ideas, his communicational realities. For they are just that, the realities by which he can live and therefore must live. The third world of his existence in infinite. But he cannot live by all possibilities. He must live by those few realities which he is capable of articulating in his own consciousness. What needs answering is why some men do not or cannot live by a self-examined idea. What needs answering is why some men would abdicate their freedom of consciousness without putting it to the test. What needs answering is not why some men are different but why so many are the same. What needs answering is why the many seem to prefer the few to do their thinking for them, to live their consciousness for them. What needs answering is not why we have so little change or even why we have so much, but why so few people live on the leading edge of change, why so few are the actual source of human and social evolution.

Burckhardt and others have argued that even consciousness in the modern world is shifting from the individual to the group and that soon the accomplishment of even the most rudimentary social tasks will require a team, for consciousness will lie "in" the aggregate and not in the individual.[18] Perhaps! If so, what could we conclude? That somehow we have learned better how to institutionalize people than how to humanize them? Or, what's a society for?[19]

EVOLUTION AND REVOLUTION

This leaves us to account for "revolution."

Like "communication," perhaps too much has already been said about "revolution" for the term to retain its usefulness for intellectual purposes. Nonetheless, as with communication, there are certain phenomena still to be accounted for; and these perhaps can still best be indexed by the term "revolution."

There are, however, preliminary problems. There is first the problem of accounting for what occurs when there is an over-sufficient rate-of-change (a rate of communicational-reality diversification or variation which exceeds the metaphoric reach of individuals). Second, there is the problem of accounting for what occurs when there has been an over-sufficient redundancy in the communicational channels by which one supports his conscious existence.

I view the present "student revolution" in the U.S. as an instance of the former and the white "backlash" in southern U.S. as an instance of the latter. But these are perhaps over-reaching generalizations.

The point is that there are two rather distinct types of "counter-evolution."[20] There is, first, that which occurs when one epistemic community reacts against crossing over into the mainstream of the larger society because its constituents have insufficient metaphors for articulating the greater diversity of variation of the larger society, or because the *rate* of evolution in that larger society exceeds the *rate* at which those articulating metaphors can be invented and mutually defined. This is not a case of culture-lag, but of (communicational) reality-lag.[21]

Second, there is that "counter-evolution" which occurs when one or more epistemic communities react against the *encroachment* of the communicational realities of the larger society into their consciousnesses because some long period of redundancy in their communication channels (i.e., of homogeneous confirmation of the metaphors by which their institutions are defined and by which the members define themselves) has temporarily incapacitated them from articulating what is imposed from without with what is more familiar within.

Both are instances, albeit extreme, of "counter-evolution." But I believe that one should look for "revolution" only when he is satisfied that the phenomenon he observes is not an instance of "counter-evolution."

Even so, one is faced with the dilemma that counter-movements are invented and energized in the same way that movements are invented and energized (including the evolutionary "movement" of the larger society). And this is via some metaphor, some communicational reality, which has the faith of its subscribers that this is a reality by which they can and therefore should define themselves into contacts and relations with others. A new metaphor (or an old one revivified) acts like a catalyst for the emergence of diverse and variant epistemic communities, of social institutions. What is knowable is how this happens, not when it will or when it should happen.

There can, therefore, be no predictive "science" of social evolution.[22] For, as long as substantial numbers of people remain conscious of their freedom to translate the third world of their existences into communicable metaphors, human society will exhibit some of the openness of that third world. Further, science itself is based on a metaphor as precarious as many others, and to demonstrate that society is fully predictable would be to demonstrate that science is therefore irrelevant. Only a man who was free to create the metaphor for doing so could ultimately "understand" society. Yet, if there were but one more man who had the same freedom, there would be no *necessary* reason why he should agree. And so new metaphors would evolve out of the imperfect articulations of insufficiently or imperfectly in-formed knowers; and society would remain open, if only to that degree.

Revolution, then. In my view, revolution is ubiquitous. Revolution occurs whenever one person invents or construes a guiding metaphor to be mutually-contradictory and mutually-exclusive of the major guiding metaphors of the day, and then denies or subverts the source of any further articulating metaphor by "closing his mind" to the third world. Revolution occurs when one person, finding himself unable to articulate one guiding metaphor with another in his own consciousness, closes his third world in order to mount the hubris of a "right" or an "is" which he is convinced is not a metaphor but an absolute truth. Revolution*s* occur when the man or his passionate inarticulations enlist or otherwise gain followers, who attempt, in one way or another, to make true in fact that which is true by mutual commitment.

Revolution is typically bloodless and nonviolent, often even imperceptible. We ought not to assume that history consists *only* of those radicalizations of political order which one man's metaphor and a pack of "true believers" have attempted to bring off or have succeeded in bringing off in fact.[23] There are radicalizations of aesthetic order, radicalizations of spiritual order, radicalizations of the order of knowledge—radicalizations of the order of guiding metaphors in every sphere of social life. And there are always radicalizations of the order of individual life. Without these, society would certainly become fixed, closed, changeless,[24] for they germinate and energize every movement, whether those movements consequent in revolutions or not.

A revolution in fact is a man trying to force the world to fit his idealistic metaphors of it, rather than the other way round. He denies himself in order to infuse society with a metaphor which blocks his own growth.

What a painful paradox, indeed, is the sight of revolutionary man, who must refuse to believe in himself in order to believe in his metaphor. More painful still the paradox of the counter-revolutionary man, who must resist the evolution of society because he has neither the courage nor the capacity to grow himself.

SYSTEM CRITERIA

But we cannot be done with it at this point. It is not enough merely to describe. The final question, when the study is of man, is not, What is going on? but where do we stand with respect to the criteria by which we are going to understand and evaluate the consequences of what is going on? That is, what are the system criteria?

Heschel has written, "is it not conceivable that our entire civilization is built upon a misinterpretation of man? Or that the tragedy of modern man is due to the fact that he is a being who forgot the question: Who is man?"[25] He is asking the question of system criteria.

Given that certain things have happened and are happening; given that conditions will likely lead to more of the same; given these, where do we stand? Do we concern ourselves that human society comes out the way it should? Do we concern ourselves that humans evolve in the direction they *could*, if society were as it should be? Or is it enough simply to "do" science the way *it* is "supposed" to be done? Is it enough, when the study is of man, to assume that what we see is what must be?

There can be little doubt that the guiding metaphor of science is one of the most powerful metaphors ever invented by man. But have we something other than science itself to demonstrate that "it" is the ultimate and final metaphor? In a typically excellent essay, Louis J. Halle has written, "I know of no natural law which says that the evolution of the universe must be toward a humanly desirable end, thereby warranting the inquirer in equating truth with his own desires."[26] Quite so! And we need reminding of this from time to time. Yet one's system criteria cannot be assumed away, as those who would counterfeit science assume by a posture of "objectivity." I know of no *natural* law which says that the evolution of the universe must *not* be toward some humanly desirable end.

There are those zealots, as there are following any great metaphor, who press the guiding metaphors of science beyond their relevant limits. It is all too easy to forget that the concept of an evolving universe is *also* a metaphor created by humans. A metaphor which denies its maker is, if so pressed, a dangerous metaphor.[27]

Even assuming that the future is not going to produce a humanly desirable end, mankind faces an immediate and present danger in those who presume to see only "what is there" because they have a powerful new methodology for doing so which is denied to the imperfectly, or the insufficiently, in-formed.

I mean no muddleheaded sentimentalism or "humanism." Scientific descriptions of man without accompanying system criteria are either dangerous or irrelevant.[28] There is no *necessary* connection between the way men conceive of themselves and their ultimate "nature," whatever that unknowable condition might be. There is no *necessary* set of institutions by which man *must* live or die, including science. The criterion which is given in a description of what *is* is no *necessary* description, whatever its guiding metaphor.

What is needed is evolutionary man, men who are both able and willing to think for themselves as well as on-behalf-of the institutions to which they are in-formed. Muller writes, "The only possible virtue in being a civilized man instead of a barbarian, an ignoramus, or a moron is in being a free, responsible individual with a mind of one's own . . . The best society, accordingly, is that which is most conducive to the growth of such persons."[29]

That's putting one's system criteria on the line.

I agree.

Notes and References

1. I think this is no necessary contradiction of Bertrand Russell's comment that most people would rather die than think—and in fact usually do so. He

meant, we can assume, a particular kind of thinking. That people will talk to themselves and to each other until they arrive at an acceptable explanation of why something did (or did not) happen the way it did is, indeed, the source of all human sociation. That they sometimes do not thereby arrive at mutually acceptable explanations is, by contrast, the occasion for many a black eye or bloody war. *Cf.* Lewis A. Coser, *The Functions of Social Conflict* (New York: Free Press, 1956).

2. It was Voltaire who said that history is nothing but a bag of tricks played on the dead, and Hegel who said that what history teaches us is that we haven't learned anything from it.

3. *Who Is Man?* (Stanford University Press, 1965), p. 8. *Cf.* Fred J. Polak, *The Image of the Future. Vol. I: The Promised Land, Source of Living Culture* (Leyden: Sythoff, 1961), whose thesis is that it is our image of the future that guides our present. Goethe called these images "illusions," and saw them, as I see them here by a different name, as the very foundation of social conduct (in *Fragment ueber die Natur*).

4. I felt particularly enlightened, rightly or wrongly, by the following: Herbert J. Muller, *The Uses of the Past* (London: Oxford University Press, 1952); David H. Fischer, *Historians' Fallacies* (New York: Harper & Row, 1970); Robert A. Nisbet, *Social Change and History* (London: Oxford University Press, 1969); Charles Van Doren, *The Idea of Progress* (New York: Praeger, 1967); Patrick Gardiner (ed.), *Theories of History* (New York: Free Press, 1959); Karl R. Popper, *The Poverty of Historicism* (New York: Harper Torchbooks, 1964); and Frederick J. Teggart, *Theory and Processes of History* (rev. ed., Berkeley: Unversity of California Press, 1949).

5. One cannot but wonder how different our grandest interpretations of humanity might have been had it not been for the unpopularity of a term with the intellectual elite, or its popularity with the non-intellectuals—or vice versa. Hundreds of thousands of dollars and untold man-hours have undoubtedly gone into the study of the learning of nonsense syllables, which learning, as far as I know, is not a very popular activity amongst non-intellectuals. But by contrast, there have been no more than occasional studies of how people *actually* learn to care for someone or to hate or to fall in love with or fall out of love with someone, all of which are, in my observation, rather commonplace events of everyday life. Intellectuals seem to be sceptical or disdainful of those concepts which might be so commonplace as to be of concern to non-intellectuals, and to grasp enthusiastically and unblinkingly at those which, if nothing else, serve to keep the dividing line between the two "classes" clear and present.

Communication, being something which everyone *does*, is therefore a most unlikely candidate for admission to intellectual jargon. But Information Theory (for example), which has very little relevance for human communication, became a part of the intellectual jargon almost overnight. The minds of intellectuals seem often to be turned on or turned off depending upon the status of the jargon being used. What does one do with his baggage of "social structures" and "social functions" and "historical cycles" and "preconditions for revolution" when a self-styled and influential revolutionary literally screams in our ears that he wants to make revolution "just for the hell of it"? How can one fashion a "theory" of revolution on the principle of some occasional screwball making revolution just-for-the-hell-of-it? Or that societies evolve out of the process of people just talking to each other?

6. This basic framework is set forth more fully in my "Communication— *Sine qua non* of the Behavioral Sciences," in D. L. Arm (ed.), *Vistas in Science* (University of New Mexico Press, 1968), pp. 48–77. I have not dealt with the technologies of "communication" in this essay for two reasons. First, our understanding of these technologies follow from the exigencies of private or common metaphors; it does not precede them. Second, except at the level of techniques (e.g., of thinking), what are referred to as "communication" technologies are rather technics for the acquisition, generation, distribution, storage, etc., of *event-data*. There is no piece of hardware which, e.g., "transmits" the stuff of consciousness. The "information" of consciousness is of a different order of reality from the "information" of, say, Information Theory. There are technologies for the latter, but not even the "hardware" of the human sensors is capable of accepting anything but "pure"—i.e., informationless, in the human sense—event-data. A human must be the *de facto* creator of what comes into his consciousness—i.e., that which *he* can articulate.

Before McLuhan's restatement of Harold Innis's thesis got vulgarized by the world's word-merchants and knowledge-brokers, what was clear was that all technologies may change patterns of human behavior and that communication technologies might also alter the human himself. It was certainly no news that a medium (e.g., a messenger) might in certain cases be, for someone, a sufficient message. But no medium is a message for all people in all circumstances any more than a given message is a message for all people in all circumstances.

What the popularizers did not and seem not yet to understand is that it is not our technologies which affect the *substance* of our communication, but the metaphors by which we talk about them.

7. A "classic" treatment of reality constructs is George A. Kelly, *A Theory of Personality* (New York: Norton, 1963); *cf.* the chapter "On Being Communicated-With," in my *Communication and Communication Systems* (Homewood, Ill.: Irwin, 1968).

8. I have dealt more thoroughly with this relationship between communication and control in my "Communication and Change: Some Provocations," *Systematics*, 1968, 6 (3), pp. 190–200.

9. This term is Burkart Holzner's, in *Reality Construction in Society* (Cambridge, Mass.: Schenkman, 1968).

10. Cf. Elizabeth Sewell, *The Human Metaphor* (University of Notre Dame Press, 1964). It is, similarly, Louis J. Halle's thesis (in *The Society of Man*, New York, Delta Books, 1969) that "the nominal is always more real to us than the real."

11. I hope it is clear that this concept of intercommunication differs significantly from the conventional (and even popularly "scientific") notion of communication as being something which someone *does* to someone else— which is implicit in all variations of the ubiquitous model, $A \rightarrow B = X$ (i.e., A communicates something ($\rightarrow$) to B with X result), a linear, unidirectional, algebraic model which is undemonstrable by any critical empirical test.

12. This is no more than an estimate, of course. But it is widely and often independently proffered by psychiatrists (McCulloch), child development specialists (Winnicott), learning theorists (Piaget), and others.

13. Walter Bagehot (in *Physics and History,* Boston: Beacon Press, 1956), has put this side of the matter eloquently: "The great difficulty which history records is not that of the first step, but that of the second step. What is most evident is not the difficulty of getting a fixed law, but of getting out of a fixed law; not of cementing, but of breaking the cake of custom; not of making the first preservative habit, but of breaking through it, and reaching something better" (p. 39). *Cf.* Sören Kierkegaard (in *Concluding Unscientific Postscript,* tr. D. F. Swenson & W. Lowrie, Princeton University Press, 1944): "There always lurks some such concern in a man, at the same time indolent and anxious, a wish to lay hold of something so really fixed that it can exclude all dialectics . . ." (Book I, chap, 2, p. 35).

14. This may *sound* like Spencerian "social Darwinism," but it is not; it is the *absence* of inherent constraints on man's consciousness, and not their presence, which characterizes human diversity. *Cf.* F. Dovring, "The Principle of Acceleration: A Non-Dialectical Theory of Progress," *Comparative Studies in Society and History,* 1969, 2 (4); and A. I. Hallowell, "The Protocultural Foundations of Human Adaptation," in S. L. Washburn (ed.), *Social Life of Early Man* (Chicago: Aldine, 1961, e.g., p. 253): "What should not be over-looked is the potential that exists for transcending what is learned—a capacity for innovation, creativity, reorganization, and change in sociocultural systems themselves."

15. In the "Alpbach Conversations" (*Beyond Reductionism:* *New Perspectives in the Life of Sciences,* ed. Arthur Koestler and J. R. Smythies, London: Hutchinson, 1969), Koestler reverts at least once to what he calls "axe-grinding" when he says, ". . . our main predicament is not individual aggression but individual devotion to totems, flags, creeds and causes; and that the egotism of the group, the aggression of the group feeds on the altruism of the individual and not on the aggression of the individual" (p. 353). What I have suggested here is quite in agreement with the *direction* from which we ought to be looking at this predicament. But I would argue that what the hybris or the egotism of the group feeds on is not individual altruism, but individual denial or abregation of the "openness" of the third world—i.e., the world of communicational realities. To the extent that men admit of this third world, they must live at the boundary of the synchronous and the diachronous (as suggested by Durkheim and Radcliffe-Brown).

16. John N. Bleibtreu, in his book, *The Parable of the Beast* (New York: Macmillan, 1968, p. 164), recognizes "the evolutionary role of excommunication as a stimulator of diversity" (*cf.* the notion of "schismogenesis"). For man, however, excommunication may be voluntary or involuntary. Those who would pursue their own "thing" apart from the mainstream (of art, of religion, of science, of whatever social institution) must either excommunicate themselves to some degree in order to persist in the singlemindedness of their purposes or face excommunication by those whose more conventional realities are being threatened. To "belong" to a social institution, one must subscribe minimally to its forming and energizing metaphors.

17. As variously discussed in L. L. Whyte, *et al.* (eds.), *Hierarchical Structures* (New York: American Elsevier, 1969). From this point of view, so-called "resistance" to change is altogether as functional and as necessary to the "evolution" of the system (e.g., the Yin and the Yang) as is "change." *Cf.* Konrad Lorenz, "A Talk with Konrad Lorenz," *The New York Times Magazine,* July 5, 1970.

18. See Jacob Burckhardt, *Force and Freedom: Reflections on History,* ed. J. H. Nichols (Boston: Beacon Press, 1964), e.g., p. 150. The issue is not the Rousseauian contention (in *The Social Contract*) that man has depraved himself while making himself sociable (or even Kant—of "unsocial sociability"), although Rousseau's suggestion that the complex mutual inter-dependence of civilization jeopardizes equality is not irrelevant. The issue is not how consciousness gets narrowed, but where in fact, as a phenomenon, it resides. At the same time, one cannot ignore those who have argued that consciousness is a social product (I think of Mead, whose argument I find most compatible). Even so, the question of what man does with it once he achieves it must be answered. Can it be answered by saying "it's in committee"? This whole Jacobin tradition (that the individual actualizes himself only by "losing" himself in the collectivity) is epitomized in the work of Teilhard de Chardin.

Cf. R. B. Fuller's proposition that over-specialization in species leads to extinction (in "An Operating Manual for Spaceship Earth," in Wm. R. Ewald, Jr., *Environment and Change,* Indiana University Press, 1968).

19. Some of the backgrounds for raising this question, a corollary of the question, what are people for? are brought together in my papers, "Communication and the Human Condition," *Proceedings* of the VII Semana de Estudios Sociales, *Mass Communication and Human Understanding,* Barcelona, Institute de Ciencias Sociales, 1970; and "On Human Communication and Social Development," prepared for the 1st World Conference on Social Communication for Development, Mexico City, March 1970, and reprinted in *Economies et Sociétés* (Paris), 1970. Heschel writes (*op. cit.,* p. 98): "A person is responsible for what he is, not only for what he does."

20. Wilbert Moore is careful to distinguish (in his systematic treatment of *Social Change,* Englewood Cliffs, N. J.: Prentice-Hall, 1963) between *change* and mere *activity.* There is much human bumbling about which, I suppose, ought to be classified as mere activity, though it seems to me impossible to distinguish the two except in retrospect. One bored student at a middle western university didn't have anything else to do, so he decided to go out and sell rocks. Was this mere activity or did it fit into some pattern of change? In any case, the notion of metaphor as the source of both stability and change would at least satisfy Talcott Parsons' oft-cited dictum that one process must account for both (as in *Societies: Evolutionary and Comparative Perspectives,* New York: Free Press, 1951).

21. Youth are typically counter-evolutionary (not revolutionary). They are more typically so in the more advanced or "open" societies, and they are more typically so when the level of their own affluence opens more personal choices to them for deciding on a life style. I think the problem is not so much what Erikson calls "identity crisis" as what Fromm has characterized as a "fear of freedom," with, in the case of the more affluent, democratic nations, a

measure of boredom. Too, we should not forget that it is youth who do things "just for the hell of it," and it is in one's youth where being out of fashion is not just a matter of embarrassment but of personal if not social catastrophe. Is it possible that youth have the same "identity crises" where their futures are literally given in the conditions of their birth?

Most so-called "revolutionary" youth of our time revert to old clichés, to metaphors out of the past. And they do so in an a-historical way. This appears to me generally to be better described as "passionatism" than as "idealism," in the same way that a spoiled child wants not to alter the world to a self-examined ideal, but just to see if he can get it the way he wants it.

Stephen Spender (in *The Year of the Young Rebels,* New York, 1969) would seem to disagree. Yet it may be that we sometimes attribute motives and attributes to our "young rebels" which we had hoped to have or to see in a society about which we cannot ourselves be fully optimistic.

Cf. also the excellent review by Walter Laqueur ("Reflections on Youth Movements," in *Commentary,* June 1969, 33–41), and the commentary by Edmund Stillman, "Before the Fall," *Horizon,* Autumn 1968, pp. 5–15. The issue may be neatly summed up in D. Gabor's conjecture (in *Inventing the Future,* New York: Knopf, 1964, p. 152) that the more permissive the society, the less permissive must be the education which makes the individual fit to live in it." But then perhaps any explanation is an over-explanation.

22.   This is neither a new nor a non-controversial position (see references in note 1, above, for detailed explorations of various views that have been set forth). Also the conception of science as being essentially "predictive" is a popular social science misconception. (But *Cf.* Kenneth E. Bock, "Evolution, Function, and Change," *American Sociological Review,* 1963, 28 229–37.)

It seems to me in no event to be a matter for the kind of pessimism expressed by Nicholas Berdyaev (in *The Meaning of History*): "None of the problems of any given historical epoch whatsoever have been solved; no aims attained; no hopes realized . . ."; or by Fourier (as quoted in Frank E. Manuel, *The Prophets of Paris,* Harvard University Press, 1962, p. 215): "Civilization is therefore a society that is contrary to nature, a reign of violence and cunning, and political science and morality, which have taken three thousand years to create this monstrosity, are sciences that are contrary to nature and worthy of profound contempt" (one is reminded of Quesnay); or by Vico or Oswald Spengler or Brooks Adams. But the Mandevilleian resolution (as in *The Fable of the Bees*) is hardly a satisfactory one. What I argue here is not that society is either doomed or saved by the fact that there can be no traditional "science" of social evolution (à la Calvinism), but that the condition of the openness of man's communicational ecosystem puts it out of reach of science as we know it today. *Cf.* R. Lichtman, "Indeterminacy in the Social Sciences," *Inquiry,* 1967, 10, 139–50.

Heschel (*op. cit.,* p. 174) says we should be ashamed: "We are better than our assertions, more intricate, more profound than our theories maintain. Our thinking is behind the times."

23.   Historians, for example, seem typically to assume that history is some acceptable explanation of large numbers of people involved in some essentially political, if not military, enterprise. This leads them to look for an explanation

of these "grand" events at the level of the abstraction and usually in terms of radicalizations only of the political order. (Why men's politics are considered more vital to man's ultimate well-being than his art or his business is more often assumed than explained.)

This raises certain problems of method. In his valuable inquiry into the "causes" of modern revolution, Crane Brinton (*The Anatomy of Revolution,* New York, Norton, 1938) isolates some (essentially) political similarities amongst the four revolutions studied with respect to their onset. This leaves me unsatisfied. I want to know as well: is there no instance in history when the same set of conditions obtained but when a revolution did not follow? And is there no instance in history when, in the absence of these same conditions, revolution did occur?

Hannah Arendt is, of course, more equivocal (in *On Revolution,* New York, Viking Press, 1963), and Eric Hoffer less so; *Cf.*: ". . . worthwhile revolutionary activity," writes Alex Comfort, "like education, the rehabilitation of criminals, or psychiatry is inherently non-directive . . ." (in *Nature and Human Nature,* London, Weidenfeld & Nicholson, 1966, p. 212). Another view on the present situation is Lewis S. Feuer, *Marx and the Intellectuals* (New York, Anchor Books, 1969), especially his essay on "Neo-Primitiviam: the New Marxism of the Alienated intellectuals."

Jean-François Revel, in his new book, *Ni Marx Ni Jésus* (Paris: Robert Laffont, 1970), offers a different kind of explanation when he suggests that Americans have now outlived the usefulness of the metaphors provided by Jesus and by Marx, and that a social revolution based upon some new or some overriding metaphor is now shaping up. The crucial role of the metaphors by which we live (and die) is also emphasized by Fred L. Polak (*op. cit.,* p. 70): "The death of the tragic spirit in our time foreshadows the death of our culture;" and, elsewhere (p. 233): "Those times which create positive images of the future, shall live forever; those times which end by destroying constructive images of the future, must irrevocably perish—having condemned themselves to death."

The casual equating of revolution with violence is, in my opinion, a most unfortunate mixing of metaphors. One might argue, with Georges Sorel (and many others) that violence is the "great purging, cleansing act in society." But with the ancient Greeks and even contemporary primitives, we might also recognize that it may be the vicarious experience and not the real one which cleanses best.

Finally, one could take the position—which very few historians seem to do —that revolution is simply the tangible (or intangible) growth pangs of filling in a life or social space which has already been opened up. The "hippies," for example, although the fad is now largely obsolete, had more of what they said they wanted than those who preceded them in history. Perhaps the "revolution" had already occurred, and what they represented was but one of several modes that people might follow in personally re-articulating a fractionated society.

All of these views should perhaps be compared and contrasted with Anthony F. C. Wallace's "Revitalization Movements" (*American Anthropologist,* 1956, 58, 264–81).

24. Except, of course, as genetic variations might account for such changes (*Cf.* C. D. Darlington, *The Evolution of Man and Society,* New York: Simon & Schuster, 1969). It seems to me, however, in spite of the beautiful metaphors of Darlington (and others) regarding the comparability of biological and social genetics, that men have demonstrated, through their institutions, the social capacity to neutralize the effects of genetic variation. *I.e.,* if people are genetically different, they must also be unequal in *some* sense. But our political and social communicational realities for at least the past 400 years have more and more effectively denied those inequalities. We seem to be following Condorcet ("Our hopes for the future condition of the human race can be subsumed under three important heads: the abolition of inequality between nations, the progress of equality within each nation, and the true perfection of mankind," in *Sketch for a Historical Picture of the Progress of the Human Mind*) rather than Comte (social progress is in the direction of inequality), Saint-Simon (the ideal society will be based on inequality), or de Tocqueville (who recognized that increased and increasing equality was the primary fact of history, but who questioned whether that is necessarily a better state of affairs for men). We should note, as well, W. H. Thorpe's contention that culture changes faster than gene pool (in *Science, Man and Morals,* Cornell University Press, 1965).

Between the seed and its flowering, there may be some *necessary* relationship. But the relationship between a man's genes and his ultimate conscious and social existence is at best but a *sufficient* one. I rest with T. H. Huxley when he says (in *Prolegomena to Evolution and Ethics,* 1893, reprinted by AMS Press, New York, 1970): ". . . the evolution of society is in fact a process of an essentially different kind from that which brings about the evolution of the species . . ." Even so, as Slijper suggests, this may be the very condition that will lead to the expiration of the species.

25. *Op. cit.,* pp. 5–6.

26. In "Ariadne's Thread," Books and the Arts, *The New Republic* (December 22, 1962), p. 20. The argument, it seems, is inexhaustible. Wrote Emmanuel Mounier (in *Personalism,* London, Routledge & Kegan Paul, 1952, and, I think, typifying the most opposed view): "Whoever argues from necessities of nature to the denial of human potentialities is either bowing down to a myth or trying to justify his own fatalism" (p. 6).

27. Herbert A. Simon titles his most recent little volume *The Sciences of the Artificial* (MIT Press, 1969). The metaphor runs amuck! It is the scientist who creates the "artificial" in his laboratory, not the fellow who makes love or asks forgiveness for doing so. When what the scientist studies (and creates in the form of *his* communicational realities) is *real* but, by mutual exclusion, what people think or do is "artificial," it's time to call the travel agency for tickets to the next crusade.

There are those who claim that the universe, including man's societies, are simply evolving, that to be "scientific," one can only observe the inevitable from his lofty detachment. But if a man of that faith were about to be run down in the street by a wild and careless driver, I do not think he would say to himself, "That automobile is evolving toward me; and I know of no natural law which says that what is going to happen should be desirable to me;

therefore I have no *rational* justification for moving out of the way." He would undoubtedly have rather clearheaded system criteria about certain men, including himself. It's only that abstraction, that slippery communicational reality—*man*—which he feels free to talk about without specifying system criteria.

28. Having abandoned once useful system criteria, science finds its goals increasingly unclear. "What, actually, would it *mean* if one understood the origin of the universe? And what would it *mean* if one had finally found the most fundamental of the fundamental particles? . . . the pursuit of an open-ended science . . . is . . . an endless, and ultimately tiresome succession of Chinese boxes," writes Gunther S. Stent, *The Coming of the Golden Age: A View of the End of Progress* (Garden City, New York: Natural History Press, 1969, p. 113).

If there is to be a "science" of man, it will have to find relevant goals or criteria for what *might* be or *should* be. "Our business as rational beings," writes Muller, "is not to argue for what is going to be but to strive for what ought to be . . ." (*op. cit.,* p. 70).

29. *Op. cit.,* p. 71 Muller is quick to acknowledge that the thesis of Karl Popper's book, *The Open Society and its Enemies* (London, 1945, New York: Harper & Row, 1964), is a "cogent argument" for the same view.

For Whitehead, adventure is the key to civilization and thus, for him, man's highest achievements will therefore evolve out of discontent.

THE DRAMA OF REALTIME COMPLEMENTARITY

EDMUND F. BYRNE

Indiana University, Indianapolis

WERE I TO exploit a cue that Alistair M. Taylor provides in his essay on evolution and revolution in society, I might begin by saying that my own remarks will be *complementary* to his. For, although those aspects of the human scene that I have chosen to emphasize remain peripheral to his expressed interests, our views are not, I think, incompatible, especially when interpreted as divergent estimates of systems boundaries. However, a discussion carried on between us at this level might well resemble a quarrel between movie cameramen about a light reading or lens aperture or, perhaps more to the point, about when to catch the panorama and when to zoom in for a close-up. I do not question his claim that evolution and revolution may be considered complementary models of societal change.[1] Nor would I seriously question the appropriateness of drawing upon general systems theory to develop a level-theory of societal change.[2] But were I to proceed along these lines I would merely contribute to the sort of inter-personal equilibrium that is painless for all concerned but clarifies nothing. This would be regrettable, since on a normative level our differences are such as might require one to re-examine his most cherished presuppositions about the academic or intellectual enterprise. On this level, it may be noted, our disagreement proceeds from different views about the procedural ground-rules, the scope and perhaps even the very goals of societal analysis.

In a word, though we are both participants in and both observers of the human scene, Professor Taylor chooses to write primarily as a disinterested observer, I rather as a concerned participant. From this latter perspective, it seems to me that his attempt at objective analysis of societal change is inadequate—not because he tries to go beyond an evolution/revolution dichotomy in the name of quanta, systems and complementarity, but because, while seeking so meticulously to legitimate his views on society by an appeal to theoretical physics, he allows himself little time to tell us what possible bearing any of this might have upon the existential, or historical, problematic of deciding between or committing oneself and/or others to evolution or revolution.[3] Instead, he almost seems to be saying that if we can just develop the right conceptual tools in time, we might yet be able to avert such obstacles to human progress as "world wars, gas chambers and genocide, etc." as well as "thermonuclear holocaust."[4] In particular, had he chosen to deal more explicitly with the sociological and ecological implications of societal change, he might not have been able to discuss evolution and revolution without acknowledging the worldwide problem of social dualism and marginalization : the almost systematic shunting aside of millions of people whose labor, and hence whose very existence, has been rendered industrially superfluous.[5] These lacunae, in turn, point to some overarching methodological questions about scientific objectivity : what it really entails and whether and to what extent it is appropriate in dealing with problems that concern us all as human beings.

I fully appreciate the importance of objectivity as an ideal, but am persuaded that it neither is nor can be nor ever should be any more than this, at least not when human beings are at stake. In particular, I am of the opinion that questions such as that of evolution and revolution on the level of human society are to a significant degree a function of internal and external power politics and that, as a result, they cannot be totally abstracted from those great questions of personal and public responsibility that the events of the twentieth century have taught us to regard as so excruciatingly realistic. Accordingly, in my judgment, when issues such as these are "on the floor," if one strives too strenuously for disinterested objectivity he runs the risk (an occupational hazard among theoreticians) of giving the impression that what is is indeed what ought to be, thereby at times transmuting the innumerable horrors of man's inhumanity to man into a patient Martian's research report on what is left of earthlings. On the other hand, to allow oneself to become embroiled in the particularities of human events, to quite blatantly succumb to making value judgments, to take committed stands on controversial issues before, as they say, all the data are in, is to put oneself at the mercy of whatever gods preside over the fickleness of historical inevitability. Admittedly, when faced with what may be called a micro-problem, it is often better to be safe than sorry; but when it comes to the macro-problem of human survival, no one is safe and few may live to be sorry.[6]

What is tacitly at issue here is whether anyone can really do anything significant about anything significant that happens. The ancient Stoics had one answer to this, bonzes burning in the streets of Saigon have had another, and revolutionaries of all kinds have had yet others. To be sure, all of them are eventually swept into oblivion by the plodding thoroughness of geological time; yet because of them, in spite of them and regardless of them, changes have occurred in human affairs. Some of these changes have indeed been so significant that, even before the Industrial Revolution had really taken hold in the world, a German philosopher was willing to surmise that the best-laid plans of mice and men may well gang awry but are coopted by "the cunning of Reason" to get it all together anyway.[7] Few today would allow that Georg Wilhelm Friedrich Hegel had thereby really succeeded in putting history in a basket. But, then, for all the subsequent efforts of a Darwin, a Marx and a Freud, no one else has succeeded very well either—at least not on the level of ideas.[8] On the level of action, on the other hand, there are signs that, in spite of the so-called "population bomb," *homo sapiens* may yet experience the elimination of many and the control of the rest by a very few. These few, these technocratic few, constituting at least the beginnings of an international and supranational power elite, will continue to be in conflict with one another to decide who shall control the controllers.[9] But, at least partly on the grounds that might makes right, they will be essentially in agreement with one another that they and they alone are capable of deciding what is good for all.[10]

Somewhat along these lines lie the differences—in method if not in motive—between my approach and that of Professor Taylor. To spell out these differences I will consider the question of societal change from three different, but interrelated, points of view. First, from the viewpoint of *social theoretics,* I recognize advantages to thinking about social change in terms of systems, levels, and complementary modes of transition, but also certain disadvantages that derive from the predominantly Western elitism that enables one to see "modernization" while remaining blind to "marginalization" (Part I: Thought and Reality). Secondly, from the viewpoint of *societal cybernetics,* I acknowledge the technological potential for regulating social change (especially through applications of systems theory, cybernetics and real time science), but am wary of the inherent dangers of technocracy (Part II: Science and Society). Thirdly, from the viewpoint of *intersocietal dialectics,* I contend that when societal cybernetics is allowed to proceed without criticism, goal orientation and even periodical redirection, there inevitably results a system imbalance which in its turn engenders frustration, desperation and rebellion on the part of those left out as "unfit" (Part III: Evolution and Revolution). Whence, in my judgment, the exceptional urgency in our times of restoring to theoretics its once vital function as societal critique.[11]

I.   THOUGHT AND REALITY

Lest this be misunderstood as representing some sort of anti-intellectualist manifesto, let it be noted at the outset that action, if it is to be truly human, must be goal-directed, and goal-directedness presupposes thought.[12] Accordingly, every society that seeks to survive as a society into the future must face the task of developing appropriate and adequate ways of thinking about and dealing with social change. This task is not always described or carried out in the same way from one age to another, from one society to another, or from one sub-society to another. But generally speaking, it calls for devising symbols or myths whereby the members of a society might learn to evaluate all social change against the primordial value of societal stability.[13]

It is quite widely assumed that one may characterize a given society as primitive or modern according to its commonly held attitudes toward social change. The primitive society, on this view, thrives on ideological justifications of the status quo, whereas the modern society is continually changing and hence is in need of justifying such change to itself.[14] These extremes of a continuum of attitudes toward social change might be thought of dichotomously as emphasizing either synchrony or diachrony; but they cannot be handily equated with such modern political tags as those of conservative or progressive (or radical). For as the common myth has it, all moderns of whatever political stripe accept and favor social change, disagreements being rather over the means, the rate and the scope of such change. As is implicit in Alvin Toffler's recent study of what he calls "future shock," there may not be as many moderns around as one is sometimes led to believe; or, to put it somewhat differently, there may well be something of the primitive in all of us.[15]

This element of primitivism in modern thought about social change is manifest in a form of theoretical bias that derives from attributing fully human significance to sub-human systems and that, accordingly, results in totemism. Interpreting totemism as a primitive form of objectivism the modern exacerbation of which is scientism, I shall contend that scientistic totemism, like any other, is metaphysically void if taken dogmatically, methodologically suspect even if taken only hypothetically, and notoriously ambiguous and indecisive if taken programmatically. Especially is this the case when it comes to theories of social change; for no matter how stable a social system, so-called, may appear to be at any given moment in history, it is a very delicate balance when considered from the perspective of geological time. Classical (in essence, Aristotelian) analysis of society tended to emphasize the stability of a society, and hence in its more contemporary dress tends to find some sort of systems theory to be quite congenial.[16] This static approach translates a substantialist bias into the assertion that there must be *something constant* that changes when there is social change. Hence, *ex hypothesi*, society is essentially stable : what change does occur is usually change *within* society, rarely change from one society to another. Some recent approaches to social change, being more dynamic or, perhaps better, dialectical in their bias, discredit stability as an arbitrary and a-historical claim, take conflict as a constant dimension of human interaction, and hence see their chief problem not as one of accounting for "deviance" and "disruption" but rather of accounting for social stability as such. A position which I take to be somewhat intermediate between these theoretical or at least methodological extremes is stage-theory, which seeks in the notion of stages (or levels, as in Taylor's theory) a way to account for radical transformations through time without sacrificing formal or even organic continuity.[17]

Given these major differences in theoretical posture, there is perhaps no notion in contemporary thought about social change that is more theory-laden than that of society. Often viewed as having developed out of and even as standing in opposition to more primitive, personal or organic sets of human relationships each of which is identified as a community (*Gemeinschaft* as opposed to *Gesellschaft*), a society (often considered to be co-extensive with a nation if not with a government or state) is thought of as a sophisticated, complex, impersonal, bureaucratic and industrially oriented set of positions or roles that persons happen to fill. Thus conceived, a society might be defined as a supra-communitarian, hierarchically differentiated set of positions or roles controlled by a government or state that corresponds at least approximately to the consensus demands of technological power.[18] This particular way of defining society is not in itself biased in favor of "modernization," or technological development, but writers who would subscribe to it seem generally persuaded that such modernization, however brought about, is at least inevitable if not intrinsically progressive.[19]

What is here suggested is that a theory, in its broadest (and etymological) sense is *a way of looking at* a certain set of recurrent events, data, phenomena, or whatever. In this vein, "evolution" and "revolution," especially as applied to social change, refer only indirectly to social change as such and directly to a theory about or interpretation of events, data, phenomena that give content to the concept of social change. As alternatives, one might, for example, prefer to conceptualize social change cyclically, dialectically, millenaristically, fatalistically, or in any number of other ways. If, however, one does choose to theorize about social change either in terms of "evolution" or in terms of "revolution," he is bound in by long-established usages that ascribe social continuity to the former, discontinuity to the latter.[20] Moreover, as a result of a long history of ideas about events and of events attributed to ideas, both "evolution" and "revolution" contain built-in evaluations that make them almost irremediably normative and even prescriptive with regard to what is good or evil for human society in general or, as is more often the case, for some concrete human society in particular.

Now, to go one step farther, what leads some people to appeal to evolution as the mechanism of social change while others appeal to revolution? Moreover, so as not to overlook the normative dimension of this question, why do some favor and support "evolution" while others favor and support "revolution"? Surely one of the relevant factors here, and perhaps the most decisive of all, is that of power-differential, i.e., the extent to which one does in fact control or at least share in control over the social system in question, and hence over one's destiny insofar as this latter is dependent upon that social system.

On the contrary, it might be objected, if the factor of power be given so central a role in one's theory of social change, the society that changes becomes little more than an *ad hoc* ("temporary" or "transitional") set of rules, roles and relations that obtain among a group of people who are forced by circumstances to live together and hence somehow to make the best of it.[21] But who are thought to be "living together" depends in part upon how narrowly one associates "together" with physical as opposed to psychological boundaries, geographical as opposed to sociological contiguity. As Professor Taylor has so graphically suggested, people's physical contiguity extends beyond "impleted" space in virtue of means of communication (including transportation) to include "expleted" space.[22] But human (as compared, say, to ant) society can also be analyzed in terms of psychosocial contiguity; and from this point of view the problem of societal boundaries (so important for a systems approach) becomes quite as difficult to solve in theory as it often is in practice. In either case, not infrequently, truth is but the offspring of power.

What all this means from the viewpoint of the powerless can for the moment be illustrated by a comparison between today's Third World and the North American Indians of the nineteenth century. Much was demanded of an Indian tribesman who was forced by events to think beyond intratribal or even intertribal struggles for power in such wise as to take into account white expansionism and "manifest destiny." Even under comparatively stable circumstances, societal maintenance requires a delicate balance between centripetal and centrifugal forces: the latter increases with pressure from within; the former with pressure from without. But as that external pressure becomes so great as to render societal maintenance from within impossible, outsiders who were heretofore peripheral to the old society now become the center of the new.[23]

This, in greatly expanded fashion, is what is happening on a massive scale all over the world today. Numerous pockets of these "decentralized" rejects are found in the unenviable role of desperate rebels struggling often quite literally for bare survival. Their so-called and self-styled revolutionary responses to being first discarded and then directly oppressed are, to be sure, occasionally from strength, but usually (at least under present conditions) from weakness and frustration. Hence the many and varied attempts on the part of various leaders of the Third World to unite scattered forces of opposition in order somehow to hold their own if not to catch up with the Great Powers. Whatever may come of all this in, say, the century ahead, it seems clear that neither for the supranational power elite nor for the devolutionary outcasts that it indirectly creates are present social systems considered sacred or even insurmountably stable. On this point, the history of the world just since 1945 provides enough instances of major social transformations to warrant some like expectations for the future.

On the assumption, then, that the dynamic view of society is verifiable, though not to the exclusion of the static view, I should now like to draw some corollaries from my general thesis that a society is a delicate balance.

First, it is impossible to establish a pattern for societal change that would be so definitively declarative of what is necessarily the case ("valid," perhaps, in Professor Taylor's terminology) as to transcend entirely the bias of one's own position in some historical society.[24] Whether one appeals to the Tao or to a social contract, to synchrony or to diachrony, to tradition or to progress, to natural selection or to a mandate given or taken back, to a synthesis of opposites or to complementarity, one is inevitably, inescapably and insurmountably drawing upon metaphor—however sophisticated its origin—quite literally to "make sense" out of the human drama in which each of us manages to have at least a walk-on part.

Secondly, when one does appeal to some such metaphor as revelatory of the pattern of social change, he may do so in any number of ways, most if not all of which can be co-present (whether manifest or latent) at the same time. Even assuming that the mode of making such appeal is verbal, i.e., linguistic in a narrow sense, this verbal appeal occurs in what might be called concentric circles of context. And depending upon which concentric circle(s) the critic or analyst chooses to consider or to ignore, and which to emphasize or to underplay, he might attribute such an appeal to some emotional state (e.g., fear), to some motive (e.g., desire for power, prestige or whatever), some interpersonal need (e.g., to humor one's guests), some taboo or ritual of one's clan, tribe or in-group (e.g., to appeal whenever possible to science for one's imagery, never to poetry), one's social class or stratum (e.g., "noblesse oblige"), or one's era in time (e.g., "the space age"). There is no one method that could handle all these concentric circles well, but anything from psychoanalysis to systems analysis to language analysis can be relevant or applicable.

Thirdly, the critic or analyst of such an appeal, who could, of course, be identical with the appealer, might want to explicate the latter's hierarchy of values, his set of practical priorities, his views about how best to justify—or rationalize—what he believes or wills to believe, and, finally, his opinion as to how well he has succeeded or can succeed at developing such a justification. With regard to views about social changes, one's view may be described as absolutist if, on the metalevel of criticism, it is considered to be beyond doubt or correction; hypothetical, relative or pragmatic if it is considered to be tentative, heuristic, exploratory, or the like.[25]

Fourthly, any such appeal, with or without justification and critique, is a social action, a performative, inasmuch as it is intended to have an influence upon the present and future behavior of others.[26] Thus, taking the example of Professor Taylor's essay, one can readily discern his intention to have us conceptualize about social change complementarily afer the fashion of general systems theory; and he goes an extra mile, so to speak, to provide us with suggestions about how we might in fact do this. But he does not tell us explicitly what concrete actions we might carry out on the basis of such complementarity-thinking.

One might like to say that on this level we are dealing with a "middle-of-the-road" approach to social change; but the message is not clear—not because of coding or static, but because the metaphor utilized becomes highly ambiguous when searched for clues to action. The subatomic world can be handled in terms of equations that allow for either a wave or a particle interpretation of matter. However valuable all of this might be to the nuclear physicist, what does it have to do with societal change? Does it encourage evolution or revolution in society? If both, as Taylor would apparently have it, then what are the practical consequences of such a position? Should one push for evolution today, revolution tomorrow—or vice versa? Does it permit one to say, let others revolt, evolution is good enough for us; or, on the other hand, let others wait for evolution; as for us, we revolt? In fact, it says nothing about any of these burning questions. Rather it seems to be saying that the important thing about social change is to have an adequate theory in terms of which to conceptualize it without falling back on mere analogies.[27] Further implied, it seems, is the not uncommon belief that for at least one segment of society—the theoretician—impartial and objective observation is the only appropriate kind of action to take with regard to social change. But this, in turn, is precisely to propose that one should do nothing, at least not if one has a similar station or role in society. Others, however, do take such observations and use them as they will.[28] So if the "observer" is not opposed to such third-party utilization of his observations, then he tacitly approves or at least tolerates it—in which case he is not entirely as impartial as his image would lead one to suppose.

Fifthly, it is important to know for which society or societies one's theory of change is a model. Some theories—especially those that appeal to evolution and/or revolution (including that of Professor Taylor)—are presented as being universal, that is, as applying to all societies of all times and places. Others are limited both spatially and temporally, e.g., to Western Europe during the nineteenth century or to Peru in 1969, and still others pertain only to a limited segment (or class or caste) of a given society over a limited period of time.[29]

Sixthly, the very notion of a society becomes fraught with ambiguity once it is asserted that a society can and does change levels. Professor Taylor has graphically shown us that the "impleted" and the "expleted" spaces of today's societies are far greater than were those of primitive societies. But in what sense does this imply that something identifiable as "society" has changed? What is the constant through time? It is not clear from Taylor's analysis why the supposed continuity from one level of organization to another level should be thought of as anything more than a useful literary device to provide a vastly oversimplified "over-view" of extremely complex and generally unpredictable interactions between different groups of people over space and time.

Seventhly, it still seems widely assumed that one can indeed talk meaningfully and even facilely about an isolated "society." This, in my opinion, is debatable. Taking the notion as referring to an organization of people geographically separated from other people who are organized among themselves, it is certainly not difficult to view one society as being distinctly different from and largely independent of another society. But once these societies enter representationally into contact with one another, the assumption that there are two societies rather than one begins to be something of an abstraction. And once their *interrelationship* becomes formalized and stabilized, the "twoness" borders on pure fiction.[30] This, however, is just what does happen as better means of communication, transportation and domination are developed.

As eighth and final point, one risks a rather immense distortion of the real state of affairs in the world today if he telescopes modern society into the narrow confines of today's technocrats and their toys.

It may indeed be the case, for example, that a sub-group of Americans has become supportive of or actively engaged in space exploration. But there is another, and much larger, sub-group of Americans who suffer from chronic malnutrition; and it is not easy to prove that the former sub-group does not owe something of its success to the ability of these others to endure deprivation. Moreover, there are large sub-groups of people in what we still call "other countries" who are starving while the particular technocratic masters who happen to control their fate capitalize on the natural resources (including cheap labor) available in their land.[31]

It is perhaps edifying to know that a power elite from various countries—shall we think of them as the "fittest" or simply as the "beautiful people"?—are moving on, perhaps "taking off," towards L_n.[32] But many human beings in the world today not only are not advancing to a higher level but are in fact retrogressing to a lower one. Primitive tribes, whether in North or South America; in Africa or Australia or wherever, once had at least the satisfaction of some control over their fate; but now those proud peoples of the past are, according to one euphemism, "self-employed"—the outcast slag of a technocracy that has no further need of their comparatively inefficient energy output. They may be found by the thousands, sometimes by the tens of thousands, clinging helplessly and hopelessly to one another, sometimes on the outskirts, sometimes in the decaying center of that megalopolis that appears so prominently on Professor Taylor's itinerary to the world ahead.[33] Not unlike the slaves who died while building the Pharaoh's pyramids, these lost millions are like the fecal matter left along the road on which the great evolving beast is passing on its way to "progress." Disorganized and disinherited from the wealth of their own lands, they are quite at the mercy of an imperialist power elite, be it, say, the Communist Party in Czechoslovakia, or American oilmen in Saudi Arabia or Venezuela.

Such exploitative arrangements, resulting in multinational and supranational centralization of control, have come to be a predominant characteristic of this century. It would seem, therefore, that an adequate assessment of societal change would at least make mention of it.

II.   SCIENCE AND SOCIETY

Be not misled by the note of gloom, if not of doom, that hangs over these remarks about societal control in the world today. Societal control is not necessarily in and of itself oppressive. That it has taken and continues to take this form is due, among other things, to the fact that only in recent times have science and technology begun developing the technical means to handle global (not to mention galactic) problems on a truly global scale. Thus, with the help of ever more subtle and sophisticated tools, once seemingly impossible projects are now being seriously planned and one day will undoubtedly be carried out: for example, to irrigate the desert regions of the earth with water from polar ice; to eliminate many if not all physical and mental disorders through genetic and chemical control; to provide precise, thorough and instant information on any subject to anyone who needs it anywhere in the world through tele-vision, video and microwave, lasers and ever more "ingenious" computers.[34]

With particular regard to computers, it now seems that we are even moving beyond the age of cybernetics (preprogrammed control or guidance systems) to an age of real-time control. Born out of the challenge to detect and destroy the multiple warheads of approach-ing ICBM's, real time science has grown up solving "on the scene" a great variety of problems that develop without warning under the ocean, out in space, or wherever. The new systems—both hardware and software—that are being utilized for continual *ad hoc* problem-solving may be viewed as the technological implementation of a pragmatist philosophy of action as well as the basis for a post-positivist philosophy of science. Science, on this view, can no longer be totally identified with the pre-established harmonies that are built into theoretical systems, since it also includes the highly complex, and now computer-assisted, search for answers at the very moment when the question arises.[35]

This problem-oriented conceptualization of science does not lend itself well to definitive formulations of what science says and still less of what science itself is all about. A discussion of the old distinctions between pure and applied science or between science and technology would be relevant, to be sure, but inconclusive. For what is at issue goes quite beyond the reach not only of verbal gymnastics but also of all classical philosophy of science.[36] This is due to (1) a new appreciation of the primacy of practice or, better, the continuity between theory and practice and (2) a new appreciation of the independence of the social sciences both from physicalist and from formalist pontificating.

With regard, first, to a theory/practice continuum, it is especially important to note that the new technology extends not merely man's arms, legs and senses but his brain, not just his capacity to do work but his capacity to think. There is, accordingly, ample reason to draw new models of thinking from the realm of the computer, as many have already done.[37] Secondly, the declaration of independence on the part of the social sciences is perhaps not yet strong enough to be called a movement but it is advanced enough to be taken quite seriously. The sometimes unduly anti-scientific stance of the existentialists has been drawn into a wider perspective of humanistic psychology.[38] The sociology of knowledge has provided us with at least the beginnings of a sociology of sciences loosely associated with studies of the relationship between science and society.[39] The ecological approach to environmental control has led a significant minority of scientists and engineers to move beyond narrow technical proficiency to a socially conscious awareness of their skills and of how they use them.[40]

In short, in these days of growing concern over such problems as environmental pollution, population explosion and thermonuclear annihilation, it is no longer considered quite so bizarre to maintain, as did one writer a few years back, that science is a sacred cow.[41] Nor is the progress of science any longer taken to be an obvious and unqualified blessing. From the vantage point of the social implications of science and of the corresponding social responsibility of scientists, all science, and perhaps especially physical science, has begun to be more widely recognized as a thoroughly human cooperative enterprise that introduces into the world changes that are often as threatening as they are profound and fundamental.[42] Adherents of the rigorous but narrow philosophy that science is just a consistent set of law-like statements contend mightily that man's role in science is, if more than incidental, then largely obstructive.[43] But men with a broader and inestimably more realistic outlook are now insisting that to be at all fruitful our analysis of science must see it as a system of interlocking components all aimed at preselected research goals, hence must relinquish uni-dimensional models in favor of a meticulous reconstruction of all the components that enter into that system as well as the links between them. This reconstruction, in turn, would show that the most important component of all is the prosthesis/cyborg : man-interfaced-with-machine.[44]

As a result of this modern transformation of science into a largely institutional enterprise, the individual who "goes into science" generally finds that he is little more than a functionary in a complex system of roles that are closely tied to the great engines of industry. His own personality and need for responsibility having thus been reduced to largely irrelevant appendages, it is even more of a shock for "the working scientist" to be told that his role in society at large is no longer looked upon as either obviously or necessarily valuable. Writers of science fiction, especially in the United States, continue to tell him what he wants and perhaps needs to believe about himself as scientist, namely, that he as an individual is or at least can be of crucial importance to the well-being of his people. But negative feedback from the world that Big Science has built is increasing so rapidly that many are now ready to presume the scientist guilty until proven innocent.[45]

In a context such as this, one might think it somewhat insensitive of Professor Taylor to have drawn up a theory of social change that dances to the tune of subatomic particles. But like most academicians today (here I must no doubt include myself), he has been conditioned to try to say something that will be of lasting and permanent value, what is called a "contribution" to a given discipline. And since, apparently, he does not find any bedrock of certitude in the social sciences,[46] he turns to what the high priests of the Age of Science have revealed to us from out of their holy of holies. That their formulas have literally nothing to do with human beings as such is of little moment; what matters is that their formulas are *sacred*. Thus, just as institutional theologians of every faith are constrained to show that their own reasoned opinions can somehow be traced to the appropriate holy writings, so we find Professor Taylor legitimizing his views about social change by a quasi-magical appeal to the inner sanctum of the physics laboratory.[47]

Like all other primitives, many moderns tend, sometimes in spite of themselves, to return to Mother Earth for foundations. Having thus involved themselves in a highly sophisticated form of totemism, they rely on this "concrete science" (as Claude Lévi-Strauss might interpret it) to structuralize the "two-confusing-for-words" details of everyday life in human society. But, as Lévi-Strauss has argued, this kind of thinking is not primitive in the sense of original but is in fact decadent. For instead of facing directly the socio-cultural conflicts out of which the delicate balance of a given societal "arrangement" has come into being, one instead traces that arrangement to sub-human—usually animal or vegetable—beings and then in time comes to think of the latter as somehow causally responsible for such arrangements.[48] In this way, one comes to regard a circumstantial structure as an inevitable consequence of natural laws; and, like the citizens of Bouville in J.-P. Sartre's *Nausea,* can hardly imagine that one might be directly, even if only partially, responsible for the maintenance of that system.[49]

To be sure, the totems of our day are no longer as simple as trees, or streams, or animals. Man's interests and needs have led him in the past few hundred years to concentrate his attention on the innards of inanimate objects. Much has been learned and much has thereupon been applied both for weal and for woe. With regard especially to the latter, we have found ourselves becoming more and more ingenious in our ability to destroy one another, with men doing the dirty work and women doing the cheering. As we move from catastrophe to catastrophe, we continue to bemoan the low yield of the so-called social sciences, make a plea for them to catch up with the natural sciences, and then burn another candle at the shrine of our favorite up-to-date totem, be it atom-smasher, television, computer, laser beam or whatever, thereby inducing in ourselves a false consciousness more elaborate than anything ever created out of primitive lore. And thus do we make ourselves oblivious of inhuman forces that are directly responsible either for establishing or for maintaining the delicate balance that we call a society.[50]

In short, on the level of theoretics, appeals to theories from physics are of limited value in dealing with the complexities of any human society. But on the level of *societal cybernetics,* unceasing development of man's experimental and technological capabilities constitutes an indispensable condition for human well-being on this planet. Setbacks along this line have been numerous; and disasters, including even total annihilation of all living species, are always possible as a result of misdirection or exploitation of this very technology. But setbacks and disasters just as staggering would inevitably result from somehow preventing any and every scientific discovery from having any effect upon society.

In effect, then, the very dangers we fear complement the benefits we anticipate to provide a kind of projection of the present-day parameters of global freedom.[51] For by virtue of knowledge attained and utilized, men steadily become ever more capable of controlling their own environment and one another as well; and in our own day it even becomes meaningful to envision an electronic elite controlling a worldwide society of the whole human race. Such a notion, perhaps, smacks of science fiction; but it is far closer to becoming a feasible option than the uninformed might dare to imagine. In this context evolution is now not only a process to observe and conceptualize but also, and in a far more immanent sense, a product of human ingenuity. In the language of Julian Huxley, evolution has become conscious of itself in and through men who know; even more than that, such men have become the principal engine of evolution on this planet. With the help of those marvelous machines —everything from the wheel to the electronic computer—that extend their organic information systems literally in all directions, they are planetizing Bacon's claim that knowledge is power, thereby bringing into being what Teilhard de Chardin called a "mega-synthesis" and what Marshall McLuhan thinks of as an electronic "global village."[52]

From this point of view, it becomes clear that, in spite of the lack of political power of the average "working scientist," the leading intellectuals, especially scientists and technologists, can have a great impact upon a society. Precisely to the extent that they assume their research findings to be disinterested, they are subject to being "coopted" for anything but disinterested purposes, most of which have to do with bringing still more wealth and power to those who already have more of both than most.[53] But just because it is to the intellectuals that power-brokers turn and they whom they reward so handsomely for their efforts, there is established in society a symbiotic relationship that saves the truth of Bacon's power-theory of knowledge : knowledge in and of itself is societally powerless; but between two otherwise balanced societal forces, superior knowledge will tip the balance in favor of its possessor.[54]

Until as recently as World War II, the knowledge that was most in demand tended, in general, to be knowledge that makes possible increased control over making things, so as to make them more efficiently, precisely, rapidly, economically, etc. Since World War II, however, the knowledge that is most in demand is nothing less than knowledge that makes possible increased control over the very process of knowing—gathering, interpreting and controlling the dissemination of knowledge, now understood as information—so as to know more efficiently, precisely, rapidly, economically, etc.[35]

Ideally speaking, control of media here means so regulating their use that one learns through them whatever one wants to know and keeps others from learning through them whatever one does not want them to know. The importance of thus controlling the media of communication has no doubt been recognized by power elites at least since the invention of language.[56] Thus, the same Plato who bemoaned the ignorance of those in the cave was also careful to indicate the need for strong censorship in that "ideal" society that could be governed only by a philosopher-king (*Republic*). On a broader scale, by means of everything from taboos to weapons of war, in-groups of whatever size and functions have sought to safeguard the secrecy of their arcane lore; and, inversely, by means of everything from fast runners to orbiting satellites they have sought to penetrate the secrecy of the out-group(s).[57] It has even been argued, and rendered at least plausible, that control over printing after the fifteenth century was a *sine qua non* condition for the establishment and maintenance of the modern nation-states of western Europe.[58] Be that as it may, few doubt that control of the press has long represented an important factor in societal maintenance. But all the rules of this game have been changing now that print communication is being superseded by such electronic media as radio, television, communication satellites, micro-waves, and the computer.

The so-called "underground" press still allows an alienated Left some means of expressing and to some extent of constituting a group identity; and the "overground" press, in general, helps the establishmentarian elements of society to maintain their posture as guardians of the public weal. But even politicians have come to realize that he who controls visual images on television can jolly well forget about symbolic images on the printed page.[59] Similarly, the industry that built a sizable empire out of the auditory inventions of Alexander Graham Bell now finds itself threatened by obsolescence unless it can muster up enough forces to prove its right to control over micro-wave communications—a right in itself as defensible as the right of carriage-makers to have controlled the automotive industry that threatened to end much of their usefulness.[60] But nowhere is the struggle for control more intense than in the area of that revolutionary extension of the brain itself, the computer. For by means of ever more sophisticated computers, men can not only process data with incredible ease and rapidity; they can also organize data into information and establish realtime control over as many systems or processes as imagination, ingenuity and the state of the art can simulate with model and machine. Whence the awesome impact of computer sciences, cybernetics and realtime science upon the whole process of transmitting information and, consequently, upon alignments of power insofar as these depend upon rapid access to reliable and appropriate information.[61]

As the restricted but highly significant example of the ongoing struggle for control of media suggests, it is well recognized by the power elites of today that now more than ever before societal control for the future depends upon control of technology. How the latter will work itself out in the years ahead can be only inadequately anticipated, as even futurists admit. But to the extent that the seeds of the future are indeed contained in the present, there is reason for believing that the human race is heading towards vast and unprecedented social upheaval. Why this seems likely remains to be discussed; but the discussion may be anticipated by two interconnected sets of observations about science and society.

In the first place, narrowly technical knowledge and skills do not of themselves qualify anyone to give advice on the social uses of science; yet on such matters almost no other qualifications have been sought by governmental decision-makers.[62] Some members of the technical community doubt that their advice carries any political weight, especially when they appeal to values or ask for priorities other than efficiency or profit.[63] Others feel that the impact of technology upon developed societies is already so great as to make traditional political structures and parties obsolete.[64] In either case, technicians are consulted about technology consistently enough to lead some writers to identify the resulting *modus operandi* as a technocracy.[65] However, due in large measure to a belated but growing realization that our technology may be leading us into social and ecological disaster, it may become somewhat more acceptable in high places to look to the social sciences and perhaps even the humanities for recommendations about life in a post-industrial if not leisure-oriented society.[66] This new willingness to search beyond the machine-men for clues to societal guidance is further precipitated by a somewhat startled realization that the old values that have held our society together are being discarded and that the hard sciences have little to offer by way of a replacement.[67]

However traumatic all this may seem from inside a technological society such as ours, finally, it is in many respects a phenomenon arising within and limited to a highly technologized society. For, as Herman Kahn has noted almost casually :[68]

> It seems quite likely that outside of the 20% of the world that is expected to live in postindustrial societies by the year 2000, the other 80% of humanity is likely to be deeply preoccupied with various kinds of reactions that resulted from the process of more or less forced Westernization and then withdrawal.

Poor grammatical tensing and matter-of-factness aside, this statement cannot be adequately handled within the limited confines of societal cybernetics, but calls for the broader context of intersocietal dialectics. And it is on this level that there appears, in its most dramatic form, the realtime drama of evolution and revolution.

III. EVOLUTION AND REVOLUTION

On the level of societal cybernetics, men have begun to assume control over the evolutionary process, especially by means of technology. But it is not all men who exercise such control. Indeed, an increasingly large segment of the human race has little or nothing to say about or even to do with this advanced edge of the evolutionary process. Quite the contrary, for every advance that gives some human beings still greater societal control, other human beings are disengaged, alienated, exploited and sometimes just brutally eliminated. Should they prove unwilling to play the externally imposed role of dregs along the evolutionary way, they thereby would actualize a polarity latent in any social system that is not responsive to the needs and claims of all affected human beings.

This being said, we are at once faced with an immense problem that has already been anticipated just by referring to a society as a delicate balance. As Hobbes, Machiavelli and other analysts of societal power relationships saw, a society tends to be what some are powerful enough to make others accept, and it remains that only so long as and to the extent that such acceptance, for whatever reasons, perdures. From this point of view, a given society might need to be viewed first, say, as a vector sum, then perhaps as a component of a vector. As internal forces shift, as new force is introduced either from within or from without, so does a given society change.[69] Something of this sort occurs where there is a transfer of the so-called seat of power from one political party to another, as a result of an election, a *coup,* or whatever. But a far more profound vectorial transformation results from the "taking over" of a given society by outsiders. For, in such a case, whatever may have been the internal organization of the newly subordinated society, the set of all roles and relationships within that society tends to become a function of the set of those of the now dominant society. A tribe that has been colonized is no longer the tribe that existed before colonization; neither is a business taken over by a corporation nor is a corporation merged into a conglomerate the same; nor, finally, is a conquered nation the same as it was prior to having been conquered. Titles and perquisites may remain unchanged; but, in terms of available power, yesterday's rulers become today's errand boys.[70]

More generally, all those disenfranchised by the power-brokers of societal control and unable to accept "natural selection" as a sufficient explanation for their degradation have no alternative but to leave the territory now under the control of others or to regain control of that territory themselves.[71] Thus might those deemed or treated as "unfit" and hence "inferior" disturb or even prevent the expected survival of the purportedly fittest; and, to the degree that they are successful, the so-called unfit might well reactivate societal values that lay dormant so long as they merely endured oppression.[72] From the viewpoint of *intersocietal dialectics,* then, evolution takes the form of a gradualism that appeals to those with a vested interest in the *status quo,* whereas revolution takes the form of an action-oriented program of liberation deemed necessary precisely because of the power-brokers' quasi-theological insistence upon their own historical inevitability.

It is, of course, customary to use the word "reactionary" to describe the attitudes and actions of those who defend an established social system against attempts to change that system in any of its essentials (however defined). This usage, though understandable enough, owes its meaning to a very narrow conception of the dynamics of a society through time. Within this limited and ahistorical context, those in control of a society may indeed do little more than "react" to an attempt by others (be they described as "revolutionary" or whatever) to alter or even overthrow that society. But if such dialectical confrontation is viewed within a wider historical context, one comes to see, I think, that it is often the revolt or rebellion against the established system that is "reactionary."

Taken superficially, it is a truism that one who rebels does so in reaction against powerful agents whose influence upon him he interprets as oppressive. More profoundly, rebellion is reaction in the sense that it proceeds from weakness rather than from strength, from frustration and even despair rather than from hubris and love of innovation.[73] When viewed synchronically or ahistorically, such reaction to oppression tends to be interpreted acontextually as an unwarranted and unprovoked attack upon a system that others have found to be advantageous in their own regard. But when viewed diachronically, in the light of historical context and origins, such an uprising can appear to have been not only inevitable but even remarkably belated in view of grievances previously endured without openly belligerent protest.

To restate the matter somewhat differently, once a given set of human relationships is taken for granted, any attempt to deviate from these relationships or, more traumatically still, to change them, tends to be viewed as "radical"—not in the etymological sense of "getting to the roots of the matter" (radix = root in Latin) but in the polemical sense of extreme and irresponsible. If, on the other hand, this set of human relationships is not taken for granted but is seen to be (as, in fact, it is) a contingent product of historical events, then any corresponding attitude or action will be radical precisely in the sense that at least in intention it uproots the system from comparatively shortsighted claims to historical inevitability.

In political terms, such radicalism has often been identified with the Left, and there are good reasons for so doing. But here one must be on guard against that facile labeling that would explain everything but understands nothing. People in the so-called free world have been taught to associate the Left with Communism. But among so-called Communist nations there is a growing spirit of reaction against any institutional rigidity, however ideologically orthodox, that is inappropriate to new conditions. In this setting, it is precisely the Establishment that identifies itself with a revolution—not a present and continuing revolution, to be sure, but a once-upon-a-time revolution that is for all practical purposes over and done. Many states, of course, still trace their origins to some revolution of the past two hundred years, especially France and the United States. States thus established as recently as the twentieth century are likely to be closely identified with and even to derive their mythology from a revolution. This is obviously the case, for example, in Russia and even more so in Cuba. In both instances, however, the younger generation, which has not itself lived through the revolution eulogized in its elders' mythology, tends to reject that revolution as but another excuse for unbending authoritarianism.[74]

The point here, quite simply, is that revolution as an ideology and revolution as a deliberately engineered process of radical change must be carefully distinguished. Protagonist control may be legitimated in any terms, and those terms associated with revolution are no exception. This is not to say, of course, that the notion of revolution has no utopian function among the oppressed antagonists of an established societal order. They both can be and at times are the agents of revolutionary societal change, just to the extent that they are successful in their efforts to achieve power for some group or groups heretofore at least partially excluded.

The function of societal antagonists differs according to the type of social system against which they are reacting. The most obvious form of opposition to oppression is that which arises among one group (a clan, a tribe, a race, a caste, a nation) that has been forcibly subdued and brought under control by an alien group which superimposes itself from without .Whether the interface of superimposition is an occupation army, a puppet government, a colonial administration or a more subtle control of the internal economy through international monetary controls and various trade restrictions, the effect is essentially the same : a group of people who once determined their own destiny within a narrower world-context now find themselves to have become exploited has-beens in a world that outreaches their learned ability to cope.

If a revolution is a radical change in form of government, then for a society once relatively independent suddenly to become subservient to more powerful others from outside is most assuredly a revolution. Nor can those subdued be expected to react favorably to such a major deterioration in their status. They may (1) despise themselves and strive to make themselves as inconspicuous as possible; (2) adulate and seek to imitate and be accepted by those who control their lives; or (3) despair of ever being fully accepted into the ranks of the controllers and in their despair turn to active antagonism against their societal protagonists.[75]

This reaction pattern on the part of the oppressed is perhaps most clearly exemplified in the response of underdeveloped or even primitive indigenous peoples to their being continually exploited at the hands of more powerful people who have come to conquer and control. Less easily seen in this light, because more assiduously glamorized and romanticized in the West, is colonization against which many peoples have in recent years rebelled. Still less obvious is the fact that some subgroup(s) of people within one and the same group might truly be oppressed and impoverished, at least in comparison to others in that same group. This is perhaps less true of primitive tribes than of modern nation-states.[76] But in either instance the reactions of the oppressed can range from at least overt acquiescence to militant rebellion, the latter extreme being more likely to occur (though less likely to succeed) in a modern nation-state. Between the extremes one notes such "movements" as the student movement, the black movement, the women's liberation movement.

In all these instances what is important to note is that rebellion constitutes an antagonist's reaction to a consistent pattern of actions on the part of a protagonist who, in turn, responds to the reaction by counter-reaction. It is accordingly inaccurate and mystifying, as the Marxists would say, to claim that a protagonist group "does nothing" until the antagonist group rises up against it. The rebellion would not—probably could not—have occurred without a considerable amount of "doing" that has already proven detrimental (if not virtually genocidal) with regard to the antagonists. Whether such actions on the part of the antagonists are detrimental to themselves or to the human race as a whole is, of course, another though by no means unrelated question. But however that question is answered, there is good reason for saying, on the level of intersocietal dialectics, (1) that the evolution of protagonists may be a direct cause of the devolution and even extinction of others and (2) that a revolution by antagonists may be a direct cause of devolution on the part of erstwhile protagonists. This societal version of Newton's third law has been operative as far back in time as our information can take us, and differs today only in the magnitude of its scope and possible consequences.

With regard to the scope of intersocietal dialectics today, it is becoming more and more realistic to talk in terms of global confrontations. The so-called world wars that blighted the first half of this century were truly worldwide in the sense that they affected every continent to one degree or another. Since 1945, some formerly hostile power elites have been learning to compete and even cooperate with one another as privileged peers across international boundaries, usually to their mutual advantage but not necessarily to the advantage of the masses of people in the countries thus represented.[77] There does not seem to be any one pattern that may be said to characterize all these contemporary intersocietal situations except that they are precisely that, inasmuch as they spill over the boundaries of a host country or countries and involve one or more supranational interests that are powerful enough to control others for profit.[78]

What shall we say, secondly, about the possible consequences of intersocietal dialectics today? It seems in general that the thrust of evolution is, as ever, a two-edged sword that cuts a clear path to "progress" for the few who are powerful enough to wield it, while dismembering the others who happen to be—or, what is perhaps more regressive yet, choose to place themselves—in their way. It is of no particular benefit to species now in danger of extinction—the trumpeter swan, the bald eagle, the crocodile, and, only slightly more remotely, all the nonhuman flora and fauna of this planet—that simians begat anthropoids or that anthropoids begat homo sapiens or that homo sapiens begat the industrialist or the industrialist is now begetting the technocrat or that the technocrat may one day beget instruments of worldwide or even galactic control. Nor is such convergent evolution (quite different, to be sure, from that envisioned by Teilhard de Chardin!) of much benefit to those innumerable branches of the human species—almost all non-whites and a growing number of superfluous whites—that have simply been in the way. For the protagonists of evolution have been no more mindful of bison hunters than of bison, no more mindful of fishermen than they were of the streams and lakes and oceans upon which those fishermen have depended. In other words, what is often thought of as pollution of the earth is inevitably a pollution of man as well.[79] Problems once confined to ghettos finally attract the attention of "the public" once they have spilled over into the suburbs. So also on a global scale, the narrowly conceived and stupidly executed technologization of this planet has now come to endanger not only all subhuman flora and fauna and all less developed or at least unscrupulous humans but even the protagonists whose will to power has in recent centuries functioned as a real but ambivalent agent of evolutionary advancement. In the span of a single generation it has come to be almost humanly impossible to live and work in New York City. In a finite number of generations—and for some experts the number is quickly counted—the same may well be said of every once habitable metropolis on this planet.[80]

The obverse side of this coin is that in sharpening the action edge of the sword of evolution, protagonists have in spite of themselves sharpened the reaction edge as well. The colonialism that was so much a part of the industrialization of western Europe came crashing down after Europe's colonial empires reached their paroxysm of futility in two world wars. Now the neo-colonialism of the super-states, flying the banner of "modernization," is moving steadily towards heretofore undreamed-of levels of control. But that very control is breeding dissent and opposition and open rebellion of many kinds and in many parts of the world.[81] The basic question of proprietorship and its legitimation is written small when students anywhere in the world ask aloud to whom the university belongs. On a global scale, men in all parts of the world are being compelled by the multiplying hints of impending doom to ask : to whom does the world belong? It is a mark of our heritage that this question is framed in a context of ownership; but the very asking of the question can perhaps lead men out beyond proprietary abuse towards a shared responsibility for recuperation, maintenance and ecologically balanced development of the limited resources available to us on this planet. And chief of all these resources, as even economists are being challenged to take seriously, are human beings.[82]

As Erich Fromm recently put it, a "revolution of hope" is possible only in and through a humanization of technology.[83] But neither past nor present policies with regard to "modernization" lend much support to the notion that a humanization of technology might come about by way of some sort of historical inevitability. If it is to come at all, it will somehow be the result of a deliberate and concerted effort on the part of all who have recognized other alternatives as intolerable and have somehow translated this recognition into political—or, perhaps better, supra-political-power. It is indeed conceivable that present-day socio-political relations can be transformed into what Amitai Etzioni calls "the active society." But, as this socio-political analyst intimates, the structure within which each power group must operate is rapidly becoming international and to some extent already global in scope. Accordingly, so long as the efforts of social outcasts in South Africa to control their own destiny are not effectively combined with those of others in the same region or continent, if not beyond, the supranational power elite who contend primarily only among themselves for control of knowledge, communications and technology will continue to confuse the weakness of the politically disorganized as a mandate to continue ruling as in the past.[84]

It would seem, then, that the very survival—to say nothing of the prosperity—of the great majority of human beings on this planet is to a great extent a function of utility as defined by the powerful few.[85] But as machines continue to substitute for and supersede not only the physical but even the intellectual labor of human beings, the latter will, in terms of work, be of no further use save to provide "services" for one another.[86] At this point, their continued existence may be tolerated only so long as they are willing to stay out of the way and not try to gain what some call "a piece of the action." "Short-run" unemployment is still so identified in developed societies with "long-run" economic growth that its presumably temporary victims are afforded at least minimal compensation. But extended unemployment of the kind that is often the lot of people in the "developing" societies leads inevitably to widespread hunger and despair, and the latter tend to find expression in rebellion. But a despair born of hunger (spiritual as well as physical) and reared on oppression may seek its maturity in violence.[87]

The long-range consequences of this increasingly dangerous world situation might well be a century not of genocide as such but of paupercide : the indirect elimination of the non-productive and the direct elimination of the counter-productive in many if not all parts of the world. If this be a kind of paranoia, then let it be said in defense of paranoia that it is supported by a large body of substantive data whose implications become clearer with each passing decade. In spite of such ultimately insignificant interferences as those brought about by kidnapping, airplane hijacking, rooftop sniping or the bombing of buildings, the opposition of outsiders against insiders seems inherently futile; and, if pursued in any given area to the point of major confrontation, could well lead to reprisals so effective and encompassing as to put the ovens of Auschwitz back into the Middle Ages of Population Control.[88] For, while one-half of the world's population is centered in East, South and Southeast Asia and another one-fifteenth each inhabit Latin America and Africa, control over and development of the world's resources is centered predominantly in North America, Europe and the USSR.[89] Moreover, the power-gap here suggested is widening steadily rather than narrowing.[90]

As is well known, living conditions in the United States and in India were closer together 150 years ago than they are now, partly because the latter country's economic growth has been absorbed by its increasing population. What is yet more sobering, however, is that India is not likely to "catch up." Given its present per capita annual income ($50 as compared with $2,000 in the U.S.A. in 1960), if both were to double over a decade or so, India's per capita income would then be $100, that of the United States, $4,000. In the meantime, the United States produces enough food to provide each of its inhabitants with 3,200 calories per day, twice that of India, which however has more than twice as many people and less than half as much land, not to mention its incomparably fewer roads, automobiles, and other modern conveniences. To mention just a few key indicators of technological development, in the United States per capita consumption of energy in 1959 was almost 8,000 kilograms of coal equivalent, in India it was 150 kilograms; in the United States, 89 million long tons of steel were produced in 1960 (27 percent of the world total for that year), and 373 million metric tons of oil were produced in 1963 (29 percent of the world total for that year), whereas in India production of both of these sources of energy was then and still is negligible.[91] As these sample figures indicate, the United States and India are almost literally worlds apart; yet it is unlikely that the former will prove very helpful in the latter's efforts towards economic development, since its economic policies respond primarily if not exclusively to internal demands for protection against competitive imports.[92]

Yet another aspect of the growing international power-gap is the tendency of highly developed countries to exploit the mineral resources of under-developed nations not yet in a position to take advantage of such resources themselves. A good example is that of oil. The currently prominent area of northern Africa called "the Middle East" would be a case in point, as would be the now somewhat quieter area of Indonesia.[93] But it will be easier for our purposes to consider only the example of Venezuela.

About 13 percent of the world total of oil is produced in Venezuela : 149 metric tons in 1960, representing three-fourths of the total production of oil in all of Latin America.[94] Venezuelan oil accounted for 23 percent of all exports from Latin America between 1953 and 1957 and close to 28 percent in 1957, as well as almost one-fourth of the annual average per capita rate of growth of the Latin American GNP during the years 1950-1957.[95] As a result of this signal contribution to modern productivity, the per capita consumption of energy in Venezuela in 1959 was about 2,500 kilograms of coal equivalent—twice as much as Argentina, a little over half as much as the United Kingdom and about one-third that of the United States.[96] Yet, however superficially impressive, these economic ratings are deceptive. For at least one-third of the GNP in Venezuela is due directly and far more indirectly to oil production, and this is entirely in the control of Dutch and American companies whose official welcome depends on little more than the payment of royalties to the local government. Out of a total population of over 7 million people, only 2 percent (about 60,000) of the active labor force are directly employed by the oil industry, compared to at least 30 times that many in agriculture, which, however, accounts for only 7 percent of the GNP.[97] Declared unemployment in 1960 stood at 13.7 percent, without even counting the many who were underemployed—figures which represent demographically a growing class of superfluous outcasts referred to elsewhere (in Peru) as the marginal unemployed.[98] In other words, behind the impressive façade of Caracas, poverty is rampant : one-fourth of the people receive two-thirds of the national income, and over 200,000 immigrants from the countryside wait in makeshift hovels called "ranchos" for their hopeless lives to ebb away.[99]

In view, then, of all the foregoing, it is the case in Venezuela that—as the stock colonialist movie would have it—"the natives are restless." And how this all came about makes yet another inglorious chapter in the story of technocracy's indifference to human misery for which it is at least in part responsible. Until 1960 there was no organized guerrilla movement in Venezuela. Then in October of that year the Betancourt government arrested three leaders of the MIR (Movement of the Revolutionary Left), in whose weekly newspaper *Izquierda* on October 14 there had appeared a controversial editorial. Whereupon some 2,000 students of the Central University protested the government action as being unconstitutional. Violently suppressed by government forces, some of them then fled to the hills, and thus began the guerrilla movement that since February 20, 1963, has been known as the FALN (Armed Forces of National Liberation) and has since become predominantly agrarian in its membership.[100]

The editorial in question insisted that a popular insurrection could not succeed; and a week after its appearance a spokesman for MIR was still insisting on his organization's nonsubversive intention to work toward elections scheduled for 1963. The Betancourt government, however, rationalized its violent behavior on the grounds that the editorial had advocated violent overthrow of the government, and it is the latter interpretation, quite without foundation, that has found its way into a United States military manual prepared by SORO (Special Operations Research Office) and entitled *Venezuela: U.S. Army Area Handbook for Venezuela*.[101] Thus indoctrinated, some 200 American military advisers have since made it possible for the Venezuelan guerrillas to share with Peruvians and Vietnamese the dubious distinction of having experienced the heat if not the light of napalm.[102]

On occasion, of course, the outcasts of this earth learn to articulate not only their misery, as in the American Negro spiritual or "blues," but also their anger. This is, for example, the extraordinary significance of the writings of Frantz Fanon, best known for a work entitled *The Wretched of the Earth*.[103] A position such as that of Fanon, however, is utterly unintelligible to those who pride themselves on proclaiming the secular heaven which at least the fittest will survive to see. Thus Zbigniew Brzezinski, a Polish exile now devoted to the Pax Americana, can say:[104]

> . . . today the differences between the two worlds are so pronounced that is is difficult to conceive a new ideological wave originating from the developed world, where the tradition of utopian thinking is generally declining.
>
> With the widening gap dooming any hope of imitation, the more likely development is an ideology of rejection of the developed world. Racial hatred could provide the necessary emotional force, exploited by xenophobic and romantic leaders. The writings of Frantz Fanon—violent and racist—are a good example. Such ideologies of

rejection, combining racialism with nationalism, would further reduce the chance of meaningful regional cooperation, so essential if technology and science are to be effectively applied. They would certainly widen the existing psychological and emotional gaps. Indeed, one might ask at this point : who is the truer repository of that indefinable quality we call human? The technologically dominant and conditioned technetron, increasingly trained to adjust to leisure, or the more "natural" and backward agrarian, more and more dominated by racial passions and continuously exhorted to work harder, even as his goal of the good life becomes more elusive?

The incredible hubris manifested in this statement by yet another apologist for the ancient ideology of might-makes-right constitutes in itself a neo-primitive artifact of great anthropological significance. For, in its barely hominoid grunts of superiority are recorded for any surviving intelligent beings (if not from the planet earth then perchance from beyond) the evolutionary dialectics that, *ex hypothesi,* shall have brought an end to *homo sapiens* at that very moment in geological time when members of the species were beginning to discover in many ways and from many sources that people are more important than prestige, more important than power, and even more important than progress. But to end on an only slightly more optimistic note, some future ecologist exploring the planet earth might possibly be able to date the establishment of a homeostatic system from the disappearance of man.

Notes and References

1. By speaking here of "societal" rather than "social" change I deviate from the usage of those for whom the theory of *social* change is a major branch of sociology; but at the same time I thereby acknowledge a strong sympathy for Amitai Etzioni's arguments in favor of what he calls macro-sociology. Although highly dubious about the ontological status of what Etzioni calls the "emergent properties" of a macro-system, I nonetheless find the adjective "societal" methodologically useful as shorthand for the conceptualization of a society·as made up not only of micro-individuals or even roles but of mirco-societies which act either collectively or representationally upon other micro-societies and/or upon the macro-society and are themselves so acted upon. I shall use "social" to refer generically to any society or societies without regard to internal complexity, "societal" precisely to specify such internal complexity.

Finally, to speak of the confrontation and interaction of societies each of which is itself internally complex, I shall use the word "intersocietal." See Amitai Etzioni, *The Active Society: A Theory of Societal and Political Processes* (New York: The Free Press, 1968), pp. 41–59.

2. Etzioni's observations on the various meanings of "sytem," especially as referring sometimes to a supra-unit (sociologists and anthropologists) and sometimes to an inter-unit ("general system" theorists), is especially instructive in this regard (*ibid.*, pp. 65, 123–5, 129 fn. 41). See also David S. Landes, *The Unbound Prometheus: Technological Change and Industrial Development in Western Europe from 1750 to the Present* (Cambridge: Cambridge University Press, 1969), pp. 229–30.

3. He does suggest in his "Concluding Observations" that one might distinguish "societal attitudes and forms of behaviour" in terms of positive and negative feedback [p. 53] and adds that "both forms of feedback are required" of the United Nations in its efforts to build a better world [p. 54]; but such observations point only to a methodological conculsion: "General systems theory provides both a fresh conceptual framework and new analytical tools for the study of societal attitudes and behavioral patterns in the international political environment" [p. 54].

4. Taylor, [pp. 49, 55]. See also [p. 22], where he calls attention to "the population explosion, diminution of non-renewable resources, and environmental pollution."

5. See, however, Taylor, [p. 22], fn. 40. Robert Redfield, whom Taylor quotes [pp. 30–31], seems to deny that any such thing is happening; but Taylor himself seems prepared to allow the contrary when he notes that a sociocultural system might "quantize from a more complex to simpler level of organization" [p. 35]. These matters will be dealt with in some detail further on.

6. A good example of the sort of concerned analysis to which I here refer is Karl Jaspers, *The Future of Mankind,* trans. E. B. Ashton, Chicago and London: Phoenix Books, 1963. See also Gunnar Myrdal, *Objectivity in Social Research,* New York: Random House Pantheon Books, 1969; Ernest Becker, *The Structure of Evil: An Essay on the Unification of the Science of Man,* New York: Braziller, 1968; Philip Slater, *The Pursuit of Loneliness: American Culture at the Breaking Point,* Boston: Beacon Press, 1970.

7. See J. N. Findlay, *Hegel: A Re-Examination* (New York: Collier Books, 1962), pp. 252–4, 334–5; Sidney Hook, *From Hegel to Marx: Studies in the Intellectual Development of Karl Marx* (New York: Humanities Press, 1958), pp. 36–41; Herbert Marcuse, *Reason and Revolution: Hegel and the Rise of Social Theory* (Boston: Beacon Press, 1960), pp. 232–4.

8. Successful or no, however, there have been many attempts. See in this regard W. Warren Wagar, *The City of Man: Prophecies of a World Civilization in Twentieth-Century Thought,* Baltimore: Penguin, 1967.

9. This problem, to be dealt with at length below, is well stated by S. N. Eisenstadt, *Modernization: Protest and Change* (Englewood Cliffs, N.J.: Prentice-Hall, 1966), p. 20; Amitai Etzioni, *op. cit.,* pp. 549–613; Ruth C. Lawson (ed.), *International Regional Organizations: Constitutional Foundations,* New York: Praeger, 1962; *Factors Affecting the International Transfer of Technology among Developed Countries,* Washington: U.S.

Government Printing Office, 1970; George Thayer, *The War Business: The International Trade in Armaments,* New York: Avon Discus Books, 1970.

10. This undoubtedly pessimistic view is perhaps most strongly stated by Herbert Marcuse, especially in *Soviet Marxism: A Critical Analysis* (New York: Random House Vintage, 1961) and *One-Dimensional Man: Studies in the Ideology of Advanced Industrial Society* (Boston: Beacon Press, 1964). See also Lionel Rubinoff, *The Pornography of Power,* Chicago: Quadrangle, 1967; Mary Warnock, *Ethics Since 1900,* 2nd ed., London: Oxford, 1966.

11. "Critique," as here used, is not unrelated to the usage of Herbert Marcuse and the tradition of the sociology of knowledge as reviewed in Gunter W. Remmling's *Road to Suspicion: A Study of Modern Mentality and the Sociology of Knowledge* (New York: Appleton-Century-Crofts, 1967); but its methodological implications are better expressed by Gerard Radnitzky, *Contemporary Schools of Metascience,* 2nd rev. ed., 2 vols. in one (New York: Humanities, 1970), especially Vol. II, pp. 126–85; its pragmatist overtones by Harold Sackman, *Computers, System Science, and Evolving Society: The Challenge of Man-Machine Digital Systems,* New York: Wiley, 1967; and its tactical optimism by Etzioni's *The Active Society.*

12. This tenet is articulated in psychology by Charlotte Bühler and Fred Massaryk (eds.), *The Course of Human Life: A Study of Goals in the Humanistic Perspective* (New York: Springer, 1968) and in sociology by Alvin Toffler, *Future Shock* (New York: Random House, 1970). The related Weberian notion of social action, reformulated by Talcott Parsons and others, has been thoroughly expostulated by Alfred Schutz, *The Phenomenology of the Social World,* trans. George Walsh and Frederick Lehnert, Evanston: Northwestern, 1967.

13. See Florian Znaniecki, *The Social Role of the Man of Knowledge,* New York and Evanston: Harper Torchbooks, 1968; E. O. James, *The Nature and Function of Priesthood; A Comparative and Anthropological Study,* New York: Barnes & Noble, 1955; Claude Lévi-Strauss, *The Savage Mind,* Chicago: University of Chicago, 1966.

14. Whether any people, primitive or otherwise, are particularly open to social change surely depends in part upon their evaluation of the consequences of such change. Accordingly, Wilbert E. Moore's analyses in *Social Change* (Englewood Cliffs, N.J.: Prentice-Hall, 1963) and in *Man, Time and Society* (New York and London: Wiley, 1963) are undoubtedly more balanced than Eric Hoffer's claim in *The Ordeal of Change* (New York: Harper & Row, 1963) that people tend to resist all change. The history of technology, for example, provides innumerable examples of different responses to the same call for change, e.g. that of Citroën as compared to Renault in France: Landes, *op. cit.,* pp. 446–51; see also pp. 122–3, 352–3.

15. Toffler, *op. cit.,* pp. 289–326. See also David W. Ewing. *The Human Side of Planning: Tool or Tyrant?* New York: Macmillan, 1969; Ronald Lippitt, Jeanne Watson, and Bruce Westley, *Planned Change: A Comparative Study of Principles and Techniques,* New York: Harcourt, Brace & World, 1958.

16. See Alex Inkeles, *What is Sociology? An Introduction to the Discipline and Profession* (Englewood Cliffs, N.J.: Prentice-Hall, 1964), pp. 18–46;

Talcott Parsons, *The Social System,* Chicago: Free Press, 1951; Joseph H. Monane, *A Sociology of Human Systems,* New York: Appleton-Century-Crofts, 1967.

17. The concept of power, long a valuable tool for political analysts (e.g Bertrand de Jouvenel, *On Power: Its Nature and the History of Its Growth,* New York: Viking, 1949), is now finding its way into sociological literature, e.g. Richard A. Schermerhorn, *Society and Power,* New York: Random House, 1961; Amitai Etzioni, *op. cit.,* pp. 313–86. Related to the latter are the favorable evaluations of conflict by such sociologists as Lewis Coser (*The Functions of Social Conflict,* New York: Free Press, 1956) and Thomas Schelling (*The Strategy of Conflict,* New York: Oxford Galaxy, 1963). Stage theories are considered outmoded by Inkeles (*op. cit.,* pp. 30–33); of somewhat limited applicability by Landes (*op. cit.,* pp. 229–30), by Wilbert E. Moore (*Social Change,* pp. 33–44, 113–7), and by Joseph Spengler, "Social Evolution and the Theory of Economic Development," *Social Change in Developing Areas: A Reinterpretation of Evolutionary Theory* (eds. Herbert R. Barringer et al.; Cambridge, Mass.: Schenkman Publishing Co., 1965), pp. 256–62.

18. This view of society suggests, among other things, that any major change in the industrial-technological power base of a given society would result in a corresponding change in the social control and hence in the structure of that society. See the works of Sackman and Landes, cited above, as well as J.-J. Servan-Schreiber, *The American Challenge,* trans. Ronald Steel, New York: Avon, 1969; Victor C. Ferkiss, *Technological Man: The Myth and the Reality,* New York: Mentor, 1970.

19. As Wilbert E. Moore points out (*Social Change,* pp. 40–2, 89–112), modernization and progress are by no means interchangeable concepts. Human values, in particular, tend to be undermined by merely technological advancement. See Eric and Mary Josephson (eds.), *Man Alone: Alienation in Modern Society,* New York: Dell Laurel, 1962; Erich Fromm (ed.), *Socialist Humanism: An International Symposium,* Garden City, New York: Double-day Anchor, 1966; Landes, *op. cit.,* p. 6.

20. Taylor, it will be noted, takes these usages as the starting point for his appeal to complementarity [pp. 4, 11, 23, 58].

21. The relativism here suggested is perhaps positivist in flavor, but would be accepted, for example, by a Sartre (see, for example, his *No Exit*) as well as by Amitai Etzioni: "The confines of social life are frequently composed of other people in the same predicament" (*op. cit.,* p. 3).

22. [p. 35 and fig. 3.]

23. See Peter L. Berger and Thomas Luckmann, *The Social Construction of Reality: A Treatise in the Sociology of Knowledge* (Garden City, New York: Doubleday Anchor, 1967), pp. 92–128; Vittorio Lanternari, *The Religions of the Oppressed: A Study of Modern Messianic Cults,* New York: Mentor, 1965; James Mooney, *The Ghost–Dance Religion and the Sioux Outbreak of 1890,* Chicago and London: Phoenix, 1965; Schermerhorn, *op. cit.,* pp. 53–69.

24. Although Taylor twice speaks of general systems theory as being "valid" [pp. 29 and 52], he does not show that such validity excludes either "analogy" or "Western ethnocentricity," both of which he clearly wishes to surmount

[pp. 28 and 55]. Nor is it likely that he could show this, if we may believe Ernest Becker, *op. cit.*, pp. 364 and 366.

25. In lieu of entering here into the much disputed question of "objectivity" in the social sciences I refer the reader to Ernest Nagel, *The Structure of Science* (New York and Burlingame: Harcourt, Brace & World, 1961), pp. 447–546; Richard S. Rudner, *Philosophy of Social Science* (Englewood Cliffs, N.J.: Prentice-Hall, 1966), pp. 68–83; Lucien Goldmann, *The Human Sciences & Philosophy,* trans. Hayden V. White and Robert Anchor (London: Cape, 1969), pp. 35–84; and, finally, I make my own Alex Inkeles' observation that "the critical question is not so much what is a man's ideology of research but rather what is the extent of his contribution to knowledge" (*op. cit.,* p. 105).

26. J. L. Austin, "Performative-Constative," *Philosophy and Ordinary Language* (ed. Charles E. Caton; Urbana: University of Illinois Press, 1963), pp. 22–54.

27. I do not suggest that Taylor is immune from analogies, but rather that he turns not to the biological but to the physical sciences for his own favored analogy, viz., that of complementarity. Nor do I suggest that the latter notion has no relevance to problems on the human level, but only that what is thereby gained in prestige is lost in specificity. Ultimately at issue, however, is Ludwig von Bertalanffy's claim that "in sciences that are not within the framework of physico-chemical laws, such as demography and sociology . . . exact laws can be stated if suitable model-conceptions are chosen"—*Problems of Life: An Evaluation of Modern Biological and Scientific Thought* (New York: Harper Torchbooks, 1960), p. 200.

28. This well-known thesis of Eric Hoffer (*The True Believer; Thoughts on the Nature of Mass Movements,* New York: Perennial, 1966) is well exemplified by the way in which evolutionary theory has been used to rationalize racist policies and procedures—once, even, as Taylor reminds us [p. 8], by Charles Darwin himself. See Donald T. Campbell, "Variation and Selective Retention in Socio-Cultural Evolution," *Social Change in Developing Areas, op. cit.,* pp. 20–5.

29. That proponents of general system theory hope to surmount such limitations is in itself no guarantee that they have or in fact can succeed. See Lévi-Strauss's critique of Sartre, *The Savage Mind,* pp. 245–69.

30. See Claude Lévi-Strauss, *Tristes Tropiques: An Anthropological Study of Primitive Societies in Brazil,* trans. John Russell, New York: Atheneum, 1964.

31. Aníbal Quijano Obregón, "Tendencies in Peruvian Development and in the Class Structure," *Latin America: Reform or Revolution?* (eds. James Petras and Maurice Zeitlin; Greenwich, Conn.: Fawcett Premier, 1968), p. 327; Landes, *op. cit.,* pp. 34–5; Ronald Segal, *The Americans: A Conflict of Creed and Reality,* New York: Bantam, 1970; *Hard Times,* March 30–April 6, 1970.

32. I refer here, of course, to what I take to be Taylor's designation for the ultimate integrative level [p. 34].

33. Oscar Lewis, *The Children of Sanchez: Autobiography of a Mexican Family* (New York: Vintage, 1961), pp. xxiv-xxv; Petras and Zeitlin (eds.), *op. cit.,* pp. 58, 196, 313–14, 319, 326–9, 367; Jagdish Bhagwati, *The Economics*

of Underdeveloped Countries (New York and Toronto: World University Library, 1966), pp. 18–19; Landes, op. cit., pp. 499–503; Christian Science Monitor, Oct. 3, 1970. The former U.S. Secretary of Labor Willard Wirtz once spoke of "a separate nation of the poor, the unskilled, the jobless" as "a human slag heap." See Erich Fromm (ed.), Socialist Humanism, op cit., p. 452. See also Etzioni, op. cit., p. 11.

34. Nigel Calder, op. cit., pp. 207 ff., 246–9; Edward A. Feigenbaum and Julian Feldman (eds.), Computers and Thought. New York: McGraw-Hill, 1963; Indianapolis Star, Jan. 4, 1970; Christian Science Monitor, Feb. 16, May 1, June 6–8, 18, and 19, 1970; To Improve Learning. Washington: U.S. Government Printing Office, March, 1970; Mary Elizabeth Stevens, Automatic Indexing: A State-of-the-Art Report, National Bureau of Standards Monograph 91, Washington: U.S. Government Printing Office, March 30, 1965.

35. Sackman, op. cit., pp. 206–300; see also definitions of "real time," "realtime processing," and "real time science." p. 617.

36. Ibid., pp. 505–50; John E. Smith, The Spirit of American Philosophy, New York: Oxford University Press, 1963; Radnitzky, op. cit., pp. iii–xl.

37. See, for example, Donald G. Fink, Computers and the Human Mind: An Introduction to Artificial Intelligence, Garden City, New York: Doubleday Anchor. 1966; Kenneth M. Sayre and Frederick J. Crosson (eds.) The Modeling of Mind: Computers and Intelligence, New York: Simon and Schuster Clarion, 1968; Alan Ross Anderson (ed.) Minds and Machines. Englewood Cliffs, N.J.: Prentice-Hall, 1964.

38. See Rollo May (ed.), Existential Psychology, New York: Random House, 1961; Hendrik M. Ruitenbeek (ed.), Psychoanalysis and Existential Philosophy, New York: Dutton, 1962; James F. T. Bugental (ed.) Challenges of Humanistic Psychology, New York: McGraw-Hill, 1967.

39. A useful bibliography will be found in Gerard DeGré, Science as a Social Institution: An Introduction to the Sociology of Science, New York: Random House, 1955. More recent works include Don K. Price, The Scientific Estate, Cambridge, Mass.: Harvard University Press, 1965; Bernard Barber and Walter Hirsch (eds.), The Sociology of Science, New York: Free Press, 1962.

40. René Dubos, So Human an Animal, New York: Scribners, 1968.

41. A. Standen, Science Is a Sacred Cow, New York: Dutton, 1950.

42. Abraham H. Maslow, The Psychology of Science: A Reconnaissance, Chicago: Regnery Gateway, 1969 (including relevant bibliography).

43. This view, treated at some length in Vol. I of Radnitzky, op cit., is defended in various ways by Israel Scheffler, Science and Subjectivity, Indianapolis: Bobbs-Merrill, 1967; Mario Bunge, Intuition and Science. Englewood Cliffs, N.J.: Prentice-Hall Spectrum, 1962; C. C. Gillispie, The Edge of Objectivity: An Essay in the History of Scientific Ideas, Princeton, N.J.: Princeton University Press, 1960. See also Nagel, op. cit.

44. The man-machine system, variously thought of as "prosthesis" or "cyborg," is discussed by Alvin Toffler, op. cit. pp. 186–91. See also Sackman, op. cit., and Ferkiss, op. cit., pp. 209–10.

45. This view is exemplified by such dramatic treatments as Friedrich Dürrenmatt, The Physicists, trans. James Kirkup, New York: Grove, 1964;

Bertolt Brecht, *Galileo*, trans. Charles Laughton, New York: Grove, 1966 (see especially p. 16 ff.). See also Lois and Stephen Rose, *The Shattered Ring: Science Fiction and the Quest for Meaning*, Galesburg, Ill.: John Knox, 1970.

46. Taylor, [pp. 19, 48–9, 52, and fn. 60].

47. The point here is not that such notions as complementarity are inappropriate to social analysis, but that (1) their legitimation is in no way dependent upon the authority of the physicist and (2) might well derive from other sources. Floyd W. Matson seems to suspect the first point: *The Broken Image: Man, Science and Society* (New York: Braziller, 1964), p. 150; Philipp Frank spells it out in detail: *Philosophy of Science: The Link Between Science and Philosophy* (Englewood Cliffs, N.J.: Prentice-Hall, 1957), pp. 163–88, 207–59. For the second poin tsee Peter L. Berger, *The Sacred Canopy: Elements of a Sociological Theory of Religion*, Garden City, New York.: Anchor, 1969; Alan W. Watts, *The Two Hands of God: The Myths of Polarity*, New York: Collier, 1969.

48. Claude Lévi-Strauss, *The Savage Mind, op. cit.; Totemism*, trans. Rodney Needham, Boston: Beacon, 1963; René Dubos, *A Theology of the Earth*, Washington, D.C.: U.S. Government Printing Office, 1969.

49. Jean-Paul Sartre, *Nausea*, trans. Lloyd Alexander, New York: New Directions, 1949.

50. Norbert Wiener, *God & Golem, Inc.: A Comment on Certain Points Where Cybernetics Impinges on Religion*, Cambridge, Mass.: M.I.T., 1964; Everett Knight, *The Objective Society*, New York: Braziller, 1960; Roszak, *op. cit.*, pp. 205–89.

51. Referring not to what all men can do, but to what our species can do in virtue of the capabilities of some. See Edmund F. Byrne and Edward A. Maziarz, *Human Being and Being Human: Man's Philosophies of Man* (New York: Appleton-Century-Crofts, 1969), pp. 327–33.

52. Remmling, *op. cit.*, pp. 118–27; Gillispie, *op. cit.*, 74–82; Sackman, *op. cit.*, 583–94; Calder, *op. cit.*, pp. 271–353.

53. Lewis A. Coser, *Men of Ideas: A Sociologist's View*, New York: Free Press Paperback, 1970. This conflict-theory work should be compared to Bernard Barber's *Science and the Social Order* (New York: Collier, 1962) and to two works written under wartime conditions: Hans J. Morgenthau, *Scientific Man versus Power Politics*, Chicago and London: Unversity of Chicago, 1946, Phoenix, 1965; Noam Chomsky, *American Power and the New Mandarins*, New York: Random House Pantheon, 1969.

54. Whence the growing interest in avoiding what is called "technological surprise": *Science and Technology: Tools for Progress*, Report of the President's Task Force on Science Policy (Washington, D.C.: U.S. Government Printing Office, 1970), p. 38; *Christian Science Monitor*, April 25–7, 1970. See also Etzioni, *op. cit.*, Part Two, pp. 132–309, p. 77.

55. Marshall McLuhan, *Understanding Media: The Extensions of Man*, New York: Signet, 1966; Henry Jacobowitz, *Electronic Computers* (Garden City, New York: Doubleday Made Simple Books, 1963), pp. 4–16, 93–8; Norbert Wiener, *Cybernetics*, 2nd ed. (Cambridge, Mass.: M.I.T., 1961), pp. 1–29.

56. E. O. James, *op. cit.*, pp. 38–39.

57. See, for example, David Wise and Thomas B. Ross, *The Espionage Establishment,* New York: Random House, 1967; Stafford Beer, "Managing Modern Complexity," *The Management of Information and Knowledge* (Washington, D.C.: U.S. Government Printing Office, 1970), pp. 41–62.

58. Marshall McLuhan, *The Gutenberg Galaxy: The Making of Typographic Man,* New York: Signet, 1969.

59. Jesse Kornbluth (ed.), *Notes from the New Underground: An Anthology,* New York: Ace, 1968; Ronald Segal, *op. cit., pp.* 139–40; Dale Minor, *The Information War,* New York: Hawthorn, 1969.

60. "The Supersonic Seventies," *Business Automation,* XVII (1970), 55–60; Joseph Goulden and Marshall Singer, "Dial-A-Bomb: AT&T and the ABM," *Ramparts,* VIII (1969), 29–37; *Christian Science Monitor,* Jan. 29–30, 1970.

61. In addition to observations in Calder, *op. cit., passim,* one might consult the *Inventory of Automatic Data Processing Equipment in the United States Government,* now being published annually by GSA through the U.S. Government Printing Office.

62. See, however, Toffler, *op. cit.,* pp. 400–5.

63. Compare Sackman, *op. cit.,* pp. 530–2, 544–5, 556, with Peter F. Drucker, *The New Society: The Anatomy of Industrial Order* (New York: Harper Torchbooks, 1962), p. 63.

64. Calder, *op. cit.,* pp. 273–88.

65. Terminology especially dear to Roszak, *op. cit.*

66. See *Science and Technology: Tools for Progress,* Report of the President's Task Force on Science Policy (Washington, D.C.: U.S. Government Printing Office, April 1970), pp. ii, vii, 14–23, 47.

67. That non-scientists are little better prepared to fill the gap may be seen from Paul Kurtz, *Decision and the Condition of Man* (New York: Delta, 1968), pp. 282–3; Nicholas Rescher, *Introduction to Value Theory* (Englewood Cliffs, N.J.: Prentice-Hall, 1969), pp. 11–27; Arnold A. Rogow, *The Psychiatrists* (New York: Putnam's, 1970), pp. 9–30.

68. *The Management of Information and Knowledge, op. cit.,* p. 27.

69. What follows might also be formulated according to the tension-management approach, provided that macro-dynamics be made an integral part of one's theory, somewhat as Etzioni is doing. See Arnold S. Feldman, "Evolutionary Theory and Social Change," *Social Change in Developing Areas: A Reinterpretation of Evolutionary Theory* (ed. Herbert R. Barringer et al.; Cambridge, Mass.: Schenkman, 1965), pp. 273–84.

70. The distinction between external and internal elites (Etzioni, *op. cit.,* p. 114) is relevant here, as are Chinua Achebe, *The Arrow of God* (London: Heinemann, 1964) and *Things Fall Apart* (Greenwich, Conn.: Fawcett Premier, 1969); Toffler, *op. cit.,* ch. 7.

71. I allude here not only to the maxim, "Love it or leave it," but more importantly to the notion of territoriality as reported by Robert Ardrey, *The Territorial Imperative: A Personal Inquiry into the Animal Origins of Property and Nations* (New York: Atheneum, 1966)—but without accepting conclusions drawn from either.

72. Friedrich Nietzsche, "Human, All Too Human," n. 224, *The Portable Nietzsche* (ed. and trans. Walter Kaufman; New York: Viking, 1954), pp.

54–56; *Christian Science Monitor*, Nov. 18, 1970.

73.   Compare Eric Hoffer, *The Ordeal of Change* (New York: Perennial, 1967), pp. 3–6.

74.   Maurice Zeitlin, "Political Generations in the Cuban Working Class," *Latin America: Reform or Revolution? op. cit.*, pp. 264–88; Alex Inkeles and Raymond Bauer, *The Soviet Citizen*, Cambridge, Mass.: Harvard, 1961; Robert C. Tucker, *The Marxian Revolutionary Idea*, New York: Norton, 1969; Daniel and Gabriel Cohn-Bendit, *Obsolete Communism: The Left-Wing Alternative*, New York: McGraw-Hill, 1969.

75.   Albert Memmi, *The Colonizer and the Colonized*, trans. Howard Greenfield, Boston: Beacon, 1967.

76.   Freud's theory of a primal horde, though given at least symbolic value by Herbert Marcuse and repeated in a way by Jean-Paul Sartre, does not seem borne out in recent studies by Lévi-Strauss (cited above) or, among others, Robert Redfield: *The Primitive World and Its Transformations*, Ithaca, New York: Cornell, 1953; *The Little Community and Peasant Society and Culture*, Chicago: University of Chicago Press, 1960.

77.   What James Burnham foresaw in 1941 as *The Managerial Revolution* (reprinted, Bloomington and London: Indiana University, 1960) is rapidly leading, for example, to a United States of Europe; and this, in turn, is but another major step towards a stabilized world economy: J.-J. Servan-Schreiber, *op. cit; The Indianapolis Star*, Nov. 8, 1970 (Sec. 1, p. 26), Sept. 10, 1970 (p. 31); *Christian Science Monitor*, Sept. 15, 1970, Sept. 25, 1970. But the growth of international investment has as one of its side-effects an internationalization of job instability, for reasons that may be learned either from following business news or, in terms of background, from J. P. Cole, *Geography of World Affairs*, 3rd. ed., Baltimore: Penguin 1964.

78.   As Sartre tells Memmi with regard to colonialism, perhaps "system" is already a better word here than "situation": *The Colonizer and the Colonized, op. cit.*, p. xxv. See also Etzioni, *op. cit.*, pp. 124–5.

79.   Kingsland Crowe, *World Wildlife: The Last Stand*, New York: Scribners, 1970; Lila Freilicher, "An Ecology Reading List," *National Catholic Reporter*, Sept. 25, 1970; Landes, *op, cit.*, pp. 4, 33–8.

80.   William H. Whyte, *The Last Landscape*, Garden City, New York: Doubleday Anchor, 1965; Jacques Ellul, *The Meaning of the City*. Grand Rapids, Mich.: Eerdmans, 1970; Jean Gottman, *Megalopolis*. Cambridge, Mass.: M.I.T., 1961; Joseph Dietch, ". . . But Who Wants to Live There?" *Christian Science Monitor*, Dec. 18, 1969. See also *ibid.*, Nov. 24 and 29, 1969.

81.   Robert L. Heilbroner, *The Future as History*, New York: Harper & Row, 1960; Harper Torchbooks, 1968.

82.   Landes, op. cit., pp. 513–4; Kenneth Boulding, "Philosophy, Behavioral Science, and the Nature of Man," *World Politics*, XII (1960), 272–9; Toffler, *op. cit.*, pp. 405, 480–2.

83.   *The Revolution of Hope: Towards a Humanized Technology*, New York: Harper & Row and Bantam, 1968.

84.   Etzioni, *The Active Society, op. cit.*, pp. 583–6.

85.   Bhagwati, *op. cit.*, pp. 205–44; Reginald H. Green and Ann Seidman, *Unity or Poverty? The Economics of Pan-Africanism* (Baltimore: Penguin,

1968), pp. 99–131; Petras and Zeitlin, *op. cit.,* especially pp. 1–144. Also useful in this connection are such United States government publications as: *Minerals in the World Economy,* Washington: U.S. Government Printing Office, 1968; *Essential United States Foreign Trade Routes,* Washington: U.S. Government Printing Office, December 1969.

86. The point here is not to discredit hopes for "post-industrial" leisure, but to note that such hopes are futile within the context of a work ethic that equates individual worth with "productive" work and thus inversely tends to view the "able-bodied" workless as worthless. See Fred Davis, "Why All of Us May Be Flower Children Someday," *Trans-Action* V (1967), 730–9; Petras and Zeitlin, *op. cit.,* p. 61, fn. 42.

87. In spite of great variations in method and motivation, most analysts of international affairs anticipate some sort of global catastrophe in the years ahead. For an understanding of this impending crisis, absolutely basic are the works of Gunnar Myrdal: *An International Economy,* New York: Harper, 1956; *Asian Drama: An Inquiry into the Poverty of Nations,* 3 vols., Baltimore: Penguin, 1968; *The Challenge of World Poverty,* New York: Pantheon, 1970. For typical statements of concern or alarm see Calder, *op. cit.,* pp. 261–3; Landes, *op. cit.,* pp. 10–2; Bhagwati, *op. cit.,* pp. 36–8; Petras and Zeitlin, *op. cit., passim,* e.g. pp. 329, 355; Cole *op. cit.,* pp. 43–54; Wilfred Cartey and Martin Kilson (eds.), *The Africa Reader: Independent Africa* (New York: Random House Vintage, 1970), pp. 257–400. See also fn. 90.

88. As Landes notes (*op. cit.,* p. 38), there is ample precedent for such paupercide in the history of European expansionism; and this has been perpetuated most recently in southeast Asia and in such resource-rich African lands as Algeria, the Congo, and Nigeria. Detailed accounts would in large measure simply elucidate J. P. Cole's observation that the white man's supremacy in recent centuries has depended upon military superiority (*op. cit.,* p. 54). For an attitudinal response see John A. Williams, *The Man Who Cried I Am* (New York: Signet, 1967), pp. 299–316.

89. Cole, *op. cit.,* pp. 74, 127, 140, 66–106.

90. Kenneth Galbraith says resignedly that "there can be no talk of the poor countries catching up with the rich": *The Underdeveloped Country* (Toronto: CBC Publications, 1965), p. 45. Bhagwati, finding recent rate-of-growth statistics ambiguous, questions not the gap but whether it is still increasing. Others are persuaded that rich countries are getting richer, poor countries poorer: Cole, *op. cit.,* p. 103; Calder, *op. cit.,* p. 259; Landes, *op. cit.,* p. 335; Heilbroner, *op. cit.,* pp. 162–3; Sackman, *op. cit.,* pp. 583–4; Petras and Zeitlin, *op. cit.,* pp. 196, 326; *Christian Science Monitor,* Sept. 25, 1970.

91. Cole, *op. cit.,* pp. 99–103.

92. Bhagwati, *op. cit.,* pp. 67, 231 ff.; Calder, *op. cit.,* pp. 263–9; *Christian Science Monitor,* Aug. 10, 1970.

93. See Cole, *op. cit.,* pp. 191, 175–7, 266; *Hard Times,* Nov. 17–24, 1969; *Time,* Oct. 19, 1970, 81.

94. Cole, *op. cit.,* pp. 89, 131.

95. Maurice Halperin, "Growth and Crisis in the Latin American Economy," *Latin America: Reform or Revolution? op. cit.,* pp. 58–9.

96. Cole, *op. cit.,* p. 129.

97. Halperin, *op. cit.*, pp. 57–8.

98. Petras and Zeitlin, *op. cit.*, pp. 195–6, 313–4, 327, 367.

99. Halperin, *op. cit.*, p. 58. See above, fn. 33.

100. James Petras, "Revolution and Guerrilla Movements in Latin America: Venezuela, Colombia, Guatemala, and Peru," *Latin America: Reform or Revolution? op. cit.*, pp. 340–3.

101. Petras, *ibid.*, pp. 337–40. See also Merle Kling, "Toward a Theory of Power and Political Instability in Latin America," *Latin America: Reform or Revolution? op. cit.*, pp. 86–7.

102. Petras and Zeitlin, *op. cit.*, pp. 349 and 352; see also Cole, *op. cit.*, p. 125; *Hard Times*, Aug. 10–17, 1970.

103. Frantz Fanon, *The Wretched of the Earth*, trans. Constance Farrington, New York: Grove, 1963.

104. Zbigniew Brzezinski, "American in the Technetronic Age," *Philosophy for a New Generation* (eds. A. K. Bierman and James A. Gould; New York: Macmillan, 1970), p. 429. For the background and context of Brzezinski's statement, see Chaim I. Waxman (ed.), *The End of Ideology Debate*, New York: Simon and Schuster Clarion, 1969; Brzezinski, *Between Two Ages*, New York: Viking, 1970.

PART THREE

THESIS: Richard McKeon
COUNTERTHESES: Joseph J. Kockelmans
Richard Cole

WORLD ORDER IN EVOLUTION AND REVOLUTION IN ARTS, ASSOCIATIONS, AND SCIENCES

RICHARD McKEON

University of Chicago

THE RECOGNITION that the issues of our times are inseparably interconnected breaks frequently through the fixity of vocabulary and obstinacy of conviction which are products of the evolved sophistication in which mankind slowly learned to differentiate processes and to distinguish objects. The analyses and programs inspired by that recognition are paradoxical and dubious, for they are frequently simple forms of admonition to unite what has been separated and to merge what has been distinguished. Failure to recognize connections is due as much to failure to make pertinent distinctions as to readiness to apply inappropriate distinctions. Transitions in the treatment of problems may be made smoothly when new separations and integrations are suggested by, and effected within, some persisting structures and distinctions. Then attention to some aspects of a familiar subject matter suggests new questions, or re-examination of accepted assumptions of knowledge opens the way to new things. When all assumptions are questioned and all things are investigated at once, no framework remains within which to pose the resulting issues except world order. The world is a construct of human arts, associations, and sciences. Human skills, societies, and accumulations of knowledge are stages of man's awareness of and reaction to the world. Both propositions are true at all stages of human history, but they become equivalent propositions, reflexively affecting each other, at stages of crisis in human history in which all fixities and doubts are transformed relative to each other in ways which raise at each turn issues of world order. There is a profound difference between the history of conceptions of world order, which is unbroken from primitive cosmogonies to modern cosmologies, and

209

the history of issues faced by men as a result of their contacts with world order, which has followed a sine-curve with bulges at ages in which all problems are problems of world order in fact if not in formulation. Many signs suggest that ours is an age which has responded to many facts which are consequences of world order to encounter issues which are puzzling unless they are recognized as issues of world order.

To the Greeks *kosmos* meant order. It was extended in their usage to mean "good order," "government," "adornment," "honor," "ruler," and "universe." We still use "cosmos," but we think that it means the system of planets, and even when we extend it to apply to analogical universes, we seldom connect it with "order," "beauty," or "government" except in moments of historical or etymological pedantry. We even suppose, against all evidence to the contrary, that the Greeks borrowed ideas from law and art to use analogically in astronomy when they developed the idea of a universe. Yet Homer used the word to portray duly ordered deeds (*eu kata kosmon*) and shamefully ordered deeds (*ou kata kosmon*), and the chief magistrate in ancient Crete was a *kosmos*. According to Diogenes Laertius (*Lives of Eminent Philosophers* viii. 48), Pythagoras was the first to call the heavens *cosmos* and earth *spherical,* but Diogenes adds that Theophrastus says that Parmenides first used the word in that sense, and that Zeno gives the honor to Hesiod. Modern readers have some difficulty interpreting Aristotle's history of the beginnings of philosophy except by giving the statement that the Ionians thought water, air, fire, the infinite, or the elements are "principles," connections and applications as cosmological or medical principles. We are not deterred in our strictly scientific interpretation of the Greek *kosmos* by the one surviving fragment of Anaximander, in which he says that the infinite is the principle of all existing things—all the heavens and all the worlds in them—and also of their destruction, "for they do justice to each other for their injustices according to the assessment of time." We suppose unhesitatingly that he is applying legal language with poetic, philosophic license to the treatment of scientific questions, without even pausing to examine the possibility (to which an examination of "cause" or "responsibility" would provide additional evidence) that later jurisprudence borrowed the proportions which constitute justice in the state from the proportions which determine the constancy of the elements and the indestructibility of matter and motion in the cosmos. Needless to say, we have forgotten the relation between *kosmikos* and *kosmetikos*, between cosmic and cosmetic, which saved the Greeks from excesses in the operationalisms they constructed to facilitate the transition from processes bearing on order to processes effecting beauty.

The history of arts, sciences, and politics might be written as variations on the theme "world order." Such an account, however, would be a history of man's disciplines and achievements in making, knowing, and doing. In a more limited sense world order, as distinguished from conceptions of world order, has been a source of dominant and puzzling issues at particular junctures of world history. Some guidance in the enumeration and examination of such issues may be afforded by examining other periods which were given their character and destiny by the issues of world order which they faced and sought to resolve in works which have been studied as monuments of world order for use and imitation.

The development of ancient Rome was such a period. Roman history was not primarily a sequence of intellectual speculation or artistic creation. It was a massive and diversified development of contacts with peoples which led Romans not only to innovations in world law and community, but also to innovations in science and technology and in the liberal and fine arts. The three varieties of innovation proceeded *pari passu,* and were interdependent parts of the world order which the Romans encountered and constructed.

(1.1) The Romans borrowed and adapted the term *kosmos,* with connotations which had been worked out by the Greek Skeptics and Stoics in the Hellenistic period, and translated it by *mundus,* which likewise meant toilet ornaments and farm implements as well as universe. Cicero defined *mundus,* or the universe, as the common home *(domus)*, or city *(urbs)*, or state *(civitas)* of gods and men *(De natura deorum* ii. 154; *De legibus* i. 23; *De finibus* iii. 64). The contacts with peoples in the Roman Empire and the development of Roman law, together with the discovery of the relation of both to natural law, are consequences and reflections of Roman contact with world order.

(1.2) The Romans encountered cultures as well as communities, and they examined the history of past cultures, their own and those of others, and contemporary cultures, particularly the Greek. In a strict sense they invented the Humanities, the Liberal Arts, and Encyclopaedias. Indeed, they thought the three inventions were a single innovation, for *enkuklios paideia* was the name of the nine liberal arts enumerated by Varro and of the body of information assembled by Pliny the Elder in his *Natural Histories*; and Aulus Gellius argued that the term *Humanitas* was a translation of *paideia* and not of *philanthropia*. (In the Eastern Empire Themistius used *philanthropia* in the sense in which Cicero used *humanitas,* and the Byzantine encyclopaedias were not ready reference works of factual information.) The Greeks had developed the fine arts and had philosophized about their nature and consequences; but the Romans invented the Humanities. The Greeks had developed grammar, rhetoric, poetic, dialectic, and mathematics; but the Romans invented the liberal arts. The Greeks had developed philosophic and learned encyclopaedias; but the Romans invented informational encyclopaedias.

(1.3) The Romans encountered science and philosophy as well as peoples and cultures. They adapted Greek speculative and theoretic sciences and Greek analytical and demonstrative methods to practical requirements : they invented technology and the communication arts by extending the uses of rhetoric to all regions of inquiry and proof. These changes were not simplifications or degradations, but a development of scientific method : the second century A.D. was one of the great periods of scientific achievement due to concern with world order. Ptolemy's *Almagest* or *Mathematical Treatise (he mathematike suntaxis)* is an encyclopaedia of astronomy in which the elaboration of trigonometry established astronomy as a mathematical discipline. His *Geography* makes technical use of parallels and meridians; his *Optics* contains an experimental study of refraction; his *Tables of Reigns* provides chronological tables of Assyrian, Persian, Greek, and Roman kings from Nabonasar to Antoninus Pius. During the Middle Ages the *Tetrabiblos* provided encyclopaedic information in Astrology, Physics, Alchemy, and Magic. Galen is credited with approximately 500 works (of which some 100 are genuine and extant) on medicine, logic, grammar, ethics, philosophy, and literature. He systematized and unified Greek anatomical and medical knowledge and practice. He had a wide knowledge of the history of medicine and of science, and made numerous dissections and discovered new facts in anatomy, physiology, embryology, pathology, therapeutics, and pharmacology. As in the case of the Ionians, Roman contacts with world order were systematized in cosmology and medicine in a context of practical uses and technological applications.

(1.4) The practical, the humanistic, and the scientific impacts of world order were not separate or unrelated. The embracing order of these orders was the Roman contact with world order. Our interpretation of Roman history suffers the literal-minded separations which lead us to study their law but to deny the existence of their humanities and science and therefore to fail to see the lessons for our times in the interrelations they worked out among the three, to face issues similar to those which we face.

The Renaissance was another period in which issues developed directly from contact with world order. Like the Romans the men of the Renaissance rediscovered the Greek classics and used them in ways suggested by the study of Cicero and the Roman classics. Their rediscovery of the classics was conducted in a context of geographical exploration and contacts with peoples and of re-examination of the revolutions of the celestial orbs and of the functions and motions of the animal body.

(2.1) The new focus on the fine arts was a response to, and a continuation of, the Roman concern with the Humanities, the Liberal Arts, and Encyclopaedias. The study of Cicero gave rhetoric a dominant role in application to all fields of study and investigation, as well as fuller theoretic exposition in Ramus, Nizolius, Bodin, and Francis Bacon. The terministic logic of the later Middle Ages was abandoned for an aesthetic and practical communication theory. Many of the Renaissance books in which "encyclopaedia" is used are books of education devoted to an exploration of new forms of the liberal arts. Alsted's *Encyclopaedia* was an information reference work organized according to the philosophies of Lully and Ramus; Francis Bacon laid the plans for a New Instauration of Learning which was to provide the basis for later encyclopaedias.

(2.2) The development of modern nation states and their contacts on the oceans and in new territories stimulated efforts to establish a science of law and formulations of international law. Natural law was combined with compact theory in the investigation of rights and obligations, *jus* and *lex*.

(2.3) The beginnings of modern science in the Renaissance encountered world order, as had the science of the second century, in cosmology and medicine. Galileo brought the line of development from Copernicus and Kepler to a culmination in the world order of astronomy and the world order of dynamics in his *Dialague on the Two Greatest Systems of the World* and his *Dialogues on the*

Two New Sciences. Harvey brought the line of development from Paracelsus, van Helmont, and Dubois, from Mondino and Vesalius to a culmination in his *Essay on the Motion of the Heart and the Blood* and his *Essay on the Generation of Animals.* The presence of technology in these intrusions of world order is apparent in the circumstances which bring it to attention and in the instrumentalities by which it is explored. Galileo's *Dialogues on the Two New Sciences* opens with a speech by Salviati which locates the scene and occasion of the conversation :

> The constant activity which you Venetians display in your famous arsenal suggests to the studious mind a large field for investigation, especially that part of the work which involves mechanics; for in this department all types of instruments and machines are constantly being constructed by many artisans, among whom there must be some who partly by inherited experience and partly by their own observations, have become highly expert in explanation.

(2.4) Renaissance culture is an interplay of these three contacts with world order as opposed to the metaphysical and logical orders of the high and late Middle Ages.

These samples of historical encounters with world order in past ages are not presented as insights into the philosophy of history, but as sources of heuristic devices for the exploration of other times in which encyclopaedias are reoriented to put the Humanities to human uses and to put the Liberal Arts to educational uses. The samples suggest that the impact of world order presents us with issues under four heads—(1) world order as cosmos, (2) world order as community, (3) world order as communication, and (4) world order as order of orders. The problems of world order are properly those which fall under the fourth heading, but they are difficult to state and to face because one of the confusing effects of contact with world order is an induced combination of ingrained *literal* interpretations ensconced in authoritative *analogical* dogmas. The great issues of our times are rigid ideological oppositions developed from pluralisms and multiplicities and designed to eliminate them. Consideration of several dimensions of world order and of their interplay is a means by which one may hope to reduce this intellectual cold war to proportions in which it is intelligible and resolvable. As Plato says

(*Timaeus* 31C), "And the fairest of bonds (*desmôn kállistos*) is that which most perfectly unites into one both itself and the things which it binds together." The most trivial issues which we face today have implications which extend to every aspect of the universe, and in tracing the consequences of proposed treatments of issues, we must not neglect the issues raised by the relation of the direction our analyses follow to the orientations of other practical, aesthetic, and scientific enterprises. The impact of world order is not primarily practical, aesthetic, or scientific; but concessions to literal-minded statement are inevitable. It would be premature adherence to truth, after having based the analysis of Roman encounters with world order on their response in law and politics and the analysis of Renaissance encounters on their response in the arts, to refuse to credit the myth that current encounters with world order must be understood as consequences of the advances of science and technology.

1.   ISSUES OF WORLD ORDER AS COSMOS OR ENCOUNTERED STRUCTURES

The Newtonian physics gave a mathematical adjustment of the two Renaissance encounters with world order to reduce them to a single formulation and to provide the analytic framework to relate the system of the world to mechanics. The Third Part of *The Mathematical Principles of Natural Philosophy* (1687) is *The System of the World in Mathematical Treatment*. Astronomers and cosmologists worked assiduously on the detailed application of the principles of Newtonian physics to the facts of the universe. When Laplace published his *Exposition du Système du Monde* in 1796 slightly more than a hundred years after the publication of the *Principia*, encounters with world order seemed to have been translated successfully into an idea or system of world order in which facts will find their place. Laplace also stated the nebular hypothesis in the *Exposition*. The Newtonian world system depended on the application of a single set of mechanical laws to all phenomena, sublunary and celestial. The further extension of such dynamical analyses was as promising as the extension of relations to further facts in a system of the world. Two hundred years of physicists found in the mechanics of motions of solids a starting-point for ingenious and revolutionary investigations of the motions of liquids, gases, electricity, and light. The second law of thermodynamics gave entropy and degradation a place with inertia and principles of conservation. There was convincing reason to believe that the conception of world system would continue to coordinate the two directions of inquiry—into cosmic whole and into elemental or atomic parts—and to supplement one by the other, using the spectroscope to explain the structure of the atom. However, contact with world order not only led to the alteration of concepts of world order (a normal result of observation of facts inconsistent with assumptions or theories) but also to the alteration of *orders* of concepts related to simultaneity, position, sequence, continuity, contiguity, infinity, parts, and matter which has led to reconsideration of what constitutes a fact or a theory and of their relation to each other.

The dependence of technology on science is so obvious that we tend to neglect the dependence of science on technology. What we know alters the range and nature of what we can do, and what we can do alters knowledge and facts. Albert A. Michelson developed techniques to determine the speed of light with greater accuracy than had been possible before. The Michelson-Morley experiment to investigate the motion of the earth through the atmosphere destroyed ether and contributed to the formation of Einstein's theories of relativity, which required for confirmation observation of facts about the behavior of light in eclipses which had never been observed before, because the reasons and the instruments for observing them were lacking. In 1897 J. J. Thomson performed an experiment to determine the ratio of the charge and the mass of cathode rays. His first attempt was unsuccessful as had been Hertz' performance of the same experiment. Since techniques for producing vacua had improved, Thomson was able to repeat the experiment under a much higher vacuum, and to discover that the cathode ray consists of particles with a small mass and a negative charge. He called the particle a "corpuscle" and its charge, a basic unit of electric charge, an "electron." The particles were later called electrons, and their discovery started the twentieth-century hunt for elementary particles.

The peculiarity of these issues is that they depart from the simple contrarieties which were erected into systems from the seventeenth to the nineteenth centuries and which constituted issues into oppositions of systems. The ether was abandoned because its effect on motion could not be observed, but it was not abandoned for a theory of empty space, because the deviation of light in the neighborhood of large masses suggests a dense space in which motion follows geodesic lines. It seemed reasonable to seek the mass and energy of elementary particles, but they are found to be closely interrelated and one can be transformed into the other. Since position enters in the determination of motion, it seemed reasonable to examine the position and the momentum of elementary particles. It was found that either can be observed but not both at the same time. The issues of the seventeenth century were stated absolutely as oppositions between theories—between Cartesian and Newtonian laws of motion, between momentum and force, between vortices and absolute space, between particles and waves of light. One or the other theory must be eliminated. But they were theories which extended to world order; and in that extension attention focused now on one aspect of processes, now on another, and for the whole extension the two conclusions, that neither theory is true and that both are true, are equivalent. The distance between theory and fact has diminished similarly; branches of mathematics which were developed with no conceivable application have had uses in physics, and branches of mathematics have been invented to fit facts which have no expression in pure mathematics. The extensions have pushed again to the beginnings of the cosmos, in an explosion or in a steady state, which have implications for the organization and the continuing size of the cosmos and for the organization of matter and the differentiation of the elements. The implications extend to all issues between the issues of cosmic whole and the issues of elementary parts, not because a cosmic or particle theory has formulated laws which are transferable but because the dissolution of assumptions has altered the fixity of facts and the necessity of theories and has adjusted the relation of fact and theory in paradoxical truths which have displaced self-evident truths and self-contradictory antinomies.

Physics and technology have provided a model in the twentieth century for the discovery and treatment of issues of world order. It is a model, in the first place, for response to the need to abandon the fixed contrarieties of earlier theory. Physical theory has embraced paradoxes and made them principles. This has meant the abandonment of the absolutes of past theory—the Newtonian absolutes of space, time, motion, and force—but it has also meant the preservation of the basic relations established by the Newtonian physics in a more diversified context of fact and formula. It is a model, in the second place, for use of the horns of a paradox to form alternative hypotheses which lead in their interplay to further progress. For decades the quantum physicists who base their theory on the conviction that the principle of indeterminacy is an irreducible fact of nature and those other physicists who base their theory on the conviction that the establishment of a general field equation will remove indeterminacy have each made contributions to quantum mechanics, have assimilated the knowledge acquired under opposed theories, and have moved on together to new problems. There are other issues raised by contact with world order in which outworn alternatives and contraries are used without relevance or insight, and their oppositions are set up in irreconcilable controversy rather than in related and supplementary hypotheses. The issues of world order may be opportunities for discovery and progress or they may be invitations to reiteration and destruction.

2.   ISSUES OF WORLD ORDER AS COMMUNITY OR ENCOUNTERED PEOPLES

During the period in which issues of cosmic order seemed to require nothing more than patience and intelligence in fitting innumerable facts into the Newtonian conception of world order, the issues of communal order seemed to require nothing more than education and constitutions to make communities free, equal, and democratic. The beginnings of the sequences were the same. Locke undertook to apply the scientific method of Newton to problems of human nature, human understanding, and civil government. Descartes thought of moral philosophy as the trunk of the tree of knowledge that would grow from the application of the method of universal mathematics. The writers of the Federalist papers thought that large republics were possible as they had not been before because of the advance of political science. Contact with world order raised issues not only concerning the conceptions used in the constitution of the infinite world system and of the universal democratic society but also concerning the relation of science to society.

The pattern of the stages of contact of world order in the two histories is the same : the fixed oppositions of the previous period are abandoned; new and more promising contraries are made the basis of theory and action; pluralisms, multiplicities, and relativisms are advocated and desecrated. But the actions and operations by which the transitions are effected retain a remnant of the difference between theoretic and practical. Changes in cosmic thinking are warranted by a consensus of the experts; changes in community action depend on a larger consensus. A community is an expression of plurality and unity; it depends on pluralism and conformism and, conversely or consequently, deviations and conformities are subjects for books which explain the degradation of our times and for committees which investigate the subversiveness of people who are different and the servility of people who are the same.

John Stuart Mill sketched two stages of this evolution in his essay *On Liberty*. When mankind was divided into governors and governed, freedom was advanced by providing the governed protection against the tyranny of government. When people became self-governing, protection was needed against the tyranny of common opinion and the majority; and this was to be achieved by distinguishing individual and society. We have clearly moved into a third period in which men search for new individualisms and new socialisms to protect them from the misuses and injustices of the old, and in which new contraries, like public and private, are proposed to clarify analysis and action.

The sense in which the impact of world order has influenced the last transition is apparent in the evolution of rights and obligations—*jus* and *lex*—in our institutions and our theories. Rights and obligations were first thought of in a "political" context consisting of customs and practices within states and between states. In the seventeenth and eighteenth centuries they were thought of in a "constitutional" context in which forms of government were established by peoples to realize and protect natural rights. In the twentieth century they are thought of in a "universal" and "human" context in the Universal Bill of Human Rights to extend rights which are possible and rights as they are conceived by peoples. The writers of the Constitution of the United States in the eighteenth century did not debate concerning the rule of the majority (although the Constitution they wrote has not yet put it into effect), but the dangers of the tyranny of the majority and of large states did engage their attention in great detail. Our problems are stated better in terms of the manipulation of the opinions of the majority and the minority and in terms of the opposition of needs and aspirations than in terms of the rule of the majority or the tyranny of the majority or the minority. The Four Freedoms illustrate the stages of this development. Two of them, the "of" freedoms, the freedom of speech and the freedom of worship, are examples of the traditional civil and political rights : they depend on institutions to protect the individual in his exercise of his rights. Two of them, the "from" freedoms, the freedom from want and the freedom from fear, are extensions of the economic and social rights institutionalized in political contexts in the nineteenth century : they depend on common action to achieve rights in the benefits of which individuals participate. The difficulties they encounter, including the difficulties which stood in the way of translating the Universal Declaration of Human Rights into a convention, result from the complexities and confusions of a third set of rights, the "for" rights, which have no accepted title but are sometimes called cultural or intellectual rights : they depend on mutual confidence and understanding and on the common acceptance of "goods" according to which differences of "interest" may be arbitrated.

The impact of world order on world community is distinct from the development and application of the conception of world community. World community already exists in the contacts of people and of communities, of peoples and of states; world community does not yet exist in the recognition and operation of institutions required by any conception of world state or by almost any conception of world society. The issues of the impact of world order as community should be sought therefore in the massive actions which have been taken since 1945 and which are surprising and often inexplicable without recognition of the impulse of world order.

(3.1) A large number of peoples who were not self-governing before the Second World War have achieved the right of self-government and their nations have become members of the United Nations. The right of all people to have the ultimate voice in decisions affecting their welfare is broadly and ambiguously recognized.

(3.2) For the first time in history, advanced nations have made available funds, materials, and training to assist, without requirement or direct repayment, less advanced nations to relieve hardships and to satisfy needs. The right of people to minimum subsistence, security, and satisfaction of needs is broadly and ambiguously recognized.

(3.3) These recognitions and programs devised to advance the right to govern oneself and the right to live have encountered an aspiration, even more ambiguously recognized and understood, to justice in the sense of treatment according to common precept and without exception or discrimination.

(3.4) Finally, the aspiration to get somewhere, to make progress, for themselves as well as for others like themselves, is combined with the determination to preserve their own values and to measure progress inspired by other values according to them.

The impact of world order appears in the need to readjust these rights which seem inharmonious or inconsistent when the traditional contraries of political theory and policy are used to put them into operation or to protect them. It has seemed reasonable to attach political or military requirements to technical assistance; and it has seemed reasonable to assume that assistance in improving standards of living raised no questions of justice and endangered no values. In a decade and a half we have slowly learned that these rights and aspirations are interrelated, and that one cannot be treated without the others. We have also only started to learn that the issues raised by the contacts of peoples have their counterparts in issues within peoples. Finally, contact with those issues, within one's own society and in the society of nations, uncovers the basis for the oppositions among aspirations in paradoxes within each of the aspirations.

The issues of world community are problems which draw attention to the need to reexamine the institutions and ideas on which all our freedoms and associations, our spontaneities and customs, our rights and laws are founded.

(4.1) During the same period in which a large number of peoples have established new nations, a large number of established nations, which had been independent of the suzerainty of other nations, have become subject peoples. Many forms of independence and democracy have been instituted and frequently have been defended; parallel forms of imperialism, more easily distinguishable by traditional terms, such as, "political," "military," "economic," "power," and "cultural imperialisms," have been developed and frequently have been disavowed. The issues of self-government—the paradoxical freedom of imposing obligations on oneself—arise in distinguishing, among practices called "democratic," those which facilitate popular decision concerning common good and improve popular response to common duty from those which thwart exercise of freedom and sensibility to tradition. Unless these issues are examined we shall go on repeating the bland contradictories of nineteenth-century revolutionaries and tories: self-determination is alienation, and possession of one's own is interference with the property and privacy of others.

(4.2) "In principle," due to technological advances, it is possible to provide for the minimum needs of all people. "In principle," due to rates of increase of population, it is impossible to continue present living standards of any people. The right to live is the right to continue to function as an animate being; the right to live is the right to reproduce as sanctioned by the customs of one's culture and the precepts of one's religion. Security (for living) and living-space, in all its dimensions from food-room to standing-room (for living beings) have always been reasons for aggression by the strong and submission by the weak. In the issues of world order in our times they are equally frequently reasons for revolution by the weak and subversion by the strong. World order has transformed an age-old paradox and debate into a pressing, difficult, and explosive issue. Socrates argued, in the second book of Plato's *Republic,* against the conception of justice as giving every man his due, and he proposed, in the fourth book, a better definition of justice as every man performing his proper function. The proportion of reward to function in the community was named "distributive justice" by Aristotle. Socrates refuted Polemarchus in the Platonic dialogue, but Polemarchus won the debate in the legal history of the West. We all know that Roman law and all subsequent law has recognized justice as rendering every man his due, that is, the criterion of justice is a passive property in the receiver rather than an active property in the agent. The increased possibility of satisfying needs in a technological age has given the old paradox of distribution and contribution, of satisfaction of needs and reward for work, a new turn. Distributive justice, during the stage of "political" rights, was the proportion of reward to function in the community; it has come to mean, during the stage of "universal" rights, the distribution of goods, facilities, and knowledge according to the ratio of needs in the world community. It is the *right* to participate in the benefits of the advances of science, technology, and culture

(4.3) If justice is a proportion set up in the community between men and goods, justice is also the restoration of the relation of men and goods, when it has been disturbed. It is the reparation of injuries and the resolution of disputes as well as the distribution of goods. Aristotle named this form of justice "rectificatory," and it has con-situted the business of law courts, civil and criminal, since the beginning of authoritative arbitration. During the stage of "political" rights, justice "rectified" injuries (*injuria*) according to laws (*lex*) which were formulated to express justice or rights (*jus*) and which therefore required supplementation by equity or reasonableness (*epieikeia, aequitas*). During the stage of "constitutional" rights, institutions were established in constitutions to express natural rights in laws and to provide criteria for laws and their execution. The dignity of man was embodied in a precept that recognized man as an end in himself and never *merely* a means. During the stage of "universal" rights, world order transfers the criterion from the nature of man to the community of men, and both law and justice, obligations and rights, are reduced to equity. This is the sense of the demand of the new nations for a "common justice," in which they will participate, as opposed to an established law or a natural right extended to them; and it is the element of truth expressed in the exposition of "judge-made law" by the American legal realists. It is a paradoxical issue since it arises from recognition of the implications of the dignity of man. Those implications require fuller exploration for they pass from the consideration of man as an end to the consideration of his use of means. Any judgment of his management or mismanagement of available means runs the danger of imposing a law upon him and reducing him too to the status of means. Equity departs from the universality of strict law to take the person, individual or association, into account; in so doing it recognizes that the individual establishes a universal law for his actions. "Rectificatory" justice has come to mean, not the rectification of "injuries" by the restitution of alienated "goods," nor the rectification of "injustices" by the restitution of natural "rights," but the rectification of "inequities" by the restitution of recognized "responsibilities."

(4) The ends and values of world order must take into account the ends and values of peoples, nations, and individuals. It is more than an extension of the old issues of private interest and public interest, selfish benefit and common good. If the new nations aspire to progress and to get somewhere, the community of peoples which undertakes to assist them in the realization of this aspiration may easily fall into the error of assuming that the locus of the "somewhere" which defines progress is clearly marked. It is natural to assume that which define, delimit, and give significance to material goods, between progress means "material" progress, and it is natural for donors to

add to their gifts moral reflections concerning the dangers of corruption and degradation found in the cultures of prosperous peoples and concerning the need to abandon traditional values and customs in the interest of technology and industrialization. Yet it is no less natural that new nations want to get where the advanced nations are now and to decide themselves concerning the dangers and degradations of their new prosperity. It is natural that they should want to measure their progress according to their own sets of values and customs. The issues are important because they bring to the fore problems of the relation of production of material goods and cultivation of ends which define, delimit, and give significance to material goods, between goods which are used and consumed and goods which are increased by being shared; between work and cultivation, between the occupations of labor and the occupations of leisure. They are issues in which pluralisms and the variable semantics of values are not simply the contraries to determinisms and determinable systems of values, but both are at once means and threats to the good life.

The issues presented by the contacts of peoples move from the relations of nations to the relations of groups within a nation to the choices of individual men. They are being met, well or badly; in many cases the policies of action which have been evolved are better than we have the vocabulary to express. If we can define the issues more clearly and devise new terms in which to express them, much of the pessimism concerning the present world situation will undergo an immediate translation into an optimistic orientation which will be more than a verbal change, since the new mode of expression will contain the indications of modes of action which may resolve the issues. There is no reason why peoples should not in the same processes become more interdependent and more free. They become interdependent because in sharing the same environment of world order their means of communication become more numerous and they themselves become more numerous; they become more free because alternatives of action become more numerous in more dimensions of living. The resolution of the resulting issues must be found by increasing choices and agreements, by inquiry into the possible relations of the established and the emergent. It is an inquiry which will spread through "curtains" erected between independent peoples and subject peoples, because curiosity and knowledge uncover balancing fixities of subjection in the independent, and innovations of independence in the subjected. It is an inquiry which will look for distributive justice in modes of participation in the advances of knowledge and culture, and for rectificatory justice in the reciprocities of cooperation and responsibility, and for equity in an interest in the living consequences of the values of other peoples in which we share, rather than in a dubious "understanding" of their differences or in a difficult "tolerance" of their peculiarities.

3. ISSUES OF WORLD ORDER AS LIBERATING ARTS OR ENCOUNTERED PRESENTATIONS

The three ages which have been used to illustrate the issues presented by world order—Rome, the Renaissance, and the twentieth century —made use of the same modes of expressing these issues: the organization of all knowledge for ready reference in encyclopaedias, the elaboration of liberal arts as a system of education, and the employment of the techniques of rhetoric and communication theory to present and resolve all issues. Viewed as an art which takes into account the content of expression, the medium of expression, and the recipient of expression, these three modes of expression are a single art, and in the treatment of issues of world order they are not always easy to distinguish or separate.

In Rome there was a continuing tendency to follow the conviction of the elder Cato that contact with other cultures, particularly the effete arts of Greece, endangered native Roman virtues. The opposed tendency was given form and vocabulary in the Scipionic circle. Cicero and Varro worked out its practical and theoretic structures and devices. They gave a name to the "humanities," since the study of the outstanding products of man in his arts is the means of knowing Humanity; they argued that *Humanitas* was a translation of the Greek *Paideia,* which we translate sometimes by *education,* sometimes by *culture.* They used the same word in *enkuklios paideia* and applied that term to the liberal arts. Varro enumerated nine liberal arts—grammar, rhetoric, dialectic, arithmetic, geometry, astronomy, music (which became the seven mediaeval liberal arts), architecture, and medicine. This circle of education or of learning, or "general education," is a necessary part of the education of an architect, according to Vitruvius, and of an orator, or lawyer, or statesman, according to Quintilian. The "parts" of rhetoric and the "questions" and "causes" of rhetoric provide a structure for all problems and devices for the "discovery," "judgment," and "disposition" for their resolution. Pliny the elder uses *enkuklios paideia* in application to his Natural Histories, and the Romans prepared ready reference encyclopaedias of the facts of science and history. The Roman contact with Greek culture was the invention of the humanities. The Greeks developed the arts of language and of mathematics, but the Romans invented the liberal arts as a system of education and as a comprehensive schematization of the disciplines of words and of things.

By the end of the Middle Ages the arts of the trivium and quadrivium had been run into the ground. The Renaissance revolted against the verbalism of logic and the sophistical questions and *insolubilia* in which the study of quantities and things became involved—or, more accurately, took their beginning. Renaissance scholars returned, with the help of Cicero and Quintilian, from logic to rhetoric to reconstitute the liberal arts, to rediscover the humanities, and to reassemble the encyclopaedias of information. Rhetoric and its arts of discovery, judgment, and persuasion developed modes of presentation of intimations of world order in each of these encompassing activities. The use of rhetoric as the dominant art in the reconstitution of the seven liberal arts laid the foundations of the modern liberal arts by turning from disciplines to subject matters. From the new organizations of the disciplines of grammar, rhetoric, logic, and mathematics prepared by Ramus and a host of scholars in all disciplines developed the arts and sciences of "literature," "history," "philosophy," "philology," "physics," "chemistry," "psychology." So reconceived the liberal arts were "humanistic" rather than abstract and verbal, and the discovery of man's great achievements in art, facilitated by contact with other cultures, Greek and Roman, was communicated and judged by rhetorical devices designed to instruct, to move, and to please. The rebirth was announced as the consequence of contact with other cultures, and the "Middle Ages" were named in commemoration and promotion of the historical continuity of world order. The devices of rhetoric extended the effects of that order in other dimensions—Nizolius used the "commonplaces" of the older rhetoric to expose pseudo-philosophies and discover new metaphysical principles. Francis Bacon used them in application to things rather than words to establish a new inductive logic of discovery. Descartes applied scientific method to the study of human passions for the first time to discover distinctions which rhetoricians had expounded. International law found its aids in Cicero's rhetoric and Roman law. Alsted's Encyclopaedia was a comprehensive organization of knowledge based on the philosophies of discovery and languages, and Francis Bacon's organization of the sciences was to provide the schematism for generations of encyclopaedias. Even the distinctions are inaccurate or inexact. These were not different ways of reflecting world order; in so far as they served that function they are difficult to distinguish from each other—the earliest works called "encyclopaedias" were programs of studies; the humanistic handbooks became compendia of information; and handbooks of information were conceived to be instruments of discovery for further exploration of the world order.

By the end of the nineteenth century the departmentalization of the arts and sciences according to subject matters had reached a point of accomplishment and sterility comparable to that achieved by the liberal arts in the fourteenth century when they inspired revolt against verbalism, aridity, and impracticality. The simple form of the revolt has been the denial of separation and espousal of programs of "interdepartmental" or "interdisciplinary" cooperation and "unity of science." During the last fifty years the liberal arts have been subject of serious thought and experimentation, and *enkuklios paideia* has taken the form of *general education,* which is contact with world order in the universe of knowledge, the community of cultures, and the discovery of structures of communication, sensibility, and action. It has been a period of the discovery and extension of media of mass communication and of the formulation and use of theories of communication, play, and decision making—all infant children of rhetoric. It has been an age, finally, of encyclopaedias of ready information in process of growth and revision to include all the facts, to keep up to date, and to extend from limited perspectives to the world and to world order.

It is highly probable that universities and educational systems will be profoundly altered in organization and operation in the next fifty years. It is not probable that these changes will have any significant relation to whether or not the subject matter departmentalization of the modern liberal arts is continued or arrested or whether or not the disciplinary verbalism of the mediaeval liberal arts is revived or excoriated. It will depend on discovering the relevances of disciplines and on cultivating disciplines in those relevances. The contacts of world order have been rather more effective guides in these transitions than the conceptions of world order, and the transitions of the past suggest that this transition will depend on elaboration and use of disciplines that move from field to field. The history of rhetoric as a discipline would probably suggest few solutions to world issues; the history of literature, which was once treated as a part of rhetoric (or of grammar) and which now associates with the adjective "world," has more advocacy but not much more promise. The history of devices or "places" of discovery and invention, which begin in rhetoric and have their development in literature, the natural sciences, philosophy, and the behavioral sciences, has theoretic and practical possibilities to which our operations in world order turn our attention. In this process the orientation to the problems in each of the disciplines—in philosophy, history, literature, the behavioral sciences, and the natural sciences—will respond to this reorientation. That orientation to world order is the new encyclopaedia.

4.   ISSUES OF WORLD ORDER AS ORDER OF ORDERS OR ENCOUNTERED POSSIBILITIES

World order may readily be conceived as cosmos, community, or the liberating arts. In each view of world order the new issues encountered are concealed, or reduced to paradox, by our use of the contrarieties successful in a preceding age, usually a continuation of the age immediately preceding, but irrelevant to new issues. The distinction of cosmos, community, and liberating arts, however, is itself such a structure of oppositions appropriate to segmentations but not to world order. We demand more and more worrying books about the opposition or indifference between the sciences and the humanities, the theoretic and the practical, the morality (or immorality) of duty and inclination. We repeat the preachment that the knowledge of man and of values has not kept pace with the knowledge of things and the world, and we design programs to make sciences of them or to preserve them from scientism. We accept diagnoses of the culture of our times as anti-intellectual, anti-artistic, anti-human, and become involved in debates concerning mass culture and elite culture. We recognize the power, danger, and degradation of the new means of communication, and we seek new liberating uses in formulae for universal application to realize the potentialities of machines in liberal and human education and appreciation of the arts. We express our inquietudes by moving through a succession of (to use an old-fashioned term) architectonic disciplines. Science is architectonic; science will solve all our problems; and our practical and aesthetic problems are susceptible of scientific investigation; but science in turn has many meanings—it is theoretical and practical, a consensus of the community of scientists, and an art. The customs and criteria of society are architectonic; our scientific and aesthetic problems no less than our practical problems are resolved by the meteorological forces which determine the climate of opinion of a time and place; but the bonds of society are numerous and include sanctions and exercise of power, recognition of common values and meanings, mutual understanding and common purposes and objectives. The arts of presentation are architectonic; art is needed in scientific proof and political persuasion as well as in artistic creation and organization; but art does many things—it discovers, distinguishes, creates, proves, communicates, arouses, moves, and delights.

Contact with world order raises issues for solution; conceptions of world order raise issues for debate. There are long periods in the history of man in which progress is made by debate within the framework of an accepted and living conception of world order. When both sides of the debate seem to be right, the issues must be reformulated. It is often the case that the new issues are encountered first in the actions we take rather than in the statements we make about the issues of our actions. We debate about the unity of knowledge and the need for specialization; we learn the fruitful interplay of entertaining a plurality of hypotheses and selecting one among them by verification. We debate about the relations of individual and society; we learn about the conformities of social responsibility which are the condition and consequences of individual spontaneity. We debate about the degradations of taste and the misuse of leisure; we learn about diversities of interest even in the experience of a common object and the uniformities of appraisal which can be apparent only if ambiguities of attention are removed. We debate about dependency and freedom; we learn about the possible advance of independence together with interdependence.

An encyclopaedia in an age of transition to the consideration of new issues must be more than a handy reference book to all ascertained facts and all established theories; it must also provide a guide to the issues which all men face in living, inquiring, and acting. There are facts and conceptions of human nature attested in psychology, sociology, anthropology, biology and physics, or expressed in the arts, or used in ethics, metaphysics, and theology. But the reader of the encyclopaedia has also encountered human nature in ways which help him relate characters of novels, neurotic personalities of psychiatric case studies, and virtues and imperatives of ethics. Human nature as it is encountered will never be reduced to a pattern of words or a structure of theory, not because of any imperfection or paucity in language or thought, but because encountered human nature is infinite and statements and conceptions are finite. This disparity is the source of issues and the promise of progress. There are facts and conceptions of world order attested by knowledge, community, and communication. But world order is encountered outside the context of cosmologies, world-states, and art-products expressive of humanity or mankind. Ultimately it is the framework of world order which is the source of all issues newly emergent from the context of a single world of galaxies, of intellectual constructs, of projected actions, or of structured sensibilities and realizations. The world order sets the dimensions in which the circle of learning of the new encyclopaedia must be plotted. The operation of world order is detected in issues encountered long before they are formulated and solved in science, in society, or in presentations. Their interplay opens up the related possibilities which are opportunities for speculative thought, practical action, and poetic production.

ON THE MEANING OF SCIENTIFIC REVOLUTIONS

JOSEPH J. KOCKELMANS
Pennsylvania State University

I. *Introduction*

OVER THE PAST forty years many philosophers have suggested, if they did not in fact explicity defend, the thesis that a radical and exhaustive explanation of contemporary science can be given by logic and epistemology of science. Many objections to this view have been raised over the past decades by historians, sociologists, and philosophers. The intent of these criticisms is not to deny the value of the work that has been done in logic and epistomology of science, but rather to delineate the precise limits within which such an approach of science can be said to shed light on science and to explain its real meaning. It is important in regard to the topic of this article to reflect briefly on some of these objections.

Today sociologists frequently claim that regardless of the merits of the investigations made by logic and epistemology of science, the inquiries do not actually elucidate the all-important function of the scientific community in the establishment of theoretical matrices. Historians point to the fact that logic of science conceals just those moments of science which are precisely vital to an understanding of the development in science : those moments in which an individual scientist or group of scientists rebels against the prevailing paradigm of the scientific "establishment," and tries to find new ways. Finally, philosophers of different schools and trends have focussed their attention on the fact that regardless of the intrinsic value of logical investigations (a value which they do not deny), logic of science presupposes a great number of insights which, upon further reflection, appear to be questionable or at least to require further justification. These issues refer mainly to the underlying conceptions of man, his world, and the priority of theoretico-cognitive approaches

231

to the world over all other approaches. These philosophers claim that almost all of those directly involved in logic of science are inclined to subscribe to a conception which attempts to come to an understanding of man and his world only and exclusively on the basis of a better understanding of man's science. One must realize, they say, that science is the product of a historical community, that adopting a theoretical and scientific attitude is only one of the many attitudes a human being in a society can adopt and, thus, that it is not correct to try to determine the humanity of man from the viewpoint of science alone. In their view one should argue exactly the other way around: the genuine meaning of science can be determined only after one has formulated one's conception of man and the world he lives in; but this can be done "rationally" only by taking *all* human experiences into account.

If it is true that science is only one among many articulated and organized forms of experience, then one of the most important philosophical issues in this connection will be the question of precisely how science is to be related to certain other forms of experience such as religion, art, technology, social and political praxis, etc., and consequently what the *relative* value of science is within the life of man, taken individually and socially. Finally, they say, if the first conception is correct, then man is either a pure consciousness (rationalism), or he is the most complicated *automaton* (scientism). But these views are in apparent contradiction with the most primordial experiences a human being has of himself. Furthermore, in the context of the same supposition, the world in which man lives could reasonably consist only of those aspects for which science is able to account rationally. But this again is in plain contradiction with the most basic experiences a human being has of his world. If we are to account for man's genuine experiences with regard to himself and his world, we must stop trying to force man into a framework of categories which are taken from the "objects" of man's theoretical, and notably scientific, knowledge and return to those concepts which follow immediately from his pre-scientific, but nevertheless genuinely human experiences.[1]

However, as I said previously, all of this does not mean a denial of the value of the work done in logic and epistemology of science. On the contrary, these reflections merely intend to delineate the realm in which logical and epistemological reflections on science can make a genuine contribution to philosophy of science as a whole. It seems to me that many historians and sociologists exaggerate the value of their investigations as much as many logicians have in their field. I wish to show this in connection with three closely related issues. In so doing, I shall take my point of departure from the very interesting and important discussion concerning the structure of scientific revolutions.

Since I can point only to a few general ideas in this essay, I have decided to join forces with the existing literature on the subject. I have chosen those authors who have most clearly and explicitly expressed their views in this regard. Since Kuhn's conception is known by everyone interested in philosophy of science, perhaps the best way to introduce the reader to the problems is to briefly summarize some of Kuhn's views in this respect.

II. *Kuhn's view*

In his thought-provoking book *The Structure of Scientific Revolutions,* Kuhn claims that if one does not conceive of history as a repository for anecdotes and chronology but in a more fundamental and broader way, history will provide us with an image of science quite different from the one most people now possess from a serious study of textbooks, or from reading a popularization of science or even from a study of the many Anglo-American books on philosophy of science.[2] The commonly accepted image of science is drawn exclusively from the study of finished scientific achievements as they are recorded in classics and textbooks. Thus this study is unable to shed light on the social enterprise in history that produced the achievements. The commonly accepted conception of science presupposes that as far as the content of science is concerned, the latter is adequately dealt with in the observations, laws, and theories described in the classics and the better textbooks. Furthermore, as far as methods are concerned, this conception presupposes that scientific methods are simply those illustrated by the manipulative techniques used in gathering textbook data, together with the logical operations employed in relating those data to the theoretical generalizations found in these books. It is Kuhn's conviction that one comes to a quite different conception of science by following science carefully in its historical development. But such an endeavor is fruitful only if one is willing to give up the classical conception of history of science.[3]

For quite some time historians involved in the history of science have taken for granted that science is a constellation of facts, theories and methods collected in texts. Scientists are those people who have striven to contribute something to this particular constellation. Scientific development is the piecemeal process by which these items have been added to the ever growing stockpile of scientific knowledge and scientific technique. History of science is the discipline that chronicles both these successive increments and the obstacles that have inhibited that accumulation.[4]

In Kuhn's opinion many contemporary historians have great difficulty with this concept of development-by-accumulation. They are inclined to believe that science does not develop by the accumulation of individual discoveries and inventions. They feel, also, that it is incorrect to call certain theories which were later rejected, unscientific, and believe that those once current views of nature were, as a whole, neither less scientific nor more the product of human idiosyncrasy than those to which we subscribe today. But if one subscribes to this latter view, then it follows that the concept of science includes not only those views which have maintained themselves in the historical process, but equally the bodies of belief which are incompatible with those held today. If this is the case, then scientific development can no longer be seen as a process of accretion.[5]

Obviously this new conception of the history of science is founded in a large number of connected reflections and has led to many unexpected results. For our purposes it is not important to go into detail; it seems more in keeping with the topic of this essay to turn to those aspects of science which contemporary history of science has been able to bring to light. Kuhn mentions as a first result of this new approach the insight that methodological devices, by themselves, are insufficient to dictate a unique substantive conclusion to many kinds of scientific questions. It is not difficult to show that in the early developmental stages of almost all sciences there has been continual competition between a number of distinct views of nature, each partially derived from, and all at least roughly compatible with, the dictates of scientific observation and method. These various trends and schools are not differentiated by some failure of method (they were all "scientific"), but by incommensurable ways of seeing the world and of practicing science in it. True, there can be no science without a commitment of the scientists in regard to admissible scientific methods, but these methods alone do not determine a particular body of admissible scientific beliefs. In addition, there is always an apparently arbitrary element (in which personal as well as historical contingencies play a part) which is a formative ingredient in the beliefs accepted by a given community of scientists at a given time.[6]

In order to be able to give an accurate description of the kind of development which actually did take place in science, Kuhn makes a distinction between "normal science" and periods of "scientific revolution." In the case of normal science the scientific community commits itself firmly to certain conceptions concerning the fundamental entities which compose the universe, the basic way in which these entities interact between themselves and with man, the methods which can be employed in scientific research, and the questions which can legitimately be asked in this context. These commitments are firmly imbedded in the educational initiation which prepares the student for his professional practice. Through its research normal science attempts to force nature into the conceptual framework which is supplied by professional education on the basis of the fundamental commitments accepted by the group. In other words normal science presupposes that the scientific community knows what the world is like and much of the success of the whole enterprise depends upon the community's willingness to defend that basic assumption. However, because of the fact that arbitrary and contingent elements are involved in these basic commitments, fundamental novelties which are subversive of the basic commitments of the group emerge. The community will try to suppress these novelties, but it eventually has to face the anomalies which in turn will lead to a crisis situation in that science. In such a crisis situation extraordinary investigations begin that will lead the scientific community to a new set of basic commitments, to a new paradigm or set of paradigms which then constitutes a new basis for the practice of science. Kuhn calls these extraordinary episodes in which that shift of professional commitments occurs "scientific revolutions."[7]

Kuhn has described what he calls "normal science" and "scientific revolutions" with great competence and ingenuity. Space limitations do not permit me to follow his investigations here in greater detail. Because of the main topic of this essay, I will turn immediately to *some* of the consequences which follow in Kuhn's view from his approach to science, assuming that most of the readers are sufficiently familiar with Kuhn's ideas, or at least have easy access to them. What has been said about Kuhn's attempt in the foregoing pages is sufficient background for understanding the points I should like to make in this connection.

In the last chapter of the original edition of the book, Kuhn asks the following question: if this description has at all caught the essential structure of science's continuing evolution, why then is it that the enterprise sketched here moves steadily ahead in ways that for instance art, political theory, and philosophy do not? Why is progress a prerequisite reserved almost exclusively for the activities we call science? Kuhn suggests that we have to give up the idea that changes of paradigm carry scientists closer and closer to *the* truth. What is important is the development from primitive beginnings, through successive stages to an increasingly detailed and refined understanding of nature, and not the development of science toward a certain goal. We are all deeply accustomed to viewing science as the one and only enterprise that draws constantly nearer to some goal set by nature in advance. But does it really help us to imagine that there is some one, full, objective, true account of nature and that the proper measure of scientific achievement is the extent to which it brings us closer to that ultimate goal?[8]

Science is not a goal-directed process but a process in which the successive developmental stages are marked by an increase in articulation and specialization. Such a process may have occurred without the benefit of a set goal, a permanent fixed scientific truth of which each stage in the development of scientific knowledge is a better exemplar.[9] It is Kuhn's conviction that if we could learn to substitute evolution-from-what-we-do-know for evolution-toward-what-we-wish-to-know, a number of vexing problems might vanish in the process. Somewhere in this maze, he says, must lie, for instance, the problem of induction.[10]

It is precisely in regard to this view that Kuhn has been attacked by some philosophers of science. They accuse him of defending a form of relativism which is applicable to several other forms of man's experience, for example, art, but is certainly not appropriate for science. In a postscript which Kuhn added to the second edition of his book in 1969, he explained that as far as this issue is concerned, he does not defend a mere relativism, he wants only to escape from the naive view that a scientific theory is considered better if it is not merely a better instrument for puzzle solving than previous theories, but also because it is somehow a better representation of what nature is really like. "One often hears that successive theories grow ever closer to, or approximate more and more closely to, the truth. Apparently, generalizations like that refer not to the puzzle-solutions and the concrete predictions derived from a theory but rather to its ontology, to the match, that is, between the entities with which the theory populates nature and what is 'really there'."[11]

Although I agree with Kuhn's rejection of naive realism, I do have great difficulty in accepting his thesis that science is not oriented toward a goal. But before clarifying my view in regard to this particular point, I should like first to make some general remarks on Kuhn's approach to science in order to avoid having my discussion of it appear too one-sided.

Generally speaking, I am very sympathetic to Kuhn's approach to science. He makes it quite clear that a *merely* logical approach to science does not and cannot give insight into *all* the important aspects of science; it certainly cannot give us any insight into the social and historical aspects of science and its development. By making some of these social and historical elements the explicit subject matter of investigation, Kuhn is able to shed light on the revolutionary character of science's development. Whereas logic of science speaks almost exclusively about the minimal, logical requirements and conditions which must be fulfilled by scientific conceptions, systematizations and theories (found in normal science) in order to be scientifically acceptable, Kuhn is able to focus on science as a process of a living community and on the "principles" which guide its development in and through revolutionary phases.

The first difficulty which I experience in regard to Kuhn's position is that he believes the second approach can in principle solve all the problems which came to the fore in that very approach. As I see it, the historical and social approach to science can, indeed, point to difficulties, questions, and problems which do not emerge in a logical and epistemological approach. I doubt, however, whether these issues can be "settled" without an explicit appeal to an "ontology" of science.[12]

A second difficulty with Kuhn's position is the fact that his description of revolutionary phases seems to be incomplete and (connected with this) that it places all revolutionary phases on a par. Kuhn mainly stresses those elements in which an earlier disciplinary matrix (paradigm) differs from a later one, but it seems to me that he does not give sufficient attention to the many elements they still have in common. Among others things it is these common elements which make it possible to apply to all of them the basic insights gained in the logic of science. Obviously I do not deny the fact that the differences between Newton and Einstein on the one hand, and Newton and Heisenberg on the other are enormous. But all the differences notwithstanding, these three prominent scientists agree in almost everything which is basic and vital as far as natural science and its ideal of scientificity are concerned.

In order to provisionally clarify this thesis to which I shall return in the following pages, may I quote two passages from well-known physicists which, taken together, illustrate the point I am trying to make. Speaking of the progress of contemporary physics Louis de Broglie writes : "Like all the other natural sciences, Physics advances by two distinct roads. On the one hand it operates *empirically,* and thus is enabled to discover and analyse a growing number of phenomena—in this instance, of physical facts; on the other hand it also operates by *theory,* which allows it to collect and assemble the known facts in one consistent system, and to predict new ones for the guidance of experimental research. In this way the joint efforts of experiment and theory, *at any given time,* provide the body of knowledge which is the sum total of the Physics of the day." The second quotation is from Heisenberg : " . . . we must not be misled into underestimating *the firmness of the foundations of exact science.* The concept of *scientific truth,* on which science is based, can apply to many different forms of knowledge. Thus, *on it are based not only the sciences of the past centuries but modern atomic physics also . . .*"[13]

At any rate, revolutions in science do not always entail revolutions in regard to the ideal of natural science; they consist mainly in a denial of the validity or *universal* applicability of an earlier accepted explanatory scheme. There are several cases in which a scientific revolution did not lead to the abandonment of an earlier disciplinary matrix, but merely to a more careful delineation of the realm of its application. There are other cases in which a scientific revolution did lead to the abandonment of an earlier disciplinary matrix; in these cases science returned from a "dead end" to scientifically viable roads. There have even been scientific revolutions in which an earlier scientific matrix was abandoned because it became clear that it could not live up to the standards which we must impose on a genuinely scientific enterprise. It seems to me that in dealing with the nature of scientific revolutions we cannot just place all of them on a par.

The greatest problem inherent in Kuhn's view, however, is connected with this basic philosophical conception, one very seldom made explicit but one which on several occasions forces him into an extremely difficult position. I wish to consider here only those problems which are immediately relevant to the main topic of this essay. I have already pointed to the fact that in Kuhn's view there can be no question of a goal or a *telos* in science. All teleological conceptions of science are to be abandoned. In formulating this view Kuhn states that we must give up the idea of seeing science as the one enterprise that draws constantly nearer to some goal set by nature in advance.[14] I am willing to agree with Kuhn when he claims that we should not subscribe to a goal *set by nature,* because then we must account rationally for this goal *and* for our knowledge of it; this, however, can be done only by appealing to questionable "metaphysical" speculations. But, why should this goal be *set by nature?* Could it not be the case that man himself has set this goal which then predelineates an ideal of science whose basic elements and principles logic and epistemology of science precisely want to describe and explain?

In view of the fact that this is perhaps the most fundamental issue underlying the discussion, I should like first of all to make some introductory remarks on "scientific rationalism." In this connection I should like to defend the thesis that, whereas "rationalism" is an adequate label for our scientific pursuit, it can never become an adequate label for a philosophical position. From these reflections on scientific rationalism, I hope to turn to the typical historicity of science and the problem of induction in order finally to return to the questions concerning the structure of scientific revolutions.

III. *Three Basic Issues*

1. RATIONALISM AND MODERN SCIENCE

On several occasions Husserl pointed out that the origin and development of modern science since the time of Galileo have been guided by the basic insight that there is a rational, infinite totality of being with a rational science systematically mastering it; or formulated in another way, the idea that the infinite totality of all that is, is intrinsically a rational all-encompassing unity that can in principle be mastered by a corresponding science without anything left over.[15]

From a historical point of view it is not difficult to substantiate the thesis that Galileo, indeed, was guided by this basic idea. Educated in Pythagorean neo-Platonism it was almost impossible for him to look at science in any other way. It is well known, also, that Descartes' whole approach to science and philosophy was founded on the same conviction. Analogous remarks could be made about Leibniz, Kant and Laplace.

However, if this thesis can be defended on historical grounds, then the consequence will be that one will have to admit that modern science, *factually*, has been teleologically oriented from the very start. The people who started modern scientific development were actually guided by an ideal, a goal, that in their view, in principle at least, can be materialized. As they saw it, it is this goal which gives us a standard with which to evaluate the contributions made by each scientist and each group of scientists. Furthermore, this teleological orientation gives us a means of delineating in advance what should be understood by scientific truth and by a definitive scientific accomplishment. Finally, it is obvious that such a conception of science necessarily leads to an interpretation of the meaning of scientific revolutions which, in addition to the revolutionary elements to which historians have rightly pointed, attempts to explain the continuity of scientific development as a whole.

Although in contemporary logic of science a similar claim is not explicitly made (to the best of my knowledge at least), and although most philosophers today would certainly be unwilling to introduce a concept as elusive as "teleology," nonetheless, they seem to presuppose an ideal which is very similar to the one Galileo had in mind. If this were not so, then it would be difficult to understand why they can claim that certain principles, methods, and procedures are to be accepted in science and why others are to be rejected. It seems to me that only where one is willing to subscribe to an ideal which (regardless of the way in which one wishes to formulate and justify it) in the actual pursuit of science necessarily has the function of a goal to be accomplished, will it become understandable why a decision can be made and justified concerning, for instance, whether or not Mill's conception of induction is acceptable.

Be this as it may, if certain scientists (sociologists and historians) and philosophers today claim that science is not goal-oriented, then they object to a conviction which, within science, is as old as science itself. In this connection it is of importance, however, to state explicitly that the point which I am trying to make here does not necessarily hinge on the typical formulation which Husserl (inspired by Descartes and Kant) has given to this ideal of science.[16] I agree with sociologists and historians that this formulation, indeed, is inadequate. But it seems to me that this ideal could be reformulated in a way which would avoid all the difficulties inherent in Husserl's conception of it.

But, before attempting to do so let us return for a moment to Kuhn's position. Obviously, one can say that Kuhn does not exclude a goal set by man, but that in his view such a goal cannot account for that kind of progress in science which most scientists and philosophers admit. It seems to me that the most fundamental reason for Kuhn's reluctance to admit "progress in science" is connected with the fact that he takes his starting-point in a radical subject-object separation. True, he does not follow Descartes or Hume who took their points of departure in an isolated, individual, human consciousness. For this abstract consciousness, Kuhn rightly substitutes the community of scientists. In his view, this community settles on certain rules for its theoretical game. The group subscribes to this set of rules as long as they work. If they do not work, then a dissolution of the community takes place and a scientific revolution is its necessary consequence. Kuhn is forced to this position because he feels that he cannot admit a rationality in nature which could give the human enterprise the character of necessity. For Kuhn, it seems to be obvious that, if it is clear that the scientific enterprise cannot be guided by a rationality inherent in nature (because we are unable to account rationally for such a rationality), then it follows that the whole of our scientific enterprise is reduced to a game guided by rules to which the members of a scientific community freely subscribe. Kuhn is unwilling to admit that these rules are in any sense necessary. The community subscribes to them because they "work"; but one does not know in advance that this is the case; one learns it from the results which follow from application of the rules. Given these facts, Kuhn argues, it does not make sense to speak of an evolution toward a *telos*; the only reasonable thing we can do is to speak of an evolution from primitive beginnings.[17]

It seems to me that the basic question at stake here cannot be solved by approaching it, either from the viewpoint of "nature," or from the viewpoint of the community of scientists, both taken separately. In my opinion, we must try to approach this problem from the realm of meaning which precisely constitutes the scientific "dialogue" between man and world. That this realm of meaning possesses its typical rationality and necessity, *and* that the enterprise taken as a whole is oriented toward a clearly pre-delineated goal, is not due to man or nature taken in isolation, but to the fact that in this "dialogue" between man and world, the scientific community chooses to follow rules on the basis of rational principles which are in complete harmony with man's ideal of reason, *and* the fact that this community is willing to test the application of these rules on the "unities of meaning" (phenomena) which this "dialogue" itself brings about.[18]

I shall try to justify this view in greater detail in the pages to come; and I will take my point of departure from some reflections of Gaston and Suzanne Bachelard. Before doing so, I should like to make one general remark at the conclusion of this section : if there is any truth in my conception, then, the logic and epistemology of science seem to have made a greater contribution to our understanding of science than Kuhn seems to suggest. For, as I see it, while logic and epistemology of science do not shed any light on science as a social and historical process, they certainly do articulate and clarify our scientific ideal which as a *telos* prescribes the minimal conditions to be fulfilled by our scientific systematization.

2.   RECURRENT HISTORY

In trying to determine the typical historicity of science, I shall briefly compare Kuhn's view with the conception proposed by Gaston Bachelard.[19] As I see it, Bachelard would agree with Kuhn that it is impossible to give an adequate insight into contemporary science merely by referring to the logic inherent in science, but that, on the other hand, any systematic attempt to give insight into the logic of science is meaningful, important, and necessary from a philosophical point of view. Furthermore, like Kuhn, Bachelard strongly stresses the historical element of science taken as a human and social enterprise. He also underlines the revolutionary character of many new theories and paradigms. Unlike Kuhn, however, he draws quite different conclusions from these convictions. Whereas Kuhn seems to suggest that the pursuit of science is without an inner *telos,* Bachelard maintains such a *telos* in his conception of science as "applied rationalism." In other words, where Kuhn does not have a solid explanation of why evolution and revolutions in science take place the way they do, Bachelard has a quite convincing explanation for these phenomena in that he believes that science is guided by an ideal of reason whose precise meaning can be established. Finally, whereas Kuhn describes new paradigms merely in terms of a break with the past, Bachelard shows clearly that each new viable paradigm is *also* and *necessarily* a historical synthesis.[20]

Bachelard begins his explanation with the remark that the question concerning the objectivity of science cannot be isolated from science's historicity. An account of science's objectivity will have to take into consideration all the insights which have come to the fore in what we call "logic of science"; but it must equally well cope with the problems connected with science as an historical phenomenon. The basic question in this connection is : on what level of scientific thought does the history of thought become integrated into our own scientific activities. It is not correct to suggest that in its search for obectivity science can take it as a constant rule to start from a *tabula rasa*. Suppose that you would have to introduce someone to quantum mechanics; you could try to do so on a strictly axiomatic basis; but your student would not understand the "meaning" of quantum mechanics and thus its complete fullness; he would not appreciate the vital issues of the rational reorganization of experiences implied in this theory—if you did not first explain to him the history of wave-phenomena from Newton to Huygens on the one hand and, say, Einstein to de Broglie on the other.

Making this claim, however, is not tantamount to saying that Schrödinger's or Heisenberg's theories have been determined by this history. The new historical synthesis was not prepared by its history in such a way that it must follow with necessity, or even in the sense in which one says that this new synthesis was already "in the air." The new synthesis has equally and equiprimordially the character of a basic transformation and revolution. Even if one takes all the available historical data into account, one is still unable to say that science was propelled by historical reason in the direction of this new synthesis. If one compares any one of the classical conceptions concerning wave phenomena with that of quantum mechanics, it is clear that quantum mechanics is an obvious break in the evolution of modern science. And yet, notwithstanding its revolutionary character, notwithstanding its character as a break with the regular historical evolution within the context of a given paradigm, a theory such as quantum mechanics is and remains a *historical synthesis,* which although it did not follow from science's history with necessity, can, nevertheless, not be understood genuinely without that history.

From these reflections it becomes clear that there is an enormous difference between history of science and the history which deals with all other cultural phenomena, philosophy not excluded. For, unlike general cultural history which does not have access to any norms with which to separate progress from regress, error from truth, the inert from the active, prejudice from fruitful hypothesis, history of science appeals to such standards. In other words, in history of science one does not only try to understand, but also to judge. Political history, for example, correctly sees as its ideal the "objective" description and explanation of the facts; and it demands that the historian not pass judgment. If the historian imposes the values of his own era on the values of earlier times, one will rightly accuse him of following the "myth of progress." In the case of science, however, progress is and can be demonstrated; and its demonstration is even of pedagogical importance in that it immediately contributes to the development of our contemporary science. In short, progress is the dynamic element which propels and guides a scientific culture and it is this dynamic element which the history of science must describe. It must describe that dynamic element by judging, evaluating, and removing all possibility of returning to erroneous notions and conceptions. History of science points to errors of the past only in order to push them aside. Because of this dynamic element the historian is able to distinguish between positive and negative moments in the history of science. These moments are so clearly distinguished from one another that the scientist who sides with the negative moment puts himself outside the scientific community. The negative moments must be barred when they tend to re-appear. In such cases, Weyl once said, there is no return possible in science.[21] On the contrary, whatever remains as positive from the past, will work again in modern thought. The positive heritage of the past constitutes an actual past whose working is manifest in the scientific thought of the present time. If the expression "historical dialectic" makes any sense at all, then it is in the history of science, because here one must, time and again, reform the dialectic between a history which is definitely past and the history which is sanctioned by the active science of today and will forever be of importance for the pursuit of science. The expression "forever" refers here to concepts and conceptions which are so intimately connected with the scientific pursuit itself and so indispensable for scientific systematizations that we cannot understand ever being led to abandon them. True, these conceptions remain contingent, occasional, and somehow conventional; and without a doubt, they are formed in an historical atmosphere which is obscure. But they have become so precise, so clearly functional, that we do not have to be afraid of an educated doubt in their regard. As Becker once remarked, there are theories which are so essentially well-founded that they constitute a necessary step in our knowledge of nature.[22]

These reflections show the necessity of developing a recurrent history, a history which can be clarified by means of the finality of the present, thus a history which starts from the certainties of the present and discovers in the past the progressive formation of our scientific truth. This recurrent history which shows how science confirms itself in the story of its own progress, appears in the works of actual science only in the form of short historical introductions. And since these introductions are mostly cut short in textbooks, getting a thorough insight into the genuine meaning of science is often made difficult for the student.

This conception of the history of science implies the rationalism I mentioned above. For, the moment one begins to consider the history of science as the realization of the progress of science's own rationality, then that history begins to appear as the most irreversible of all histories. History of science shows that in this particular sector of our culture all forms of irrationalism are in principle excluded. This is not to say that the struggle between error and truth will be over some day. On the contrary, each generation of scientists must reaffirm the stand of rationalism, and time and again take up the balance of history. And then it will sometimes be necessary to enter into a part of science's history that has been abandoned for a long time. Perhaps in this connection it is meaningful to make with Richtmeyer[23] a distinction between history of science and the "story of science." History of science, which is an empirical science, is interested in an objective description and explanation of the facts; the story of science goes beyond that in that it describes that same history *as guided by a finality of reason,* that is, a finality of scientific truth as well as of technical realization. The story of science is, for the actual pursuit of science, pedagogically of much greater importance than science's history taken in the strict sense.[24]

The thesis that one must attribute the essential characteristic of evaluation and valorization to the very history of science and thus that scientific truth is by definition a truth *which has a future and a telos* can also be justified by pointing to the astonishing progress of the mathematization of the physical sciences which suggests a new motive of valorization. For, with mathematics incorporating itself into the physical sciences, the *apodicticity* which is essential to mathematics becomes an integral part of the systematization and explanation of our scientific experiences. The value of the mathematical concatenation which connects the axiom with its consequence is added to the value of the concatenation which attempts to connect the cause with its effect. But in order to understand the genuine meaning of the mathematization of nature, it is vital to avoid two obvious mistakes which have been made in history by many scientists-philosophers: the conception that this mathematization of nature presupposes that nature is inherently mathematical (Platonism) and, thus, that the goal of science is set by "nature," *and* the conception that mathematics is merely an instrument or a language (Conventionalism) and, thus, that science is not goal-oriented. I have dealt with this problem in another context and so will limit myself, here, to explain briefly why I feel these conceptions are in principle mistakes and how these mistakes can be avoided.

Platonism, as we find it in Kepler and Galileo, states that there is an eternal mathematical order in nature. This mathematical order is not something which the scientist adds to the phenomena he observes, but is something discovered by him *in* these phenomena and constitutes their very nature and cause. According to another conception, mathematics constitutes the language of natural science. In this view, mathematics is an optimal tool to express the insights of the natural sciences. Although there certainly are isolated cases suggesting such a view, "expressionism" falls short the moment these isolated cases are integrated into an encompassing conceptual framework. For at that moment it becomes clear that the intelligibility which is inherent in our own conceptual framework forms an integral part of the insights which are found in our description and explanation of observable phenomena; and this is even true to the extent that there are no scientific insights at all outside that or a similar conceptual framework. At that moment, it becomes clear, also, that the intelligibility of our conceptual framework itself essentially depends upon its mathematical character. Thus mathematics is not a means to express a rationality which is already there, but constitutes the rationality of our description of the observed phenomena. Without the insertion of our description of the results of observation and experiment into an essentially mathematical framework, these descriptions would merely express a matter of fact, but they would not make a positive contribution to our *understanding* of these facts, nor would they be fertile in regard to future observations and experiments. The numerous publications on the relationship between theory and observation have established without a doubt that the rationality of our insights in regard to natural phenomena essentially depends upon our own conceptual framework which is inherently mathematical in character. Thus, the value of a theory of science does not depend merely upon its experimental foundation, nor exclusively on the truth of certain mathematical insights used in the organization and coordination of the experiential data, but on the fact that the observed phenomena and the mathematical truths grew together into an indivisible unity, namely the physical meaning of these phenomena for the community of scientists through which this unity was brought to light. As long as a dialectic tension between these two essential elements is kept alive by the community of scientists, a scientific systematization can be said to give insight into nature.[25]

3. *Induction*

At this point it is appropriate to return to the "problem" of induction which was alluded to by Kuhn in a similar context. From the present point of view, natural science, indeed, begins with observation and experiment. The immediate results of observation and experiment are then subjected to a process of "induction." One must note, however, that observation and experiment in natural science substantially consist in processes of measurement which result in numbers. One must realize further that the process of induction applies to those numbers which are connected with certain quantitative aspects of the phenomena we observe, and leads to general quantitative relationships which are supposed to cover the same phenomena provided they occur under the same circumstances. However, the generalities which result from this inductive procedure are no more than *empirical* generalities. To this first induction, which is still of pure physical nature, we must now add a "mathematical induction." After the empirical generalities are formulated, a *creative* act on the part of the physicist is necessary, one which substitutes a mathematical and thus *eidetic* (i.e., universal and necessary) generality for the empirical generality. Only then can one look for more universal mathematical relationships from which the eidetic generality and others as well can be deduced. The margin between the empirical and the eidetic, mathematical generality which is *in principle unbridgeable* except by an infinite process of approximation, refers to the equally unbridgeable margin between the rational and the real. It is because of this insurmountable gap between the empirical and the eidetic generalities that each "law of nature," regardless of whether it is of higher or lower generality, has the character of a hypothesis, and also why natural science is in continuous need of verifying its final conclusions. It is, however, very important to note that the laws *as* eidetic generalities are universal and necessary, and *as such* are not hypotheses. What is hypothetical and thus asks here for a process of verification is merely the applicability of this universal and necessary law to the phenomena they are supposed to cover and claim, in this sense, to explain.

The most important point in these considerations is to realize that in the meaning brought to light by natural science, the empirical and the rational (mathematical) elements are one and inseparable. True, analysis can distinguish these two elements, but in the concrete physical meaning there is a continuous "dialectic" at work which molds these two elements together into a unity of meaning.[26]

IV.   CONCLUSION: EVOLUTION AND REVOLUTION IN SCIENCE

It should be clear by now that, according to the view developed in this essay, one must maintain that logic and epistemology of science alone cannot explain the genuine meaning of science in all its aspects; that the sociology and history of science are able to shed light on other equally important aspects of science; that, for an understanding of the development of science, the "revolutionary" phases are of greater importance than the phases of "normal science"; and that science is not guided by a goal set by nature. On all of these points the present view is in complete harmony with the conception set forth by Kuhn.

However, where Kuhn claims that the idea of a *telos* in science is to be rejected altogether, and that science basically is to be conceived of as a game played by a scientific community and based upon rules freely established by that community, here the present view defends the thesis that science is and remains goal-directed, and that for this reason the "rules of the game" are determined not only by the choice of the members of a scientific community, but also by factors which (to a certain extent at least) are independent of the choice of that community. According to my view, the scientific "game" is bound by the unities of meaning which the scientific dialogue between man and world brings about (namely the "facts" shown by observation and experiment *insofar* as both are conducted within the context of a scientific theory), and also by the teleology which is essentially inherent in the *scientific* dialogue as such. As far as the latter element is concerned, science does not want to give just any account of the phenomena; its account of "nature" must be rational and systematic. That is why the "game" is not only bound by the "facts" but also by our human ideal of scientific reason which necessarily implies principles, methods, and procedures of systematization and explanation which, in turn, prescribe what can and should be done. This complex of principles, methods, and procedures is obviously not prescribed by nature, nor is it dictated by *a priori* concepts and principles of pure reason in the sense of Kant, but by the very nature of man's *scientific* project as such. Man does not have to project the world scientifically; but if he does, he must do it in such a way that, although it still leaves him a great margin of freedom in concrete cases, his pursuit will nonetheless be determined by what he himself projects as the ideal of scientific reason. In a scientific project of the world, man decides to thematize the phenomena he wishes to examine in such a way that the realm of meaning constituted by the scientific thematization will fulfill certain minimal requirements and conditions intrinsic to, and essentially connected with, his ideal of scientific reason. The ideal can be briefly described by saying that each science must explain the relevant phenomena in such a way that this explanation is characterized by clarity, simplicity, exactness, and coherence, and encompasses as many analogous phenomena as possible; in a word, a science must give a *rational* insight into the phenomena covered by it.[27]

It is obvious to me that such an ideal predelineates a goal toward which science is continuously on its way, but which it will never completely reach because of the "infinitude" of the phenomena and the finitude of man's understanding. This ideal, furthermore, determines minimal conditions which must be fulfilled by any scientific thematization of the world. However, this ideal does not determine the concrete steps which are to be taken in each case in order to solve a problem or to answer a question that may arise within a certain realm of (scientific) meaning. In the final analysis, this is, as we have seen, due to the fact that the gap between the rational and the real is in principle unbridgeable. However, this does not alter the fact that the ideal certainly determines the minimal requirements which every scientist has to obey in order that the results to which these steps lead can be accepted as within the realm of the ideal of scientific reason. It seems to me that logic and epistemology of science are mainly concerned with making our ideal of scientific reason explicit whereas sociology and history focus their attention mainly on the creative element which is found in every genuine, scientific discovery and made possible to a great extent by the social and historical conditions in which a scientist or a group of scientists find themselves.

In a period of scientific "revolution," one or more scientists feels that certain generally accepted paradigms are either inadequate in regard to the phenomena they claim to explain, or, at least, unable to solve all the problems which can legitimately be asked. In developing a new scientific matrix, scientists are guided by an ideal which is the ideal of all scientists. On the other hand, this ideal does not show concretely what steps need to be taken to come to a new matrix capable of solving all the problems on hand and yet remaining in harmony with man's ideal of scientific reason. Thus, I can completely agree with Kuhn that methods and principles alone do not determine a particular body of admissible scientific beliefs and also that in all paradigms there is an element determined by personal creativity and historical contingencies. However, it seems to me that Kuhn, too one-sidedly, stresses the revolutionary character of these innovations. As I see it even these innovations have as much the character of a harmonious evolution as of a revolution.

In the second part of this essay I stated that in my view the term "rationalism" is an adequate label to characterize our scientific pursuit. In concluding this essay, I should like to return to this statement. If we take this term in its most original meaning so as to stand for that view according to which there is nothing which does not have its reason for being the way it is and, thus, that there is nothing which is not at least in principle intelligible,[28] then, we may say, it seems to me, that philosophy can never make any legitimate claim to such a label. For the meaning on which philosophy must reflect comes about *independently of* philosophy itself and always has already occurred before the philosopher can reflect on it. On the other hand, if the philosopher is to understand the world in which he lives in the way it actually did develop, then he cannot make use of empirical methods which, as we have seen, necessarily imply abstraction and idealization. Science, however, is in a quite different position. It may legitimately claim that as far as it is concerned there is indeed a totality of meaning, a rational and all-inclusive unity of meaning which can be mastered by a corresponding universal science without anything being left over. The reason science may claim this is that such a rational and all-inclusive totality of meaning is determined by science itself. Science projects its own world on the basis of rigorous principles and methods and this projection is such that it *a priori* excludes everything which cannot be mastered by these methods. It seems to me that this rationalism which is inherent in science, explains why, in a philosophical conception of science, the concept of evolution plays as important a role as the concept of revolution.

Notes and References

1. *Cf.* Joseph J. Kockelmans, *The World in Science and Philosophy* (Milwaukee: Bruce, 1969), pp. 134–168.

2. Thomas S. Kuhn, *The Structure of Scientific Revolution* (Chicago: University of Chicago Press, 1970), pp. 136–137 and p. 1.

3. *Ibid.*, pp. 1–2.

4. *Ibid.*

5. *Ibid.*, pp. 2–3.

6. *Ibid.*, p. 4.

7. *Ibid.*, pp. 4–6.

8. *Ibid.*, pp. 170–171.

9. *Ibid.*, pp. 171–173.

10. *Ibid.*, p. 171.

11. *Ibid.*, p. 206.

12. Joseph J. Kockelmans, *op. cit.*, pp. 144–155.

13. Louis de Broglie, "The Progress of Contemporary Physics", in Werner Heisenberg, *The Physicist's Conception of Nature* (London: Beaverbrook Newspapers Limited, 1962) p. 158; W. Heisenberg, "The Idea of Nature in Contemporary Physics", *ibid.*, p. 28 (my italics).

14. Thomas S. Kuhn, *op. cit.*, p. 171.

15. Edmund Husserl, *The Crisis of European Sciences and Transcendental*

Phenomenology trans. David Carr (Evanston, Northwestern University Press, 1970), pp. 21–23.

16. Edmund Husserl, *Crisis,* pp. 60-100; *Ideas,* trans. W. R. Boyce Gibson (New York: Collier Books, 1962), pp. 331–394.

17. Thomas S. Kuhn, *op. cit.,* pp. 168–173.

18. Joseph J. Kockelmans, *op. cit.,* pp. 155–168.

19. Gaston Bachelard, *L'Activité Rationaliste de la Physique Contemporaine* (Paris: Presses Universitaires de France, 1951), pp. 21–49. English translation of this chapter in *Phenomenology and the Natural Sciences,* Ed. Joseph J. Kockelmans and Theodore J. Kisiel (Evanston: Northwestern University Press, 1970).

20. *Ibid.,* p. 21.

21. H. Weyl, "On Space, Time, and Matter", in *Phenomenology and the Natural Sciences,* p. 94.

22. O. Becker, "Contributions toward the Phenomenological Foundation of Geometry" in *Phenomenology and the Natural Sciences,* p. 120.

23. F. K. Richtmeyer, *Introduction to Modern Physics* (London, 1934), quoted by Gaston Bachelard, *op. cit.,* p. 27.

24. Gaston Bachelard, *op. cit.,* pp. 21–31.

25. Suzanne Bachelard, *La conscience de rationalité* (Paris: Presses Universitaires de France, 1958), pp. 16–20; English translation of this chapter in *Phenomenology and the Natural Sciences.*

26. *Ibid.,* pp. 17–18; Joseph J. Kockelmans, *op. cit.,* pp. 131–133.

27. Joseph J. Kockelmans, *op. cit.,* pp. 155–166.

28. A. Lalande, *Vocabulaire Technique et Critique de la Philosophie* (Paris: Presses Universitaires de France, 1962), p. 889.

THEORETICAL BECOMING
RICHARD COLE
University of Kansas

I

ANALYSES OF CHANGE have always been answers to the question, "what remains the same, if anything, and what alters in a given change?" Change of accepted scientific theories is no exception. Differences in the way such change is characterized generally turn on what is supposed to remain constant when a new view replaces the old. "Revolutionists" in the history of science usually compress this constancy or do away with it altogether and "evolutionists" generally see the nature of science as enduring but science itself as changing for the better through accretion or the correction of error and hasty judgment. It is useful to think of Aristotle in this connection. Radical "revolutionists" tend to think of scientific change as substantival change in which all the essential properties of a scientific view are periodically replaced and nothing remains constant in the process save an indefinite infinite potentiality for such replacement, whereas conservative "evolutionists" see science as a natural process which develops toward its perfection according to its own internal principles.

Putting the matter in this Aristotelian way is rather suggestive : we can see some people gilding the lily by distinguishing some changes as accidental rather than natural as, for example, when science is seen as depending in part upon historical accident, or the national origin of a scientist, or a political ideology to its temporary detriment. But tempting as it is, I won't go on to use the Aristotelian machinery to spin off the possibilities. My interest lies in seeking to uncover what, if anything of significance, tends to remain throughout the history of changes in scientific theory and thus to shed light on the substance of science.

253

Since, on first blush, scientific theory seems more fickle than constant through time, it is important to adjust our vision so as not to be too taken with these, admittedly, vast changes. The seas are sometimes calm and sometimes stormy, the face of the sea is ever-changing, but the heart of it is still so much sea water, a fact we are apt to ignore in the tempest. What we are told the world is like by different generations of scientists reveals obvious differences among the generations, but, like another sort of generation gap, it may be that the current generation did not really discover sex and politics for the first time.

It is useful to remind ourselves that shifts in philosophical fashions among scientists in the last century are not without precedent in the intellectual history of mankind; the popularization within physical science of nineteenth- and twentieth-century empiricisms are fundamentally no different than a Roman giving up Democritus for Plotinus. So the view of Reichenbach and others, that the development of relativity theory and quantum physics signaled an unprecedented revolution in natural science, ought to be tempered by our memory, if we are fortunate enough to have such in our memory, that things like this have happened before. We must get clear on the way in which philosophical underpinnings function in scientific change before we are seduced into over-emphasizing the differences between nineteenth- and twentieth-century natural science.

I draw attention to this matter of philosophical underpinnings for two reasons; first, recent changes in scientific theory have been associated by frequent and profound confessions of changes of views identifiable as philosophical, and, second, because it is in the nature of competing philosophical views to so contrast with one another as to make a difference in philosophical perspective appear to constitute the greatest possible difference.

Care must be exercised, however, to distinguish between what a scientist says when he is talking philosophy of science and what he says when he is doing or teaching science. With a few (very important) exceptions scientists are impatient with philosophizing, and when they do it, it is usually pretty poor stuff evaluated as philosophy. Nevertheless, there is a point the physicist is making when, in a dinner-table conversation, he relates the dogmas of early Carnap or Hempel, and it has to do with the way he does his work.

Looked at as a philosophical adjunct to his work, his talk of "analytic," "synthetic," "cognitive significance," "about the world," "fullness and emptiness of empirical content," etc., reveals his desire to adopt a properly humble attitude towards his science. "Science," he is saying, "unlike mathematics, is pretty problematical stuff." The insight he absorbs from the philosophical view has to do with a truth which he thinks earlier classical physicists did not know, that one must take care not to be dogmatic in his work.

Another useful role such a philosophical perspective often plays has to do with its conventionalism and its tendency toward instrumentalism. Often, when one does physics, it becomes extremely difficult to attach significance to every line in a mathematical derivation. Much better to get the job done, to set down premises in mathematical form, forget your physics, and calculate. Scientists sometimes refer to this process as "turning the crank" and, indeed, it is very like a mechanical process with the conclusions flowing forth like so much unexpected water. So it is well, under many circumstances, to proceed instrumentalistically with theory and mathematics, especially if one's concern is with experimental results. A philosophy which justifies such a procedure and outlines some of its aspects is put to good use.

You will notice that I am not directing your attention to the question of the truth or falsity of a philosophical view the scientist may hold, but rather to the role it plays in his science. I do not mean to be making a pragmatist's point by doing this, i.e., arguing for truth or falsity by looking to utility in practice, nor do I mean to argue that, since this is mere practice, it does not tell on the truth or falsehood of the theory. I am reporting on my judgment of a fact; just that there is some use for philosophy in the workaday sciences, even when the philosophy is of quite low quality and reaches further than its use.

The fact that the philosophical views espoused over that cup of coffee go far beyond its use is also significant. A philosophical view, by its nature, is rather far reaching; one might say, comprehensive. It is commonplace for a scientist to appear inconsistent because of this. For example, many scientists, as dinner-table philosophers, are radical empiricists, but as scientists are down-the-road realists. I say —many scientists; I suspect I could say most of those who are dinner-table sceptical empiricists are not so in the lab or before a class.

Still, I do not maintain that all scientists make use of philosophy in this way, i.e., committing themselves to a dinner-table position they show no signs of fully believing in their actual work and at the same time, making some use of that dinner-table position in their actual work. Some, like Einstein, Heisenberg, or Skinner, are sufficiently committed to their philosophical views that it shows up consistently in their scientific work, and they are able to use much of it efficiently, too. For example, Einstein, in his adoption of what he sometimes called the operational viewpoint, cast his special theory of relativity in terms philosophically consistent with a modified philosophical operationalism, much to the advantage of science. Such juice squeezed from philosophy in the service of science had not been seen since Newton, who with his almost Democratean atomism, was able vastly to advance the cause of physics by stating things just in terms of parts of matter, their shapes, their positions, their changes of position in time, and certain innovations like forces acting as causes of accelerations.

In the hands of great theoreticians, as great theorists are called among physical scientists, adherence to a given philosophical view can make an important difference in their work. A rough rule of thumb is that explicit philosophical commitment plays a role proportional to the ability of the scientist as a theorist. But it can play some rule even among the "I get my hands dirty in the laboratory" type, be it only one of obedience to a canon of scientific humility and to a canon of quick mathematical procedure.

What are we to make of this talk about the utility of philosophy to the scientist? Remember, my aim was to sharpen vision with a view to seeking what is common throughout change in theory. The most obvious sign of fundamental change in science is change in the philosophical underpinnings. We are urged, by the above considerations, to exercise care in evaluating the role that philosophy plays, for the workaday scientist does not use everything that is a part of his philosophical credo in his science, and, in the case of those scientists who do make use of more than is ordinary of their philosophical canons, they are apt to do as well with widely varying sets of canons.

Now I agree that the place to look for fundamental change in science is at the point where philosophy and empirical science meet. But I am not willing to take a scientist's word when he claims to locate that junction at materialism, conventionalism, operationalism, or what have you. Still, the above considerations seem to entail the conclusion that philosophy is useful to science, one philosophy being, perhaps, as useful as another. This is a rather good place to rest and go on with one's affairs; it has the advantage of unsticking gluey philosophical dogmas from science and providing a less parochial picture than is usual—but it does not go deep enough. Might there be some philosophical commonality to the various useful philosophical views scientists have used?

The method I propose should be used to answer this question is to take the philosophical core out of science by peeling away the science from around it rather than depending upon what philosophers and scientists have said about it. If on observing periods of change we observe that everything has change we must suppose that the only underlying matter of science is prime matter and we have no business assigning it any determinate attributes. We must take care to keep our abstractive faculties about us, however, for, if we do not, we may find that the subtraction of all the differences between an earlier and a later period leaves nothing apparently remaining though there is something, albeit something somewhat abstract. So, in looking at the faces of our siblings, we may find Gertrude to look like Beatrice and Beatrice to look like Mergatroyd but claim that no common thread runs from Gertrude to Mergatroyd, forgetting that they have their parents in common. Or we might say a triangle is like a regular chiligon, having straight edges, and the regular chiligon is like a circle, but the triangle has nothing in common with the circle, forgetting they are both plane figures.

Suppose we contrast scholastic science with the physics and astronomy which Copernicus, Galileo and Newton developed. The medievals were not much for experiment and this for philosophical reasons, they were interested in what things did of their own accord without interference, having been convinced by Aristotle that the science of nature is the science of the natures of (albeit created) substances. If you fiddled with a created substance you were involved in the productive arts and that was an entirely different kettle of fish. Also, even though your language becomes very technical, if you were a scholastic scientist you did not—in science—suppose that natural effects were caused by things unseen or unseeable, except in the sense that their first cause was God. But that is theology and not natural science. So you adopted that kind of realist ontology in natural science that is associated with Aristotle and Aquinas. Natural objects are as they appear to us to be, colored, tangible, geometrical shapes with tendencies to rise or to fall (or suffering motions around and around). We discover this by experience. The scholastic scientist was so far an empiricist, in spite of later misrepresentations of his way of doing physics.

He was also, in the same way, an empiricist in his discovery and assignment of causes. His interest was in discovering the natural causes of change, i.e., causes for those changes which were internal to the object changed. The way in which we distinguish a natural from a violent change (as changes with external causes were called) is by experience; we *witness* constraints and movers and, by elimination, judge that certain objects are changing naturally, of their own accord, because there is no observed agent apart from the object acting on it. The procedure used to determine the causes of a natural change was also empirical; it depended upon selecting out from among the empirically observable, but essentially attributable, properties of the object that one, or that conjunction, which was invariably connected with the unconstrained change.

It bears remarking that the pattern of finding causes of natural physical changes in the object itself was preserved in medieval science even where a most prominent exception occurs, in the theory of impetus. In that theory, certain locomotions, like those of projectiles, are brought under the aegis of scientific inquiry even though there is obviously a mover involved, the hurler of the stone, for example. Why does the stone move in its parabolic path? Because it acquires impetus, which is regarded as a temporary property of the object and which is then, suitably, located within the object. So the vertical component of motion is explained by the heaviness of the object and its horizontal component by its impetus, both properties belonging to the object at the time it is traveling.

A point that must be attended to is the connection of theology with natural science as viewed by the scholastics (and by Aristotle). It is customary to regard medieval science as hopelessly entangled with God, for God is first mover, among other things. The fact that Galileo and Newton seemed to employ God in the same role is regarded as unfortunate derangements.

But, properly speaking, theology was not natural science for the scholastics; they regarded it as another science which was naturally connected with physics in that it gave the cause of natural causes as well as the cause for the existence of natural things. Eliminating theology leaves natural science untouched; it simply becomes necessary not to ask for the cause of causes and of the existence of things just as today.

There is a moral here: If you don't want theological answers, don't ask theological questions.

Another characteristic of medieval science was its teleological character, a character that is not apparently shared by later science. A natural change was understood as the process of acting out the potential of an object to be, or to be where, it ought most to be. The language used may be misunderstood, however, because when moderns say the same thing, they use different words. The notion of an end for a natural change is of that state or place of an object which, when attained, left the object without any internal tendency to change further. We make use of the same notion when we build certain kinds of machines and study biology.

An elevator which is self powered and automatically controlled by push buttons is so constructed that, barring misfortune, it moves to an equilibrium state, its proper floor, and stops there. It is not altogether wrong to think of Aristotelian science on the model of automatic machinery.

We are inclined to represent natural embryological development from zygote to adult making use of this idea of natural change. We regard the zygote as being the adult potentially, as developing towards maturity in consequence of its internal operation, and as ceasing to develop when maturity is reached.

I will have more to say about teleology later.

Galileo was more like Aristotle than is often assumed. Like Aristotle, Galileo regarded the objects of natural science to be those observables that are empirically accessible in a fairly direct way. Moreover, like Aristotle, Galileo assigned internal causes for natural motions, to the point that he was led into error in explaining the motion of the tides. The correlation between the tides and the position of the moon was known to Galileo, of course, but the idea that the moon could be the agent responsible seemed to him mysterious and occult. Instead, he developed what may be called the sloshing theory of the tides, explaining their ebb and flow as a sloshing caused by a somewhat complicated complex of circular motions of the earth. Indeed, Galileo regarded the action of the tides as outstanding evidence for Copernicus' heliocentric theory, since he thought the sloshing explanation a true one and that entailed the earth must indeed move.

The quarrel between Galileo and the scholastics was not so much a quarrel on matters of scientific procedure and philosophy as it was on the question of certitude. The scholastics appeared simple-minded and dogmatic to Galileo, given to taking their views from authorities, particularly Aristotle, and unable, unlike Aristotle, to pay attention to experience sufficiently to correct mistakes that might be made. It is true, of course, that Galileo disagrees with much of Aristotelean physics, but he respects the same sort of evidence as did Aristotle, and Aristotle, of course, did not appeal to himself as an authority to support his conclusions.

The scholastics and Aristotle are reputed to argue *a priori* on *a posteriori* matters. There is some truth in this, and it is important to see it, but we must locate it exactly. There are two modes of reasoning that might with some justice be called improperly *a priori* that might be confused. If a man accepts a claim about an empirical matter on authority he engages in a kind of relative *a priori* reasoning, for he does not go to the empirical evidence to support his claim but rather accepts it prior to supporting experiences, believing that what his authority says on such subjects must be true. His reasoning rests on a premise which is prior to the appropriate experience, i.e., that experience which provides evidence for the claim.

Another sort of improper *a priorizing*, somewhat less sinful, is prematurely to put the stamp of necessity on a claim, jumping, however short a distance, to a conclusion that such-and-such must be true. Given what Galileo knew about those phenomena that both he and Aristotle studied, it must be confessed that Aristotle seemed guilty of this error. Aristotle does read as though he is pretty all-fired certain about some things that turn out to be false. To put it another way, Aristotle seemed to believe the scientific task was a good deal easier than it subsequently turned out to be.

Galileo is sometimes regarded as the inventor of scientific humility, a view which ought to surprise his biographers who represent him as something less than a humble man. But Galileo's humility, what sort it was and how it ought to have been limited, is of more than mere biographical interest. "Nevertheless the earth does move," is alleged to have been Galileo's opinion after having officially recanted before the Inquisition. Yet by deserved reputation, Galileo is credited with the view that the proper attitude of a scientist is the open-minded one, an attitude which might appear, but is not inconsistent with his stubbornness to recant, we may say, in his heart.

Here lies a crucial part of the philosophical core of natural science. The problem for the theoretical scientist is fundamentally one of evidence, and of standards of evidence, given a possible theoretical interpretation of the facts. His practicing philosophy of scientific method must provide a means for questioning the truth of a theory, but it must also provide a stopping-point in this questioning process, for it is necessary to come to conclusions in science. "What is sufficient for the acceptance of a theory?" is a question that must be asked, for the scientist, unlike the philosopher, cannot afford, in his discipline, to remain in a state of eternal doubt.

A bargain must be struck between a scientist and nature, for nature does not yield easily to inquiry but it does yield just enough so as to encourage scientists that their pursuit is not entirely hopeless. Sceptical philosophies of science cannot be practicing philosophies insofar as they depend upon the rather clear truth that scientific claims differ from logical claims in the way they are judged to be true. Most philosophical scepticisms are able to call into doubt that which they call into doubt because the standard of truth they impose is, so to speak, of the highest. Something is to be doubted if it could be false; and anything can be false unless its truth is either of the transparent kind nowadays called analytic, or is derived from such transparencies.

Aristotle seems to have struck this bargain by limiting the province of scientific inquiry to those becomings which become in virtue of internal properties. There is a lot of change in the world not subject to theoretical investigation for Aristotle: any change clearly caused by an outside agent, for example, and a change attributable to spontaneous change. In viewing the whole flux of becoming he separated out a subject matter and gave the rest to other disciplines or to no discipline at all.

It will surprise some to learn that Plato's philosophy of science was much more sceptical, for Plato, in the *Timaeus,* is clearly committed to a version of scepticism in all natural science. He, like Aristotle, also divided objects of possible inquiry into two classes, that about which certainty is attainable and that about which certainty is impossible. He drew this line between the natural and the intelligible worlds, rather than drawing it within the natural world. It is, of course, notorious that Aristotle could not draw his line at that point, since he did not accept Plato's notion of an intelligible world.

The natural world, the world of becoming, was, for Plato, subject to a kind of inferior inquiry, in which one proceeds hypothetically

—telling, in his language, likely stories—and which does the job, such as it is and such that it can be done, of natural science.

So Plato and Aristotle each pay tribute to limitations of inquiry in natural science, but each in different ways. So does Galileo, in his way, and when we subtract away his displeasure with the method of authority practiced by his contemporaries, we find not a thorough-going scepticism, but a degree of confidence in the possibility of arriving at termini in natural inquiry coupled with an appreciation of the complexities of nature and the need to maintain high—but not impossibly high—standards of evidence.

It is possible to abstract out this much commonality so far between Plato, Aristotle, the Scholastics, and Galileo: the world of becoming, which forms the background of scientific inquiry, resists comprehensive, apodictic treatment, but there is something there which yields, albeit with difficulty. It has become clear that Aristotle drew the boundary between the treatable and the non-treatable with premature over-distinctness. It is also possible that Plato and Galileo erred a bit on the sceptical side. But all agree: something limits inquiry into becoming; and all agree: such inquiry is, nevertheless, not entirely hopeless.

We may also find agreement among these authors on the conditions under which nature yields to inquiry—so long as we are willing to make such a statement of agreement general enough. Nature yields, roughly, when she can be made comprehensible and to the extent she can be made comprehensible. Ancient and modern authors will differ on what this means in detail but this can be said: theory serves the task of making nature comprehensible and it does so by making it reasonable—rational—understandable—choose what synonym you please so long as it is vague enough to encompass the samenesses but not so vague as to be without value.

A nice quote from Aristotle puts these points beautifully. It is from *Generation and Corruption*, 316ᵃ5:

> Lack of experience diminishes our power of taking a comprehensive view of the admitted facts. Hence, those who dwell in intimate association with nature and its phenomena grow more and more able to formulate, as the foundations of their theories, principles such as to admit of a wide and coherent development; while those to whom devotion to abstract discussions has rendered unobservant

of the facts are too ready to dogmatize on the basis of a few observations.

I promised earlier to say more of teleology. I propose to keep that promise by asking whether or not the notion of teleology applies to the subject at hand, i.e., the task of eliciting the common part of science which remains while science suffers theoretical becoming.

There are two ways in which such an application might be made. First, it might be the case that some sort of teleological mode has remained at the heart of the theoretical interpretation of nature. It is true that teleological thinking is still to be found in natural science, particularly in biology. One can find it in physical science, too, if one does not expect too much of a teleological account. Fundamentally, a teleological account of a phenomenon involves viewing that phenomenon as a *development* which proceeds according to internal causes and terminates in what might be called an equilibrium state. Second, there is also, of course, that sort of teleological account which explains a phenomenon on the basis of the conscious purposes of some external agent, but this, while interesting to theologians, does not appear to be a part of the history of natural science beginning with Aristotle.

Seeing teleological accounts as simply involving developments of phenomena according to internal causes, one can identify many instances of such accounts in the sciences. Explaining the behavior of stars, for example, on the grounds of natural development beginning with a large hot gaseous mass contracting through its internal gravitation attraction until it is dense and hot enough internally to support nuclear reactions, which prevent further contraction until its nuclear fuel is used up, etc., is a teleological account. Marxist developmental accounts of economic history is an example from the social sciences. Accounts in biology which depend upon seeing an organism as a machine are teleological. Interestingly enough, the theory of evolution that postulates external or accidental causes as the cause of mutation, and therefore of the origin of new species are not teleological, as most biologists make haste to remind us.

Still, much of natural science lacks a teleological character; and it would be very difficult to say that post-Galilean science kept the teleological mode as a necessary condition for making nature comprehensible. It is best to conclude, then, that there is probably no abiding teleological requirement for scientific theory.

But there is another place where a teleological account might be appropriate. Perhaps science itself, which, as is the premise of this paper, changes, changes in accordance with internal laws of development toward a natural terminus. This is not to say—it is said in contrast to—the claim that the world is teleological. Perhaps, scientific theorizing is teleologically comprehensible.

I propose to make this view plausible by considering in some detail the history of astronomy and mechanics, at least as far as Galileo. If changes in scientific theory can be fit into the rubric of the sort of teleology which involves internal tensions tending to a natural end, it should be possible to illustrate this. A way to put the problem is: "Can science be made to appear to be a kind of drama; that kind which develops according to its internal logic?"

II

To most ancient astronomers, the facts of astronomy spoke for themselves. There is an outermost sphere of stars, fixed relative to each other, which rotates uniformly around an earth which does not move. The other heavenly bodies, the sun, the moon, and the five planets observable with the naked eye also move in circular paths around the earth. The exception among ancient astronomers was Aristarchus, who thought that the same appearances could be explained by supposing the earth and the planets rotated around the sun. Aristarchus' theory was considered fanciful and wrong, even obviously wrong, since it seemed clear the earth stood still. Ptolemy, using geometric and trigonometric methods, succeeded, so far as the evidence warranted and required, in mathematicizing ancient astronomy. His major accomplishment, by making the planets move circularly around a point which itself revolved around the earth, was in describing mathematically the observed paths of the planets. Aristarchus' theory, never much admired, was not a serious contender to Ptolemaic astronomy until Copernicus did for him what Ptolemy had done for conventional ancient astronomy.

Copernicus took Aristarchus' theory and gave it mathematical formulation. In showing that the apparent motions of the heavenly bodies could be described in a convenient and exact way by supposing the earth and the planets to be rotating around the sun, and supposing the earth to be also revolving around an axis connecting its poles, he did for Aristarchus what Ptolemy did for

Aristotle. Aristotle, depending upon the evidence of the senses, accepted as evident the fact that the heavens rotated around the earth, though not every part at the same rate. Since he was concerned with the "why" as well as with the "what" he developed a profound and subtle theory to explain these motions. Ptolemy, without concerning himself primarily with the why, had developed a mathematical description of the motions of the heavenly bodies that made possible a remarkable accurate statement of their past and future positions.

But Aristotelian-Ptolemaic astronomy had developed a theoretical foundation in Aristotelian physics such that the Copernican mathematical description, though an obvious aid in astronomical calculation, was a physical impossibility—if Aristotelian physics was correct. Aristotelian physics, supported as it was by the weight of sense-evidence, by a ponderous tradition, and by its happy compatibility with church dogma was a major obstacle for anyone who wished to defend the Copernican system on any grounds other than mathematical convenience. If Galileo was to make the Copernican theory acceptable even as a conjecture of physical fact, he had to make it physically possible; to do this he had to provide a theoretical underpinning different from the Aristotelian. In short, he had to develop a new science : Mechanics.

So far as facts are concerned, by Galileo's day there were many more of them. Largely because of the use of the telescope in astronomical observations, a large body of fact unknown to earlier astronomers had become available. A good many of these Galileo discovered for himself. It was observed that the moon had a very irregular surface, full of mountains and valleys; the moons of Jupiter were observed; Venus was seen to show phases like the moon, suggesting that it shone by reflected sunlight and making possible its accurate positioning in the sky relative to the sun. The planets, observed through a telescope, presented perceptible discs which varied in size as the planets changed in apparent brightness, making possible a way to calculate their orbits that had not existed before. Sun-spots were observed which Galileo turned to use to defend Copernicus, and many other things.

But it is possible to assimilate all these things to the Ptolemaic picture with his Aristotelian grounding, given sufficient ingenuity and dogmatism, for, in an extreme case, it is always possible to explain an observation as an illusion caused by some accidental peculiarity.

Galileo's defense of Copernicus, therefore, could not consist simply in proposing a new science of Mechanics, and in parading the new evidence which he could explain and Aristotle couldn't. The Aristotelian of his day were skilful in explaining or explaining away all the new data. Galileo had to represent his view as the plausible, reasoned one and his opponents' view as dogmatic, depending upon a tenacious adherence to tradition more dogmatic than scientific.

Here a tension arises. It is, in its way, a moral conflict—if one may use the word "moral" to describe the primary injunction of the Galilean scientific mentality; for it was a conflict between the secure and the risky—the traditional and the new, the orthodox and the heretical, even the pious and the sacrilegious. If Galileo was to defend Copernicus, he could not naively consider the motion of a falling or rising body to be along a straight line.

If the earth is spinning and rotating about the sun, such natural motions must be curvilinear, not straight. The Aristotelian explanation of natural motions depends upon attributing the property of gravity—which means heaviness and is not to be confused with Newton's idea—to bodies which naturally fall, and levity to bodies which naturally rise. To the heavenly bodies, which naturally moved in circles (so it appeared), he ascribed the property of circular mobility. He associated these properties with different material elements : earth and water had the property of gravity, air and fire the property of levity, and the heavenly bodies, which moved naturally in an entirely different way, and in a way which no terrestrial body exhibited, were made of their own, finer, more divine stuff. Consequently, by depending upon experience to tell him which motions were natural and which not, by observing the kinds of natural motions, and because of the requirement that the cause of such motions must reside in the moving bodies themselves, Aristotle concluded that the stuff of which the heavens were made was completely unlike terrestrial substances. From the fact that celestial motions continued in a regular way, unceasing and changeless (so it seemed), and from the apparently unchanging appearance of celestial objects, he concluded that these objects were made of an indestructible stuff whose nature it was to revolve in eternal circles.

Galileo agreed with Aristotle that circular motion was natural, and that it was in the nature of bodies which exhibited it to move in that way; the difference is that Galileo thought it was natural to *all* bodies. As to falling bodies, Galileo thought that it was natural for bodies of the same type naturally to cohere, so that if a piece of the earth, or moon, for example, were removed from it, it would naturally return to the earth, or moon, by the easiest path. This path, if observed by someone on the earth or moon, would appear straight, but could be described as curvilinear—if the earth or moon were in motion. The only bodies that would be at rest for both Aristotle and Galileo, then, are bodies in their natural places or bodies that are constrained from motion. But this meant for Aristotle, the earth is at rest, and for Galileo, either the earth is at rest or it is itself moving circularly. The Galilean notion of rest, so far as observation bore on it, was relative, for the word "rest" did not mean for him what it did for Aristotle. Galileo devotes a good deal of argument to the point that we must assign "rest" to a body on the basis of a reasoning which depends upon consistency with theory and observed fact.

If appealing to circular motion is no longer a way to show the heavens are made of indestructible divine stuff, then what of Aristotle's other arguments, which depend upon the constant appearances of the heavens? Galileo making use of recent observations of the moon and of the moons of Jupiter, makes it implausible to believe that these bodies could be quite the way Aristotle described them.

Given that it is possible the earth spins on its axis and also rotates around the sun, and that the other planets rotate around the sun, in the absence of evidence that it is fact, Aristotle and Ptolemy still could be right. After all, though Aristotelian physics provides a theoretical ground for the geocentric view and even if this ground is wrong, Ptolemy could be right, for the bulk of evidence and common sense was on his side. It remained for Galileo to establish evidence for three things : first, that the earth spins on its axis; second, that this spinning earth spins around the sun; and third, that the planets also spin around the sun. To establish the first of these notions without the others would not be sufficient, for it is possible on Galilean physical grounds that the earth spins but remains otherwise motionless.

On Galilean grounds no terrestrial observation could establish the diurnal motion of the earth. Such evidence as was later adduced was not only unavailable to Galileo; what is more interesting is that it could not be anticipated by him, since it was apparently inconsistent with Galilean physics. The fact that a projectile fired from north to south, or south to north could not participate in the daily rotation while in the air was denied by Galileo. The behavior of the Foucault pendulum, the daily rotation of which is very convincing evidence, could not be predicted by him; the equatorial bulge of the earth, if it were known to Galileo (and it was not), could not have been connected by him to the earth's diurnal rotation. All these observations support the Copernican system, but on Newtonian, not Galilean, grounds. Galileo had to restrict himself to evidence taken from celestial observations, and to an argument which depends primarily upon Occam's razor : that these observations are much more simply explained by positing the rotation of the earth rather than its motionlessness.

There is a moral to this. Galilean physics was soon superseded by Newtonian physics. But the story is not told by saying Aristotle tried and was wrong or Galileo tried and also failed or Newton tried and was successful. Galileo was a bridge between the Aristotelian physics of the schoolmen and Newton. What was right in Aristotle filtered through Galileo and shows up in Newton; and what was right in Galileo was assimilated by Newton and passed by Newton to his successors.

The Aristotelian notion of natural motion was a kind of law of inertia, for it attributed to the body a certain natural mode of moving. Galileo's principle of inertia was a refinement of it, and a universal extension of it. For Aristotle, there were two kinds of natural motions, which applied to two distinctly different kinds of matter. Newton changed Galileo's "circular" to "straight." The law of inertia was inherited from Aristotle, changed by Galileo, and straightened out by Newton. Though Galileo's explanation of gravitation was like Aristotle's in that "gravity" was a property of motion inherent in a body, it was extended by Galileo to apply to all like bodies : a piece of the moon would fall to the moon, and a piece of Jupiter would fall toward Jupiter, etc. It was made to depend upon the presence of other like material close by; Aristotle's principles of gravity and levity depended not upon other nearby matter but upon the existence in nature of a geometrical point, the center of the universe. Newton's Universal law of gravitation made all matter alike and, in this way, is a clear generalization of Galileo's idea.

What existed in germ in antiquity, budded in Galileo and flowered in Newton. There was a change from seed to bud to flower; it is correct to say that Aristotle and Galileo were wrong. It is also correct, given the development of physical science since, to say Newton was wrong. But it is again correct to say that the way Einstein was right was the way Newton was; the way Newton was was the way Galileo was; and the way Galileo saw what truth there is to see in these matters was the way in which Aristotle saw it. Newton is often quoted as saying he saw further only because he stood on the shoulders of giants. Vision in science is like that; it depends upon a perspective of a giant on the shoulders of another giant who sits atop still another giant.

III

What lesson is to be learned by this sample of analysis in the history of astronomy? First, there are changes and they are extremely pronounced. But there is also a continuity. Aristotle, Plato, Galileo, and the others involved in that inquiry which led to present-day science wanted to know what the world was truly like. To know this is to know the causes of those changes in the world we observe. Changes in scientific theory are occasioned by criticisms which have the following form: "the world cannot be exactly as you describe it, since there are things unaccounted for in your view which another view well accommodates."

But what shall we do with modern instrumentalist interpretations of natural science? The view that natural science proceeds toward an end of giving a true account through reason of changes in the world is denied by those who regard scientific theory as so much machinery for generating predictions and summarizing experimental data.

I have mentioned before that instrumentalism in working science is a sometimes thing. We can argue, with much plausibility on the evidence, that no scientist is consistently instrumentalistic in his work.

There is, however, a way in which instrumentalism itself, as a philosophy of science, illustrates the very point I am making. For what view does an instrumentalist have of the world? The world is characterized by disorder which necessarily resists embedding into a rational scheme of natural causes. Yet science exists, with causal language and the rest of an apparently realist ontology that this entails. How are we to account for this? We say, contrary to appearances, the theoretical language of science does not posit a scheme of things and causes which account for phenomena in the way they, on the surface, appear to try to do. Such language is to be compared to the recipes that a cook may use just because they produce tasty consequences.

This *accounts*—gives the reason why—for the appearance of realism in scientific theory while at the same time preserving what the instrumentalist regards as given: the world of experience is intrinsically disordered to the point that each part of it must be regarded as causally independent of each other part.

In short, the instrumentalist seeks to explain and reconcile two things he regards as true; the perspicuous accidentality of events in nature and the appearance of a denial of that accidentality in existing scientific theories.

It is further significant that common interpretations of quantum physics, apparently at odds with the notion that science aims at fitting the facts into a rubric of reasons, also tends to verify that notion, if they are looked at in a somewhat different way.

An interpretation of quantum physics known as the Copenhagen interpretation regards the fundamental events in the world as non-causal. But they do not mean by this that there is no way of under-standing them at all; they are to be brought under statistical laws rather than causal ones. It is not claimed to be a feature of quantum mechanics that the statistical laws are themselves subject to statistical variation, which would really make a chaos of things. Indeed, to create such a chaos would make science impossible.

The Copenhagen view sees the world as at bottom spastic, but whose elasticity is subject to law and can be studied. Indeed the point of certain thought-experiments associated with the Copenhagen inter-pretation of quantum physics is to establish, *a priori*, this point of the essential spastic nature of nature.

So experimental results are accounted for with reasons; and though Newtonian actional causality disappears, rational explanation remains at the very heart of science.

PART FOUR

THESIS :   Edward A. Maziarz
COUNTERTHESES : Arthur F. Holmes
                             George Schrader

THE INDIVIDUAL IN DIALOGIC INVOLUTION
EDWARD A. MAZIARZ
Loyola University

THERE ARE A number of questions of singular importance to each individual and to the common weal to which this article addresses itself. Some of these questions have been asked by men of all ages and at every level of the human condition. The simplest queries have taken the following or analogous form: Who am I? What is the meaning and purpose of my life? Is there a future life of which earthly life is but a preparatory stage? These same questions, however, can be, and often have been asked at another level, the level of speculative thought. One may inquire more globally into the meaning and significance of human, as opposed to or compared with animal nature; or, of one race, type, or species of man as contrasted with another. More ultimately, one may reflect upon the significance of human history itself and its relationship to divine or to diabolical realities.

These simple facts of experience—that each individual is both a questioner and a respondent unto himself—offer us some evidence from which we can begin our investigation of the individual and his interaction with evolutionary and revolutionary forces. Throughout his life, the individual places himself, his life, and the meaning of his life into question. Sometimes he may direct his questions to a friend or an acquaintance; at other times he may seek for the answers men have embodied in literature, in the arts and the sciences. Occasionally he may present his interrogations to a religious or psychological counsellor. But just as frequently, he may be found dialoguing or debating within himself about his total life style and its negative and positive aspects, seeking for some further differentiation, or, possibly, for a new integration. In some instances the individual may be found addressing his questions to a god or to some noumenal reality, to an abstraction like fate and destiny, or even tragically calling upon the universe at large.

Now it is when we regard the individual in this way—as the questioner of himself and of others—that we can envision him as partaking in and contributing to change of every kind. Development and decline in all realms and at every level depend, in one way or another, on the querulous character of the individual. For it is as a questioner that man reveals himself as dynamic, as opened out to others and to the world rather than as static and as closed in upon himself. And thus, if we disregard the particularity of the questions that concern the individual, we can concentrate upon the real issue under consideration: man's dialogic nature.[1]

In the first place, man's self-questioning nature seems to be incapable of fulfillment.[2] Even an individual who grants that he has had a clear and satisfactory answer to a particular question is as likely to reject or dismiss the answer as to accept it. He may continue to be doubtful, querulous, opinionative, restlessly searching for another answer as well as for another formula in which the question can be couched. Small and large associations of human beings display a similar flux of attitudes concerning common goals, values, and means to achieve their expectations.[3] One of the reasons for this diversification of views and attitudes, of course, rests on the simple fact that men themselves are not only the questioners but also the respondents to their own queries and puzzles. Even in those cultural situations wherein men accepted an extra-human source as a final answer to their problems—faith in a divine revelation, for example —their arguments, differences and disputes have been as multiple as their points of agreement. A great amount of human evolution and revolution, then, arises in response to the open-endedness of man's dialogic nature.

Secondly, when men place themselves into question or direct their questions at one another, they situate themselves in an inter-personal context. A questioner postulates some sort of duality or audience to whom the question is addressed. In some dialogic postures, a questioner may address another person or a group; but even when he addresses himself to an abstract entity as his fate or race, nation or church, he initiates a dialogic situation in which the person or thing questioned is called upon to be a respondent to his query. The inter-personal relationships, then, of I-Thou and of I-It,[4] as well as those of I-They[5] and I-We[6] seem to form the limits or end-points of those human situations in which man is dialogic and, possibly, of all human enterprises.[7]

In the third place, man's dialogic nature presents us with the possibility that it is by means of his questions, puzzles and problems that one man finds his distinction from another.[8] It is a moot point, of course—in fact, a portentous question itself—whether or not the basic human questions remain the same throughout the ages, and whether the difference lies only in the questioners. It seems to be the case, however, that men are differentiated not only in terms of their questions, but also by means of the disparate, divergent and contradictory answers that they find satisfactory. Contrasting and conflicting replies to questions are given from one tribe or culture to another, from one historical period to the next, and from one individual to his nearby neighbor. Each individual, then, in terms of his dialogic nature, places into question his own melange of subjects. Over and above the social and cultural frameworks within which he lives, the individual embodies other persons, gods, and the entire furniture of the world within his own personal perspectives.

In addressing itself, then, to the individual as both object and agent in evolutionary and revolutionary processes, this article will regard man in a dialogic context. There is a singular significance and importance in adopting this particular perspective over comparable points of view—man as a being, as reason, as soul, or man as the religious, the economical or the tool-making animal.[9] It has been argued, with great merit and from a variety of viewpoints,[10] that the latter perspectives on man have led man to his present technologically successful but interhumanly baneful condition. At least part of the reasons for contemporary man's frustrated and alienated affliction have been laid on the failure of religious, political and social leaders to hold the good of human being paramount over power, privilege and property. Blame has also been placed upon the attempts of dissident individuals and concerted groups for substituting a new mastery for an old slavery, an alternate privileged class for a decadent one and the ravaging of all property for property respect and maintenance.[11]

Without denying the merits of these and similar points of view, this article will propose a different perspective. The problems associated with human inter-relationships are extremely complex and no one solution to all human ills is viable. Consequently, I shall argue that despite some tokens of contrary evidence, individuals at all levels contribute to mutual processes of depersonalization and even degradation. Contemporary men may lay the blame for the lack of authentic inter-personal relationships upon new shibboleths—economic and social conditions, sexual frustrations, political chicanery—to replace the role formerly attributed to divine or diabolical powers. I shall argue strongly that the individual needs to reconstitute himself in dialogic relationship with others on the premise that "power to the people" can be more than another shibboleth only to the extent that each individual and small groups of individuals reconstitute themselves as possessing and sharing power together.[12]

This article, then, will approach the issue of evolution and revolution and their relationship to the individual within a dialogic perspective. After a preliminary section offering some historical remarks and aimed at clarifying terminology, the second part will portray the genesis of an individual human being within a dialogic context. The individual will be regarded as the object of the processes of Evolution-Revolution, but as mediated by parents and anonymous other persons. Two of the inter-personal reference frames, the I-They primarily representing tradition, and the I-It representing states of affairs or the condition of the world surrounding the child, will symbolize the forces of Evolution-Revolution as they bear upon the genesis of the individual. Part Three will deal with the individual himself as an associate member of the dialogic context within which he is born. It will describe the modes in which the individual appropriates the mediating activity of parents and anonymous others and the manners in which he maintains himself as an "I" or a self—a reference point from which he can be in dialogic relationship with other. Part Four, the concluding section, will center on those inter-personal relationships that we have designated as I-Thou and I-We. The individual will be singled out as a concentration of inter-personal power by means of which he can responsibly and effectively contribute to and profit from his own historical position in the inter-personal world.

Before proceeding to clarify the terms of this discourse, I need to say a few words about presuppositions and background knowledge. Change, movement, process, development, progress, and regress form one set of terms associated with the term "evolution." Reform, renewal, dissent, and rebellion make up a second set of terms associated with the term, "revolution." Both sets of terms taken together represent some very general and theoretical horizons under which the process of Evolution-Revolution takes place in relationship to the individual. I shall assume that genuine changes occur at the level of everyday reality. That is, I shall disregard whether or not scientific notions of space, time and change are real, and assume that genuine changes occur within inter-personal space and time.[13] On the other hand, I shall argue in a number of places, and this opinion serves as a supposition throughout this essay, for some sort of "involution" as explanatory of the genuine transformations that occur in the individual, and thus for some validity to the notion of cyclical time. The term "involution" shall be employed to designate those processes of differentiation and integration that take place within the individual at the reflective level before they are given linguistic substance or other forms of inter-personal significance. In this way I hope to present and defend with some success, that personalized sector of selfhood that enables the individual to be an individual along with others in the world.[14]

Finally, I shall not summarize or criticize particular theories of evolution, revolution, or theories of man except indirectly. Instead my aim will be to offer a positive, practical, and humanistic interpretation of the individual in his inter-relationships with others and with the world at large. That is to say, the viewpoint adopted here may be considered an attempt at philosophical anthropology and as a contribution to what Ernest Becker has called "the science of man."[15]

In effect, the task of the philosophical anthropologist is, in many ways, more difficult and delicate than that of the biologist, chemist, anthropologist or historian. I hope to elaborate a "theory of theories" about evolution and revolution as related to the individual within a dialogic context. In undertaking this objective, I shall not quibble about being more evaluative and prescriptive than factual or descriptive. The purported dichotomy between fact and value itself is an issue that is not only highly debatable in theory, but entangled likewise in the complexities of everyday life. At any rate, it is in the above fashion that I shall attempt to answer the principal question to which this article addresses itself: In what senses can it be said to be the case that the individual human being is both an object and an agent in the process of evolution and of revolution?

I. HISTORICAL REMARKS;
CLARIFICATION OF TERMS

The fundamental elements germane to the notions of evolution and of revolution as they bear upon the existence of the individual emerged over a long period of time. These elements are closely associated with the perennial problems of explaining change and stability, growth and decay, life and death. In the contemporary world view,[16] the term "evolution" and its cognates refers to a mega-theory or paradigm that explains change at every level—cosmic, cultural, social and personal—in a reflective manner through geological, astronomical, chemical, biological, psychological, socio-logical and historical hypotheses. The term "revolution" and its cognates was originally limited in scope to political movements and to the movements of stellar bodies. It has come to be employed as indicating a cultural change inspired, and to some extent, controlled by human beings. In this broader sense, revolution refers to changing or altering the options and life chances available to human beings at a given time and place.[17]

Over the last hundred years and more, these two paradigms have become, for at least many peoples of the Western world in their everyday life, a mythological framework[18] for viewing the universe and its processes in a self-contained manner. Some form of belief and commitment to Evolution-Revolution, then, is operative on the scientific as well as at the everyday level of human experience.

At the scientific level, theories of Evolution-Revolution are generally expressed in terms of statistical regularities and laws. Such theories seem to operate under the assumption that they apply primarily if not solely to mega-processes that take place in nature or in the social order, and that "a" or "the" individual is but a negligible or perhaps an unrecognizable factor or entity swallowed up in these larger processes. At the same time, if the processes of Evolution-Revolution are actually operative in the universe as described by theories, do not both processes and theories require some recognition of the "individual" in some meaningful sense of that term? Is not the individual not merely an object but an agent in these very processes? In other words, a cultural insight into theories of Evolution-Revolution acknowledges the presence of some transitory element or configuration not only as part of, but also as contributory to, and, in fact, as the very bearers of these processes themselves.

Now there is adequate evidence to support the view that uniqueness and individuality of some sort are recognized in all the sciences.[19] Scientific theory is concerned with the molecule, the atom, the cell, as well as with their larger groupings in terms of a gas, a chain reaction, or an organism. The sciences also address themselves to such notions as the person, the tribe, the state, and to such larger entities as world community and action-systems.[20] In fact, one might regard the goal of each scientist as aiming to encompass, both in minute detail and broad perspective, the complete knowledge of each individual. At least indirectly, then, the sciences give linguistic and symbolic significance to the individual, or to the species-specific entities that are the agents and not merely the objects of the changing dramas of nature and of society.

Evolution and revolution, then, will be regarded here as theoretical terms that complement and supplement each other. Evolution represents a theoretical summary and a unification of the natural history of the heavens, of the planet earth, of human societies, and of the life of the individual himself. The term is partly descriptive and partly explanatory of a present state of affairs as it has come to be, and as it can be made meaningful in terms of adaptive capacity and other factors.[21] As complementary to evolution and yet in contrast with it, revolution stands for those theoretical explanations of direct human interaction in the evolutionary processes themselves. It may well be the case that the active possibilities for revolution are pre-contained in the evolutionary processes at every given moment. Nevertheless, revolutionary changes stand for totally new directions and options unexplainable in terms of the past. In other words, evolutionary theory may be regarded as offering a necessary condition for understanding change of any sort; but only evolution combined with revolution represents both a necessary and a sufficient condition that makes any other changes, progress, growth and stability intelligible to human beings.

In effect, then, the natural history of the universe is inseparable for human understanding from the social conditioning of that same universe. And it is precisely at this point that Evolution-Revolution —viewed in a dialogic context—has a great amount of insight to offer on the role of the individual in natural and in human affairs. Formerly, theories of change and development tended to make man's position and role with regard to change more passive than active, more unquestioning than questionable. They tended to make the individual more a spectator than a contributor and a participator in the affairs of the universe. For example, as an image of the gods, man was destined to embody divine rather than human purpose, though the latter were not clearly known but only a matter of surmise. Within this divine-human framework, it seemed to be the case that divine purposes were ultimately achieved no matter what men contributed, and thus the affairs of the world were not really being managed by men in any shape or fashion. Evolution-Revolution will be interpreted here as transferring this same divine ultimacy to individuals and to groups of individuals as they live within the dimensions of an inter-personal and dialogic universe. In this way, a dialogic interpretation leaves open to men the realities of their own possibilities and responsibilities with regard to each other, without denying the amount or the degree of divine efficiency and causality.[22]

II. THE INDIVIDUAL AS MEDIATED BY PARENTS AND ANONYMOUS OTHERS

Whatever differences they may show, most attempts to define or to describe human nature have one thing in common : they offer images, models and types of an abstracted man. In comparison to these theoretical or reflective systems, the individual—Ghandi, Hitler, or one's next-door neighbor—are approximations or distortions of these models. In some cases a model is formed additively by combining all meritable characteristics to achieve a universal picture of man as the ideal man, the hero or the saint.[23] In other instances human and cultural traits may be successively removed to reveal an image of "natural man" or of man as a machine or a cybernetic complex.[24] Again, other methods may compare cultural likenesses and differences to obtain a sub-cultural residue or a cross-cultural product.[25] The above techniques and variants of them have been employed by theologians as well as philosophers, by social as well as by natural scientists, by politicians as well as by revolutionaries.

Most of these models, of course, have their utility both for the enlightenment of the individual's everyday life as well as for theoretical and other purposes. A person interested in personal growth and development, for instance, may employ any number of models as mirrors for himself or as stimulants for perfection or as standards for judging other persons. Individuals who may be directly involved in service to others may rely upon psychological and sociological descriptions and techniques in assisting others directly or in planning health and welfare programs.[26] In fact, one might argue successfully that it is a rare individual who does not himself harbor some abstracted image or model of what it means for him and for others to be human.

Despite the amount of talent and concern that may be exercised in formulating a model of Everyman, it is a task that is difficult and perhaps impossible to complete. The principal obstacle preventing the making of a universal image of man, of course, is the individual himself. *Man ist was man tut.*[27] Human life is a process beginning with birth and concluding with death, and becoming human is not a process of becoming a universal, abstracted man—not a copy, a model or an archetype—but becoming an individual human being. And it is only the individual man himself who offers both the ultimate meaning and final testing ground for any typologies of the human species.[28] A second obstacle blocking the formulation of an ideal type of man also arises from the ways in which men live their lives. All human beings admit to some forms of inauthenticity. Men freely acknowledge that intellectual abstractions and generalizations, scientific axioms and common sense maxims, role playing and human weaknesses frequently prevent one from embodying his own individuality or acknowledging that of another.

Thus, in addressing ourselves in this section to the genesis and to the growth of the individual within a dialogic context, we cannot avoid some degree of abstraction and some form of modelling. Human diaries, autobiographies and biographies themselves may represent but snippets of the richness and complexity associated with the process of becoming an individual human being. A completely different view of the individual, however, emerges when he is regarded as issuing from and as maintaining himself within a dialogic context. I shall propose that these four terms and variants of them—the I, the We, the They and the It—serve as four personalized reference points that the individual himself employs and relies upon in everyday life.[29]

In relating himself to others and to the world at large, then, the individual adopts the following perspectives or variants of them: (A) As related to anonymous others, an I-THEY relationship obtains; here I will include also such notions of God and of Devil as serving some individuals as standards of reference. (B) For a meaningful, one-to-one relationship with another human being, I will use the term I-THOU, and I-WE for a similar relationship to a community or a group. (C) Finally, the term I-IT will symbolize the individual's relatedness to things, occasionally to other persons, and, finally, to indicate how the individual summarizes and circumscribes for himself a given state of affairs (as in response to the question : How are things going with your family?). The two sets of expressions, I-THOU and I-WE, on the one hand, and I-THEY and I-IT, on the other, will also indicate a temporal context. The former set indicate process, novelty, dynamic and fluctuating relationships between and among human beings; the latter symbolize stability, tradition and factitious relationships between and among human beings, as well as between and among human beings and an overall, abstracted view of a context or a situation.

I shall not attempt a theoretical justification for this terminology here, but will let its merits stand or fall in its application throughout the rest of this article. I believe, however, if one wishes to approximate as close as possible to the actual dialogic relationships in terms of which individuals live, the above terminology is a defensible approximation. Simply put, I maintain that individuals "personalize" all their experiences. That is, all human relationships and interactive processes are interpersonal and intersubjective : religious faith, culture, art and the sciences are stamped with the images of individuals acting in dialogic relationships. When one opposes a ruling political party, for example, he talks about "them" or about "it" in an inter-personal manner.[30] If one favors a new social movement, he personalizes the movement in terms of a THOU or a WE. Inasmuch as the individual has never experienced himself and the world from the position of a stone, a god, an atom or a tree, he has no other recourse, in relating himself to non-human objects, but to employ himself and other human beings as imagery and as models.[31]

Turning, then, to the issues associated with the genesis and growth of the individual within a dialogic context, I argue that parents (and significant other persons) play a role of singular importance : in effect, their role is more revolutionary than evolutionary. Parents initiate the child into their dialogic universe. For, on the one hand, as representing the accumulation of human tradition, parents play the role of the THEY with respect to the child. By a variety of interpersonal techniques, parents invite the child to membership in the world of I, THOU, WE, THEY and IT. Parents name and address the child, they invoke and call upon the child for its own personal response. The child, however, is at first incapable of interpersonal response. Relative to the child, then, parents may be regarded as enacting the role of the THEY.

On the other hand, parents need to initiate the child into the spatial and temporal circumstances with which the child is surrounded. Parents want the child to accept THAT interpersonal interpretation of the world that governs their own everyday life. As a result, parents also make the child aware of interpersonal anonymity: they interpret IT, the immediately surrounding world, for the child, and employ all those forms of communication and self-expression that may help make the child compresent with them as a THOU within the immediate context of the family.

The revolutionary role played by parents is further substantiated when we also recognize that parents are, at one and the same time, the carriers of heredity or nature, and of environment or nurture. Within a dialogic context, both heredity and environment stand for a contribution to the genesis of an individual that is both open and closed to further development. The genetic materials may offer the individual a combination of forces and dispositional properties[32] to which the interpersonal context may not be open. On the other hand, the dialogic context surrounding the birth of an individual may offer a wide range of possibilities to which the genetic material may not be disposed.

In effect, this section will present the genesis of the individual as an interplay of the child and parents as the latter play the role of the THEY to the child in the following ways: (A) the THEY as power; (B) the THEY as image-makers, and (C) the THEY as offering interpersonal freedom to the individual.

Is it not true that others—whom we have called the THEY—not only offer a great amount of options to an individual, but frequently determine the ways in which he lives? One need think only of the countless customs, rules, and prescriptions at the levels of the family, the church, and the state, or consider the volumes of civil laws that men have proscribed—to realize that an individual embodies a large amount of rituals, customs, rules and habits.[33] Fashions in clothes determine the way a man knots his tie and the length of women's skirts. Customs in eating determine how the table is set for a meal, and the way one is to use his fork. The rituals by which men greet each other, how men think about mathematics,

God, and politics, and the future options open to each person—all of these argue strongly for the belief that almost all personal activity has been conditioned by, and/or is a direct response to the THEY.

When a human being conforms to social customs, laws and manners, no one seems to take much notice of him : he is "taken for granted." But if one uses the wrong fork at a state dinner or grows a beard or goes against an accepted form of attire, he is immediately noticed, perhaps challenged, sometimes ignored, and occasionally ostracized. Similar reactions occur not only when one fails to conform, or goes directly against what the THEY prescribe, but also when one exceeds the measure of what the THEY believe should be done in a given state of affairs. In the latter context, the individual and his performance or his outlook on life stand out boldly in the face of the anonymous others. As a result, a person in this situation may at one time regard what he has done as being bold, somewhat outstanding or even distinguished, while at other times he may be made to feel that his deeds are quite paltry, somewhat misguided or completely valueless.

The first manner, then, in which parents extend their power over the individual is positive, and somewhat dramatic : parents give interpersonal *reality* to the individual, as well as to his deeds, to his words, and to the variety of presentations of himself. This reality-conferring power of the THEY takes two forms : a positive affirmation, or, a merely permissive sufferance.[34]

The dominant role of parents as affirming or merely permitting types of behavior in the genesis of the individual is an illustration that is immediately evident. Both parties in this ongoing encounter exercise a powerful transfer of emotional energies. Consider the combinatorial complex of feelings embodied in the mother witnessing the successes of the child as it adopts itself humanly to the environment. The mother may have a feeling of benign goodness, love, joy and fulfillment at the child's success; she may embrace the child and, in this process, call forth a similar embodiment of feelings and emotions. In this, as in many similar ways, the child learns what types of emotional combinations are or are not reality-creditable. A similar reality-granting role, however, may continue to be actually exercised by the parent and passively expected on the part of the child for a long time. And a like relationship, moreover, obtains between any two persons : all of us acknowledge our mutual presentation of ourselves to each other either positively or permissively.

The power of *denying reality* to the individual and to his self-expression is the second way in which the THEY exercise power over the individual. This reality-denying power is exercised in an open and direct, or, merely in a permissive manner. A parent may neglect, not give much credit to, or totally ignore the variety of modes in which the child presents itself to the parent. These tentative realities on the part of the child may, as a result, never again surface to view, and thus they may form part of the inner and secret world that all men possess and wherein they harbor these illicit realities as a miser's gold. The emotional complexes in terms of which the reality-denying power is manifested vary, of course, from a completely emotional rejection of the child to a mere emotional negative ploy that is aimed at forcing the child to seek the reality of an emotional reunion. Here, again, we have a paradigm of the type of negative options that all men exercise over each other in everyday life.

Now the very possibility of becoming an individual is a direct result of this affirming/denying power on the part of parents and other significant persons. For its part, the child gradually learns to imitate, emulate and affix for itself some of those images of humanness that are given reality-credit by the parents. The fixation of these images permits the child to find some sense and meaning for itself, and it teaches the child that *the only reality within which it may function and behave is that which has been given reality-credit by the THEY*. And thus by imitation the child induces this affirming/denying power within itself, and exercises *its own* reality-crediting or denying ability whenever it appears likely that success may attend its efforts.

It is of singular importance, then, to recognize that the parents and others, by exercising their power of giving or of withholding reality-credit, literally *force* the child to accept the immediate power of the THEY, and, by extension, the power of other human beings as the ultimate standards of reference for "reality." Even nature is never to be judged for what it is, but for what functions it may serve and what purposes it may fulfill for human beings. Standards the child learns are real, but they are real relative to the persons who initiate or who apply them in an endless variety of situations and contexts. The "world" of nature, then, does not exist in an impersonal manner but in the ways that it is interpreted, personalized and credited with reality by one or by a random number of human beings: the THEY.

A striking illustration of the power of the THEY over the individual can be obtained by addressing our reflections to the internal dynamics and effects of breaking a law or prescription; and, as a consequence, of inquiring as to the type of reality possessed by laws. Is it not apparent that in breaking a law, a human being does not actually and literally "break a law" or "breach a custom?" It is true that he may be said to "know" of the law's existence and of the consequences that may ensue for those who do not follow the law. But, as a human being, his conflict is NOT between himself and the law, but between himself and the THEY. In other words, the law or a custom does not represent to him an abstract, written statute; rather, it represents the "other than himself" or the THEY.

If it be the case that breaches of custom or of law are really directed against the "THEY" as appropriated within the individual, then it follows that laws and customs have no existence in themselves precisely as such. Laws, rules, customs and procedures, having "no being" of and in themselves, are interpreted by the individual as interpersonal addresses to himself, requiring and demanding an appropriately automatic response. Even positive laws (reality-crediting), as "honor and respect your parents" as contrasted with negative laws (reality-denying) as "do not steal," evoke this same opposition between human power and freedom, between human spontaneity and randomness of behavior; laws demand conformity to a pre-determined choice by others over which one has little or no control. In breaking a law, then, the individual is guilty before the THEY or before "someone"—as even before the anonymous image of a policeman.

As an effect concomitant with the reality-crediting or reality-denying power of the They-Self, the latter unwittingly inducts within the child so multiplex a variety of images and models of how to be a human being that it beggars any classification. An individual human being, then, is pluralistic with options and monistic only secondarily. Each human being is essentially "multiphrenic"; he is replete with possibilities of what kinds of human being he may possibly become. The child does not simply receive a singular mother-image. This latter notion is a facile generalization of the variety of attitudes that a given mother herself may take during a given day: mother-working, mother-crying, mother-happy, mother-angry, mother as talking to neighbor, husband, salesman, friend, children in the family, and so forth. Each person who is influential in the interpersonal formation of the child feeds his own complex imagery into the child with the expectation that the child will be able not only to absorb, but to match this input with appropriate output.

It is likewise important to note that negative power may insert as many images or ways of being human as does positive affirmation. Though the primary function of negative power is to be circum-ambient—to set up territorial limits for the young individual—it also serves to fixate models or images of how men should NOT be, as when a parent says, "Don't do as I do but do as I say." As a matter of fact, the negative power teaches the child that there is more to the world than what appears; in fact, that there may be another world or area beyond its own familial Garden of Eden.

In effect, then, the negative power introduces into the child the whole meaning and function of "temptation" (from *tentare*) and hence makes man that animal who is "testy"—who subjects every-thing to test, to criticism, and who hypothesizes and theorizes about everything that may enter his purview. Thus the moral and the critical sense in man are so closely allied that man may be regarded simply as temptation—as the maker, the tester and the breaker of any boundaries within which he may find himself contained.

At the same time, and arising from the very multiplicity of contexts and the variety of standards that may be exercised upon him, the child finds creativity and initiative—*freedom* for itself. The child learns that mother-at-work, mother-at-play, mother-with-father, mother-with-guests—and a whole melange of different roles played by the mother—give rise to different types of conformity or of randomness in the child's behavior. But the freedom offered is not a clear and definable freedom for it is both determined, and thus also limited by the types of models of humanity offered to and experienced by the child. The child with an over-strong or domina-ting father-image, for example, may find its particular type of liberty embodied in the master-slave imagery so strongly excoriated by Nietzsche. Because of the initial imagery of the THEY in its own family, a particular child may be well content with a clear image of itself as a child of duty, as having a limited amount of choice, and as bonded to a maximum amount of responsibility.

But the very fact that one child, in contrast with another, may have become humanized in terms of strong father-images or of strong mother-images, does not of itself preclude the possibilities of there being a variety of imagery in terms of which a child could be raised. While it is true that many or most cultures promote and adulate the implications of parent/father/mother/child images, and so forth, there is a wide variety of images actually presented in the genesis of human beings. If a child cannot establish a mutually interpretative relationship with a parent, it may do so with another member of the family, with other children, or with strangers. But it is the very multiplicity of imagery and plenitude of options received by the child that form the inner basis of its freedom.

Another manner in which freedom is offered to the child can be traced to the variety of emotional complexes that it experiences in itself as the result of interpreting these same complexes in other human beings. Evidently, one of the prime goals of the THEY is to make the child feel a state of uneasiness, anxiety and guilt if it fails in following the paths of acculturation that have been mapped for it. But the very fact that an identical misdemeanor may give rise to diverse emotional reactions on the part of different persons, helps the child to realize that there is no unique combination of emotions that can be employed as the paradigm for its behavior.

In conclusion, our presentation of the genesis and growth of the individual within a dialogic context offers us a different perspective on the meaning of individuality and on the interplay that obtains between an individual and other human beings. The individual human being is not a lone, solitary self caught up in the ego-centric predicament. Faced with this view of human nature, one attempts to introduce "others" into the individual, as it were, from the "outside." In contrast, we have suggested that the individual—the "I" of any dialogical context—only subsists as one member of a dialogical framework. As a result, we need to answer two different questions. The first : In what senses can the individual be said to be similar and yet distinct from others? will be dealt with in the next section. The final section will deal with a second question : In what ways does the individual exercise his interpersonal power within the dialogic universe?

III.   BECOMING AND BEING INDIVIDUAL : THE PROCESS OF INVOLUTION

Thus far we have considered the individual principally as an object of the larger processes of Evolution-Revolution as mediated to him through parents and significant other persons. But the individual is not merely an object or a receptor of the life chances that are thrust upon him. Already at an early age, he exercises some measure of selectivity in the options offered him and achieves some form of integration of the projects expected of him. The individual, in other words, is not merely passive but is an active agent in his own growth and development as a human being. He develops his own style of life even when he is largely under the tutelage of others.

We need a theoretical term, then, to stand for the active agency that the individual exercises in his own growth and development as a human being. If the complex expression "Evolution-Revolution" designates the meta-processes within which individuality is a possibility, then the term "Involution"[35] can symbolize the micro-processes at which individuality arises and contributes its own share to the former processes. Involution, then, permits us to describe how the individual is at one and the same time a member of the dialogic universe and how, as a member, he serves as a unique center or reference.

No individual, then, can be regarded as being totally and completely made by others within dialogical process. While it is true that men are the makers of other men, each man, in some sense, is a maker of himself. The processes of involution that take place within the individual enable him to fashion his own interpersonal interpretation of the world, of others, and of himself. In effect, then, the individual as a subject in the interpersonal universe cannot best be described or understood in terms of one essential property of selfhood or as embodying one unique model.[36] Becoming and being an individual, rather, is a day-to-day process of interaction between human beings.

This section will describe the individual as being dialogically alive in relationship to others. First of all, the individual will be considered as an *I* in dialogical process—as the genuine center or locus of interpersonal dialogue. As such, he will be shown to be an individual by reason of four interpersonal relationships by which he is bonded with other human beings: he is dependent on others, he needs others for his own fulfillment, he offers others his own interpretation of the world, and he is an object of fascination to other human beings. But the dialogic life of the individual has another side. Dialogic individuality requires that the individual himself take on any or all of those interpersonal frames of reference—the THOU, the WE, the THEY and the IT for others, as others do for the individual. For only in this manner can dialogic interplay take place. Secondly, then, this section will show how these latter members of dialogic life are themselves embedded in the individual: the individual is, at one and the same time, "self" and "other."

In the light of our remarks about the genesis of an individual, we can now identify the first invariant element associated with being a human being. Each individual lives his entire life in dialogic interdependence with some other human beings. In the first place, and as we have pointed out in the previous section, some dialogic relationship is required for an individual to become invited to membership in the family of man. Even though it may be the case that an individual may come into the world as the result of artificial processes, other human beings still need to mediate the individual into the dialogic universe in one way or another. A second way in which interpersonal subsistence is manifested is in the mutual collaboration needed to maintain the necessities of life as well as to make the products of the arts and sciences available to others. But the third and principal manner in which the dependence of the individual on others may be seen lies in the interpersonal associations he maintains with friends and collaborators, with enemies and random acquaintances.

Human interdependence is closely associated with a second trait that characterizes the individual. Each individual is incomplete and unfinished. Similar to all the bodies in the universe, the dialogic individual is in motion and caught up in process. In everyday life, the individual finds himself being asked, solicited and even required to share his own individuality in work as well as in play, in marriage and in friendship, and in a whole variety of interpersonal situations and tasks. Dialogically considered, the individual finds completion and fulfillment through the interpersonal dimensions of the THOU and the WE, as well as of the THEY and the IT.

As a new being called into the presence of other persons, the individual symbolizes a third element of interpersonal relationships: he is a unique interpreter of other persons, of the worlds of the arts and sciences, and of the universe in which he now co-exists with others. Within an interpersonal context, men are mutually embodied and interpreted by each other, and each individual exercises his own prerogative of embodying and of interpreting others in his own unique fashion. In a similar way, each individual embodies his own appropriation and appreciation of the arts, the sciences, of social and of political life, and of human history or destiny. The uniqueness of the individual's contribution to dialogic life, however, is also due to the simple fact that no particular individual can know everything or do everything by himself. The very incompleteness of the individual's perspective, then, highlights the necessity and the benefit of dialogic life.

As sharing and coordinating interpersonal worlds with other persons and as inhabiting worlds of his own making, the individual represents a fourth element of invariance: he is, to himself and to others, a being of fascination. Each individual has his own attitude, perspective or feeling about one thing or another. The individual, in fact, is so multiple and changeable that he embodies within himself a countless variety of interpersonal responses and reactions. For example, the individual may present himself in dialogue with another person as a stage upon which the entire drama of human life is being enacted; and, in a moment, he may draw the curtains, alter the props, remodel the entire structure, and perhaps tear the whole stage down. He may at one time reject other persons as cohorts in his life's enterprise and later regard these same persons as collaborators and as friends. And, in this variety of costume, of role and anti-role, the individual displays such random and satanic ingenuity that one may well wonder whether or not the individual has any inner continuity, identifiable spatiality or continuous temporality: *homo homini ludus.*[37]

We have considered four ways in which persons relate themselves to each other dialogically. Each individual is mutually dependent on others for fulfillment, each relates to others as co-interpreters and as mutual centers of fascination. In turning to the other side of the dialogic relationship, our aim will be to show that ability to participate in dialogic life requires that each member of the dialogic situation must be capable of enacting a variety of roles. At times, the individual speaks and acts for himself, representing the "I" of an interhuman context. Frequently the individual may be called upon to merge his own center of reference with another person or with a larger group as in love, friendship, and in a variety of other relationships in which he is asked for interpersonal commitment. Occasionally the individual may be required by others to play the role of anonymity as in speaking and acting for others, or in permitting others to employ him or his talents for their own needs and purposes. In any and all of these interpersonal situations the individual is expected to move smoothly from one role to another in response to the dialogic situation at hand.

In effect, these four interpersonal frames of reference—the THOU, the WE, the THEY and the IT—make dialogic life possible for the individual and serve in preserving, fostering and developing individuality. In maintaining and developing his own individuality, in facing particular decisions; generally, in being occupied with the business of everyday life, the individual successively takes on any or each of these interpersonal roles in relationship to others. We will discuss each of these four interpersonal roles as they are embodied in the individual.[38]

Let us consider, first, the dialogic relationship of an individual to the different contexts, situations and states of affairs in which he lives.[39] If he finds himself involved in a set of circumstances that is displeasing to him and which he would like to see altered, the individual may, of course, consult a friend or an authoritative source for an opinion. If no one is available for advice, the individual must estimate the situation himself. He may decide simply to tolerate the situation. Another option open to him is to reflect upon the state of affairs and seek for some means by which he might alter his predicament. In any case, the individual shows that he has, on the one hand, some estimate of how matters actually stand, and, on the other hand, an alternative model of how things actually could be. In effect, then, the individual employs his own interpretation of how IT, the situation, could be altered.

In other words, the individual feels himself capable of interpreting any state of affairs. Each individual has his own point of view about the world as a totality and about any of its sub-systems or parts. The individual circumscribes a variety of situations into his own perspective : how the members of his family relate to each other; how the political leaders may be conducting the affairs of state; how the business of the entire universe might more effectively and harmoniously be deployed.

By encompassing states of affairs simply as IT,[40] then, the individual becomes comfortably situated in interpersonal space and time. He feels that he is an articulate part of the dialogic world. By developing this ability to adjudicate any state of affairs, the individual acknowledges that the entire world is somehow embodied within him and that it can be encompassed in his purview. By the very fact of developing his own articulation of the world, the individual has acquired dialogic presence and significance along with other persons.

The dialogic individual also embodies the role of the THEY. As representing everyone and yet no one in particular, the THEY represents anonymity and objectivity. The THEY is a dialogic accomplishment of great interpersonal value, for it encompasses first, the establishment of ordinary and of technical language required for a variety of dialogic relationships. Second, the THEY represents the accumulation of generalized customs, rules and laws governing interhuman relationships as well as the accumulation of practical and theoretical skills. Third, the THEY includes a wide variety of matters that have been interpreted as facts over which there can be no further dispute or dialogue. And finally, the THEY represents that wide range of interpersonal roles that human beings need to play when they are of service to one another.

On the other hand, the THEY relationship also serves as the source of inadequate human communication and cooperation, and of a great number of evils that men continue to foist upon one another. In wars, another human being represents "the enemy." In political and business matters, an official may avoid personal responsibility by concealing himself with the anonymous cloak of the state or a corporation. In civil and social life, many persons prefer to be anonymous and are affronted by a casual greeting or question. Finally, confrontation among a variety of professional persons— clergymen, scientists, scholars—may be carried out professionally and punctiliously and, as a result, anonymously.

If the individual were to be no more than an embodiment of the IT and the THEY, he would find it difficult, and, perhaps, impossible to contribute and to share in dialogic life. Without relating himself as a THOU to another person, or without committing himself to membership in some WE relationship, the individual endangers his own dialogic life. An individual may be of great service to others in many ways and he may discourse with other persons about a variety of states of affairs, but unless he offers himself as a partner to one or to many individuals in shared relationships, the individual tends to become an anonymous self.

Many words have been written to encapsulate or to eulogize the shared relationships of lovers, of friends, and of collaborators. Transforming union, mutual indwelling, authentic communication and a variety of other effects have been underscored to highlight the merits of friendship in all its forms. Within a dialogic context, the relationships of I-THOU and I-WE are the primordial sources of human creativity, dialogic renewal, and cooperation. We shall present five broad types of dialogical life, beginning with mutual exclusion, then to partial and to complete inclusion, then to composition, and conclude with mutual inclusion or identity.[41]

In the first place, as a type of dialogic life, mutual exclusion can help us to explore and understand man's reality-crediting and reality-denying power. We are not dealing here with human ignorance, or with the simple fact that one individual may not actually have embodied within him some information about the world or some particular aspect of world affairs. What is at stake is the more portentous fact that men do exercise their reality-denying power with regard to another individual and with regard to an entire class of individuals (Negroes, Jews, Poles, Baptists, Communists). And, for a variety of reasons, professional persons will use their reality-denying power to exclude other professional persons from their group totally, or to deny reality to the ideas, plans or proposals of their colleagues. In this process, it appears that the given person takes *his own world* as a class of events and employs it as the model in terms of which he excludes other objects, ideas, proposals or persons. Possibly an overweening traditionalist or an arch-conservative would exemplify this powerful prerogative of human beings to say "no"—to discredit the very being of existence of anything that is proposed for his consideration.

The overlapping of classes, as the second mode of dialogic life, enables one to encompass a wide range of human experiences. Among these, we can include the following as being quite definitive both with regard to their structure and frequency. (a) The first level can be discerned in the overlapping of jobs, tasks, duties and services that men perform in everyday life. By offering services to each other, by being workers, tools and instruments relative to each other, men have a set of functional and role-playing inter-relationships in which there is much overlapping of functions and within which no "job" can be perfectly defined or delineated. Personal or mutual exclusion, of course, can and does occur frequently : one worker will refuse to perform a certain menial task that may not fall under his job description; or, one may exclude another worker from any overlapping of personal or functional relationships. (b) The second level is that of the political order (which, itself, overlaps the order of labor). At this level, men are often obsessed by the dream of perfect consonance between word and deed, so that each man expects to have made available to him all the rights and benefits that no political order could possibly insure. In fact, the political order gives birth to the dream of a homogeneity of total wealth for all individuals with each retaining his own "individualistic" prerogatives and options. (c) A third level of overlapping of classes can be found in the inter-relationships associated with religious structures. On the one hand, each member of a given faith wishes to have *his own* beliefs and thus be the sole member of his own church, and, on the other, we encounter a variety of ways in which persons mutually recognize and consent to their differences. Within the religious dimension, one may also encounter the yearning for total identity and a merging with the Absolute or the One, again, with individuality retained in some options and not in others. But the level at which the over-lapping of classes is illustrated most strikingly is in the great variety of images of the WE as found in (d) inter-personal relationships of any sort. At this level, differences between persons may or may not be of significance: one can set up some overlapping of classes in the way one relates to his job, to a game, to books, to writing, to music, and to an endless number of other relationships. Hence, one can "anthropomorphize" or "personalize" any object or thing whether or not it is capable of response on its own part.

When we turn to the third type, inclusion of a lesser in a greater class, we find the above levels of WE-relationships also exemplified. At work, a man may be part of his company or factory as a larger class. The political order itself may be regarded as consisting of various levels of classes, characterized both by inclusion as well as overlapping. The "being of God" can be regarded as that total class within which all subclasses are includable, and many religions espouse a veritable hierarchy of priorities and inclusion of class with class. And, finally, even inter-personal relationships exhibit many types of inclusion. The beloved wants to be "totally included within" the lover, and vice versa. Parents frequently want to exercise some hegemony as progenitors over their children as the receivers of life. And frequently, in conversation, one argument or discussion may, in final analysis, be seen as having been a sub-class of a larger frame of discourse.

Composition, the fourth type of dialogic life, involves the joint inclusion of two distinct classes to form a larger one. In composition we have a type of human relationship that is more cogently a WE-relationship than the others. For, as a matter of fact, complete and mutual exclusion exhibits an anti-relationship where one person might wish to make a null class of the other. Moreover, overlapping of classes and inclusion of a class in a larger or greater one still involves as many or more THEY-relationships than those symbolized by the WE. Composition, or two distinct classes forming some larger one, emphasizes the fact that a WE-relationship is something that may be transitory (two friends may be separated from each other), may be quite circumstantial (two friends com-present within a larger group), or unrealizable for any number of reasons (two friends who may not be able to become attuned).

At the same time, it is in composition, in this formation of a mutual territory (that may have few or even no bounds), of actually inducting an ontological way of being together that we find human creativity at its highest and most creative level. In an encounter of this sort, we have both parties to the *noyau* being both spectator and participant, being mutually creative and receptive, totally exercising and exorcising themselves in a creative interpersonal relationship that beggars description. Composition, then, is a mode of dialogic life combining simple unity with complex multiplicity.

Identity or mutual inclusion, as the fifth mode of diologic life, is the penultimate dream of every human being. The ontological insight that I am a human being precisely to the extent that and in the measure that I am "other" than any and every other being is transcended in the mysticism of ecstasy, fusion, identification and mutual indwelling, found both in human and in divine love. The projected eschatologies of Buddhists, Christians, Marxists, and humanitarians, and of almost every form of political ingenuity and promise, is motivated by this type of mutual inclusion or identity. As a matter of fact, the extent to which various religions, political entities, philosophies, theologies or sciences try to be reductionist without at the same time being pluralist accounts for some of the finest accomplishments as well as the most dastardly deeds that are found in the annals of human history. When, however, we turn to the level of a one-to-one relationship, we find this type of identity or mutual inclusion achievable in a variety of friendships, marriages and collaborative relationships. "I in my beloved and my beloved in me" can be paraphrased in a variety of ways : the artistic, scientific or prophetic genius merged and identified with his *daimon,* the child with play, and possibly, God with what is *not* God but joined with Him.

It is in the above and similar ways, that the individual, in partaking of dialogic life, is both an object and an agent in evolutionary-revolutionary processes. As sharing himself with others, the dialogical individual needs to involute upon himself. That is, in terms of the demands that dialogic life puts upon him, he coordinates, simultaneously, two dialogic movements: (a) he spirals outward as a THOU, a WE, a THEY or an IT in relationship to others: he is "other" than himself; and (b) he curls inward upon himself as others play these same roles in relationship to him: he is a "self" for others.[42]

IV. POWER: THE EVOLUTIONARY AND REVOLUTIONAY POSTURE OF THE DIALOGIC INDIVIDUAL

The Hebrew story of Genesis and similar mythological narratives afforded our early ancestors a clear and paradigmatic presentation of man as power. As Paul Ricoeur has so meaningfully pointed out,[43] these early narratives, dealing with the symbolism of evil, agree in placing the onus for evil on any other person or thing except upon man himself. The blame may be placed on wicked gods, on the fact that the soul is exiled, on the person of Satan or some other evil demon, or on the portentous fact that evil forms a very part of the past of being itself.

While the Genesis story may be interpreted as pointing to the fact that man sinned and thus that man disobeyed God himself, the impact of the story can also be interpreted as inculcating into men the hubris of power: the possibility that no powers of heaven or of earth can exercise effective hegemony over man. As representing not himself but mankind, Adam is a character who will get the most out of a situation at least expense to himself. With great cunning, he satisfies the demands of woman and of Satan, and he has learned a great lesson: he can consider himself the equal and the adversary of God himself.[44]

The dialogic dimensions of the individual as power have been approached from a number of other points of view: human reason, human action, human practicality and human aggressiveness. Within a dialogic context, these latter perspectives on man are but indices or manifestations of that interpersonal power that the dialogic individual shares with other persons. This section, then, will begin with some comments on the creative and destructive nature of both natural (the IT) and dialogic power (the THOU, the WE, the THEY). We will then focus on some characteristics displayed by the dialogic individual as sharing power with other persons, and conclude with some remarks on the posture of the individual relative to the processes of Evolution and of Revolution.

One can regard the history of the human race, or of any portion thereof (as, the history of physics, of war, of religions) as exhibiting the history of human power. Regarded in this way, history points out to us a simple truth : man is the maker and the breaker of inter-personal boundaries and the builder and the destroyer of dialogic structures. What one generation, or one man has accomplished, another will likely transcend. What one person has embodied in a conversation, another will restate in his own, and, presumably, in a more precise fashion. This characteristic by means of which the dialogic individual is the initiator of dissent, of rebellion, and of revolution against the past, seems to be as old as human beings them-selves.

As a matter of fact, this boundary-breaking characteristic is older than the history of human power, for it traces its lineage back to mythological and religious beliefs about the origins of the universe. Nature, or the IT, begins with some sort of primordial power— whether it be called by a personal name, regarded as fate, as the One containing all the potential Many, or as some primary state of basic elements. Whatever may or may not have been the original condition of primordial power, the multiplicity and variety of its displays throughout millions of years forms the fascinating story called the natural history of the universe. And how do we explain the present state of affairs? We have myths and beliefs that unify this history as a struggle for power, as a divine creation, or, finally and quite recently, as a combinatory result of statistical-natural occurrences.

Power, then, is not indigenously dialogic or human, for human power represents a hominization of powers inherited from nature. And just as a natural object or individual needs to adapt itself to the limitations and confines surrounding it, so the dialogic individual, from birth to death, needs to articulate himself with others in preserving and furthering the dialogic community of men. If the history of nature as IT can be interpreted as creative evolution, then the history of the dialogic community can be interpreted as creative revolution. For, as has been well said, "man" is nature involuted upon himself; the individual is reflected nature.[45]

We can now turn to the dialogic individual as the axial point of the processes of Evolution and of Revolution. First of all, the individual serves as the focus of a dialectical interplay between nature (the IT) and the THEY. Hermann Hesse beautifully expresses one aspect of the evolutionary process:

> I was an experiment on the part of Nature, a gamble within the unknown, perhaps for a new purpose, perhaps for nothing, and my only task was to allow this game on the part of primeval depths to take its course, to fill its will within me and make it wholly mine. That or nothing.[46]

On the other hand, the individual has emerged from the sacred pastness of the human condition as it has been brought up to the present (the THEY). In addressing himself to a task of any sort, in learning a subject and in relating himself to other persons, the individual needs to be well advised on the HOW and the WHAT of his endeavors. What have other cultures and individuals done when faced with similar problems? What has Aristotle, Jesus or Ghandi said on these matters? Dialogically viewed, human history can be regarded as one vast chorus requesting that they live on in us, their ancestors, not merely by way of memory and example, but by way of model and paradigm.

Secondly, the individual serves simultaneously as the focus of address from his contemporaries. By offering him membership in their dialogic community, other persons offer him dialogic significance and recognition. Lacking this acknowledgement, the individual may find himself at most an alienated and imprisoned spectator of the interpersonal business being carried on by other persons.

As a member of the natural dialogic universe, then, the individual can be compared to a point of lying above and equidistant from the four points of a square. Assume that all four points are connected by straight lines. In this simile, two points of the square can represent the IT and the THEY, while the other two points can stand for the THOU and the WE. There are some advantages to representing the dialogic universe in this fashion. It illustrates the interdependence of the individual on the IT, the THEY, the THOU and the WE. Inasmuch as one point can substitute for another, the simile also illustrates the involutional relationships of the individual to others within the dialogic framework.

In conclusion, this article has argued strongly and conscionably for the importance and the significance of the individual within the natural-dialogic universe. There are many reasons for this emphasis on the posture of the dialogic individual, not the least of which stems from the dialogic power he possesses in authentically relating himself to one and to another and to many individuals in terms of a THOU and a WE. For, paraphrasing Martin Buber's[47] beautiful aphorism, we can conclude in all seriousness of truth : without IT and THEY man cannot live; but he who lives with IT and THEY alone is not a man.

Notes and References

1. I am indebted, of course, to Martin Buber for use of the notion of "the life of dialogue." See Maurice S. Friedman, *Martin Buber: The Life of Dialogue* (New York: Harper, 1960), Part Three, "Dialogue," pp. 57–100, and Friedman's "Introductory Essay" in his editing of Buber's *The Knowledge of Man: A Philosophy of the Interhuman,* in which he discusses other views of the life of dialogue. Similar points of view are developed in *A Study of Interpersonal Relations: New Contributions to Psychiatry* (New York: Science House), edited by Patrick Mullahy.

2. In *Human Potentialities* (New York: Basic Books, 1958), Gardner Murphy devotes the final chapter to a discussion of "The Human Natures of the Future," pp. 302–329.

3. The complex inter-relationships associated with membership in a small group have been admirably summarized in Theodore M. Mills' *The Sociology of Small Groups* (Englewood Cliffs, N. J.: Prentice-Hall, 1967), *passim.*

4. Both of these relationships, I-Thou and I-It, are developed in Martin Buber's *I and Thou* (New York: Scribner's, 1958). For a further development of these notions, see Edmund F. Byrne & Edward A. Maziarz, *Human Being and Being Human: Man's Philosophies of Man* (New York: Appleton-Century-Crofts, 1969), Ch. 11, "Authentic Communication," pp. 262–294.

5. The "They" as standing for "mankind," or the anonymous or impersonal others has been developed especially by Martin Heidegger in his *Being and Time* (London: SCM Press, 1962), pp. 126 *ff,* and pp. 238 *ff.*

6. The use of I-We is greatly dependent on Gabriel Marcel's analyses of such notions as continuity, presence and invocation; see, for example, his "The Ego and its Relation to Others," in *Homo Viator* (New York: Harper Torchbooks, 1962), pp. 13–28.

7. Much of this article had already been written when I found Hugh

Dalziel Duncan's excellent presentation of interpersonal relationships in terms of I, Thou, They, We and It, in his *Symbols in Society* (New York: Oxford, 1968), pp. 81 *ff*.

8. For a development of this notion of man as questioner, see "Choosing Among Projects of Action," in *Affred Schutz: Collected Papers, I, The Problem of Social Reality,* Edited and Introduced by Maurice Ntanson (The Hague: Nijhoff, 1962), pp. 67 *ff*.

9. For a presentation of some of these themes in relationship to the human condition, see Edmund F. Byrne & Edward A. Maziarz, *op. cit., passim.*

10. See any of the following as examples: Arend Theodoor van Leeuwen, *Christianity in World History: The Meetings of the Faiths of East and West* (New York: Scribner's, 1964), Ch. VIII, "Christianity in a Planetary World," pp. 399–436; Karl Jaspers, *The Future of Mankind* (University of Chicago, 1961) *passim;* Gabriel Marcel, *Man Against Mass Society* (Chicago: Regnery, 1962) *passim.*

11. For some recent books on the issues of dissent, rebellion and violence, see Theodore Roszak, *The Making of a Counter-Culture: Reflections on the Technocratic Society and its Youthful Opposition* (Garden City, New York: Doubleday, 1968); Jerome Tuccille, *Radical Libertarianism: A Right Wing Alternative* (New York: Bobbs-Merrill, 1970); Henry Bienen, *Violence and Social Change: A Review of Current Literature* (University of Chicago, 1968), *Bibliog;* Martin E. Marty & Dean G. Peerman, *New Theology No. 6: On Revolution and Non-Revolution, Violence and Non-Violence, Peace and Power* (London: Macmillian, 1969).

12. A wider framework within which my proposals are compatible, I believe, can be found in Jope John XXII's famous encyclical, *Pacem in Terris;* see Edward Reed (ed.) *Peace on Earth: Pacem in Terris. Proceedings of an International Convocation on the Requirement of Peace.* (New York: Pocket Books, 1963).

13. For an example of this approach to time, see Wilbert E. Moore, *Man, Time, and Society* (New York: Wiley, 1963).

14. I do not wish to enter the debate about the "private language problem" or similar issues. I intend to argue in section three of this paper for some sort of "Life's world" as the individual's personally appropriated life style.

15. See Ernest Becker, *The Structure of Evil* (New York: Braziller, 1968), Part IV, "The New Science of Man," pp. 307 *ff.*, and especially the "Appendix," pp. 384–395.

16. For a recent summary of Darwin's theory and its influence see Michael T. Ghiselin, *The Triumph of the Darwinian Method* (University of California Press, 1969); see also George Gaylord Simpson, *Biology and Man* (New York: Harcourt, Brace & World, 1964).

17. See Hannah Arendt's comments on these points in *On Revolution* (New York: Viking, 1963), Chapter 2, "The Social Question," pp. 53–110.

18. In *Social Change and History: Aspects of the Western Theory of Development* (New York: Oxford, 1969), Robert A. Nisbet argues that "Of all the metaphors in Western thought on mankind and culture, the oldest, most powerful and encompassing is the metaphor of growth." (p. 7).

19. These points are developed by a variety of scientists in John R. Platt

(ed.), *New Views of the Nature of Man* (University of Chicago Press, 1965); see especially the article by Clifford Geertz, "The Impact of the Concept of Culture on the Concept of Man," (pp. 93–118).

20. For an outstanding example of the development of these ideas see Talcott Parsons, *Societies: Evolutionary and Comparative Perspectives,* (Englewood Cliffs, N.J.: Prentice-Hall, 1966), Chapter Two, "The Concept of Society: the Components and their Interrelations," pp. 5–29.

21. For some comments on evolutionary forces and factors see "Philosophical Aspects of Darwinism," in Hans Jonas, *The Phenomenon of Life: Toward a Philosophical Biology,* (New York: Harper & Row, 1966), pp. 38–63.

22. For a recent attempt to unify evolutionary theory with theology see Eric C. Rust, *Evolutionary Philosophies and Contemporary Theology* (Philadelphia, Pa: Westminster, 1969), *passim.* A more revolutionary perspective is presented in Rubem A. Alves, *A Theology of Human Hope* (Washington, D.C.: Corpus, 1969).

23. Aristotle's *Ethics* and some books of the Christian bible are examples of this particular approach.

24. The works of Robert Ardrey and of Konrad Lorenz illustrate this technique; see the criticism of their proposals in Ashley Montagu (ed.), *Man and Aggression* (New York: Oxford, 1968), *passim.*

25. A survey of models offered by social psychologists can be found in Morton Deutsch and Robert M. Krauss, *Theories in Social Psychology* (New York: Basic Books, 1965); a similar summary of psychological models is given by Calvin S. Hall & Gardner Lindzey in *Theories of Personality* (New York: Wiley, 1967). A developmental and integrated view of the development of the individual is offered in Theodore Lidz, *The Person: His Development Throughout the Life Cycle* (New York: Basic Books, 1968).

26. For a fine illustration of the interaction between theoretical models and their impact on the social order, see Richard L. Means, *The Ethical Imperative: The Crisis in American Values* (Garden City, New York: Anchor, 1970).

27. Literally, "man is what man does."

28. An excellent critique of the applicability of some theories of social psychologists to human affairs can be found in Robert E. Lana, *Assumptions of Social Psychology* (New York: Appleton-Century-Crofts, 1969).

29. Throughout this essay I am proposing recognition of a dialogic universe as descriptive of the actual interrelationships found in everyday life. For a similar proposal, see *The Civilization of the Dialogue,* A Center Occasional Paper, Volume II, Number I (Santa Barbara, Cal.: Fund for the Republic, 1968).

30. The notion of a "moral" or a "corporate entity" serve as illustrations of the term "IT."

31. Though I am unable to develop this point here, I believe the traditional theory of the analogy of being is defective primarily because it abstracts from any personal, interpersonal or dialogic reference.

32. For an excellent presentation and defence of this notion, see David Weissman, *Dispositional Properties* (Carbondale, Ill.: Southern Illinois

University Press, 1965), Chapter 4, "Real Potentiality," pp. 159–193.

33. For a recent study of customs and laws, see Burton M. Leiser, *Custom, Law, and Morality* (Garden City, New York: Anchor, 1969).

34. For an excellent illustration of the employment of this type of interpersonal power see Erving Goffman, *Stigma: Notes on the Management of Spoiled Identity* (Englewood Cliffs, N.J.: Prentice-Hall, 1963). Albert K. Cohen's *Deviance and Control* (Englewood Cliffs, N.J.: Prentice-Hall, 1966), develops these same notions in a sociological perspective.

35. My emploment of this term is explained in a number of places. It is similar to the classical Christian notion of "conversion," and quite close in meaning to Martin Buber's notion of "turning" (*I and Thou*, p. 57), though at this point of his discussion Buber is arguing against the quasi-biological and quasi-historical thought of today with its dogma of process. I use it as a technical term within the dialogic context to show that the individual at times needs to speak for a THEY or a WE and not merely for himself.

36. I strongly favor anthematic or perspectival view of human nature; see Edmund F. Byrne & Edward A. Maziarz, *op. cit.,* "Preface."

37. Literally, "Man is to man a game." I have adapted this expression from Thomas Hobbes "homo homini lupus," or, "man is to man a wolf." I am favouring the "spirit of play" in a dialogic universe as opposed to the "spirit of seriousness."

38. I have used the word "embodied" in a number of places throughout this article; for a further explanation of its use, see Edmund F. Byrne & Edward A. Maziarz, *op. cit.,* Chapter 12, "Self and Others as Embodied," pp. 298–314.

39. My use of "IT" may be somewhat confusing. Briefly, I am trying to show that in a dialogic context, we do talk meaningfully about, say, "the United States," or, "the current situation in biology," or "how primitive man worshipped his gods," and so forth. Karl Jaspers used the notion of "encompassing" and of ciphers for a similar purpose; see his *Reason and Existenz,* (New York: Noonday, 1955), Second Lecture, "The Encompassing," pp. 51–76.

40. For a similar development see Hugh Dalziel Duncan, *op. cit.,* pi 110–116.

41. I have borrowed the analogy for this scheme of relationships from the algebra of classes, a summary of which is available in many introductory logic or mathematics texts.

42. I believe that the description of the "dialogic individual" given in this paragraph names some dialogic "universals" that are common to all men—regardless of their learning or sophistication.

43. Paul Ricoeur, *The Symbolism of Evil* (New York: Harper & Row, 1967); see Part II, "The 'Myths' of the Beginning and of the End."

44. Throughout this final section I am using the word "power" in an ultimate sense as a "category" that applies both to membership in the natural and in the dialogic universe as well.

45. "The ego persists by becoming ever more itself, in the measure in which it makes everything else itself. *So man becomes a person in and through personalisation.*" Pierre Teilhard de Chardin, *The Phenomenon of Man* (New York: Harper & Brothers, 1959), p. 172.

46. Hermann Hesse, *Demian* (New York: Bantam, 1968), p. 89.

47. "And in all the seriousness of truth, hear this: without *It* man cannot live. But he who lives with *It* alone is not a man." Martin Buber, *op. cit.*, p. 34.

Selected Bibliography

Alves, Rubem A. *A Theology of Human Hope.* Washington, D.C.: Corpus Books, 1969.

Becker, Ernest. *The Structure of Evil: An Essay on the Unification of the Science of Man.* New York: Braziller, 1968.

Bienen, Henry. *Violence and Social Change: A Review of Current Literature.* University of Chicago, 1968.

Buber, Martin. *I and Thou.* New York: Scribner's, 1958.

Byrne, Edmund F. and Edward A. Maziarz. *Human Being and Being Human: Man's Philosophies of Man.* New York: Appleton-Century-Crofts, 1969.

Castell, Alburey. *The Self in Philosophy.* New York: Macmillan, 1965.

Crosson, Frederick J. and Kenneth M. Sayre (eds.) *Philosophy and Cybernetics.* New York: Clarion, 1967.

Duncan, Hugh Dalziel. *Symbols in Society.* New York: Oxford, 1968.

Friedman, Maurice. *To Deny Our Nothingness: Contemporary Images of Man.* New York: Delta, 1967.

Ghiselin, Michael T. *The Triumph of the Darwinian Method.* University of California Press, 1969.

Goffman, Erving. *Stigma: Notes on the Management of Spoiled Identity.* Englewood Cliffs, N.J.: Prentice-Hall, 1963.

Goffman, Erving. *The Presentation of Self in Everyday Life.* Garden City, New York: Anchor, 1959.

Hall, Edward T. *The Hidden Dimension: An Anthropologist Examines Man's Use of Space in Public and in Private.* New York: Anchor, 1967.

London, Perry. *The Modes and Morals of Psychotherapy.* New York: Holt, Rinehart and Winston, 1967.

Lowen, Alexander. *The Betrayal of the Body.* London: Macmillan, 1967.

Maslow, Abraham H. *Toward a Psychology of Being.* New York: Van Nostrand, 1962.

Milss, C. Wright (ed.) *Images of Man: The Classical Tradition in Socioligical Thinking.* New York: Braziller, 1960.

Montagu, Ashley (ed.) *Man and Aggression.* New York: Oxford, 1968.

Montagu Ashley. *On Being Human.* New York: Schuman, 1951.

Moore, Wilbert E. *Social Change.* Englewood Cliffs, N.J.: Prentice-Hall, 1963.

Mounier, Emmanuel. *The Character of Man.* New York: Harper and Brothers, 1956.

Muller, Herbert J. *The Children of Frankenstein: A Primer on Modern Technology and Human Values.* Bloomington, Ind.: Indiana University Press, 1969.

Murphy, Gardner. *Human Potentialities.* New York: Basic Books, 1958.

Petras, John W. *George Herbert Mead: Essays on His Social Philosophy.*

Edited, with an introduction. Columbia University: Teachers College, 1968.

Platt, John R. (ed.) *New Views of the Nature of Man.* University of Chicago, 1965.

Ricoeur, Paul. *The Symbolism of Evil.* New York: Harper and Row, 1967.

Rust, Eric C. *Evolutionary Philosophies and Contemporary Theology.* Philadelphia: Westminster, 1969.

Simon, Herbert A. *Models of Man: Social and Rational.* New York: Wiley, 1957.

Simpson, George Gaylord. *Biology and Man.* New York: Harcourt, Brace and World, 1969.

Smith, Wilfred Cantwell. *The Meaning and End of Religion: A New Approach to the Religious Traditions of Mankind.* New York: Mentor, 1962.

Wagar, W. Warren (ed.) *European Intellectual History Since Darwin and Marx: Selected Essays.* New York: Harper Torchbooks, 1966.

CRUSOE, FRIDAY, AND GOD
ARTHUR F. HOLMES
Wheaton College

DANIEL DEFOE was an eighteenth-century satirist and man of letters. It took an eighteen-century man to conceive Robinson Crusoe, a solitary sailor, so self-contained and self-sufficient that he could create a life by himself, indeed a society composed of just one rational individual.[1]

Defoe was not alone. Seventeenth- and eightenth-century philosophers pictured man by nature as a sort of Robinson Crusoe, an isolated individual, self-contained and self-sufficient, needing nobody else to establish his own identity or fulfil his potential. Thomas Hobbes was one of them. But he detected seeds of conflict in the radical individualism of man's natural state, such that in the war of all against all every Crusoe fears his man Friday. Political organization therefore arises to preserve some degree of good for each individual in his relations with others, an organization that is enforced by the might of Hobbes' Crusoe, the Leviathan.

Jock Locke too, described the natural state as that of an isolated individual, self-contained and self-sufficient, a society of one rational being. Larger social relations are structured by the natural rights we share to life and liberty and private property. Crusoe's rights must be protected by the rule of reason, and his man Friday's too. Provided that is done, a man finds fulfilment by the light of reason alone.

No man is an island, not even Robinson Crusoe. He was already a mature man when he was cast ashore; he had his books and the rudimentary instruments of a culture; he both feared and loved the advent of Friday; and he had his God. No man is an island.

My point is that serious difficulties accompany the Enlightenment conception of individuality which Crusoe represents. This I take to be Edward Maziarz' view as well. He offers an alternative account of individuality with which I am largely in agreement, one which

307

draws more on recent phenomenology than on the Enlightenment model. But he does not explain in any detail the role of religious experience in personal fulfilment, even though the evolution-revolution question is perhaps most pointed in this basic area of an individual's life. I shall therefore consider some aspects of the phenomenology of religion in relation to the phenomenology of interpersonal relations which Maziarz has introduced. My purpose is not so much theological as it is philosophical, that is to say I shall not develop nor defend a religious doctrine but shall try to describe the effect of religious faith on personal fulfilment. For the sake both of clarity and of historical perspective, I shall first of all introduce some models that have been used in forming conceptions of individuality.

I.   THE CRUSOE THEORY

1.   *Metaphysical models*

How we say we achieve individuality depends on how we conceive it. That conception is often part of a larger conceptual scheme, a metaphysic, that is often coordinated and unified by means of a model drawn from one area of experience such as science. A metaphysical model illuminates our understanding of all areas of experience. It is therefore used to help us conceptualize individuality.

Aristotle, for instance, employed the model of form and matter. He conceived the individual as a member of a species, so that the universal characteristics of the human species make the individual human and give intelligible meaning to his existence. It is one thing to belong to the species, however, and another thing to be uniquely, even idiosyncratically, I. The medievals offered several explanations of individuality. (1) My individuality might be due to the fact that I fall short of the ideal of the human species or, to use the technical term, due to a "privation of form." But this implies that the individual is less than fully human in direct proportion to the degree of his individuality. Individuality, then, is bad, even sinful. Something is awry. (2) My individuality might alternatively result from some accident of my bodily existence, such that my biological heredity accounts for my uniqueness. To use technical terms again, "matter individuates." Yet if this is so, what purpose is there to my uniqueness? Purpose comes from form, and form defines the human species without differentiating individuals. My uniqueness is not essential to my humanity, only to my existing. What then is the value of my existing as *the individual I am and could be,* not as just another human? And what does it mean to find fulfilment *as an individual,* rather than as "just another" man? (3) Does religion provide the answer? For the medievals, insofar as religion is cognitive, it offers true propositions about God and about men in general, propositions that are not available to reason alone, namely the doctrines of the faith. In so far as religious experience is non-cognitive it culminates in a mystic union with the Divine that eclipses individual self-consciousness rather than establishing it. What then is the meaning of my individuality?

Duns Scotus tried to answer the question by introducing a third principle in addition to the form of the species and the material body, namely a principle of individuality, *haecceitas.* His point was that I exist by virtue of my material body and I am a man by virtue of the form of the species, but I am distinctively *I* by virtue of a unique form of me. To William of Occam, however, Scotus seemed to complicate the picture needlessly. Why not reject the theory of forms altogether and say instead that only individuals exist, that we are the individuals we are because God chose to make us that way? This seemed to him to give more value to my uniqueness than all the complexity created by Greek metaphysics.

In place of a metaphysic rooted in the Greek model of form and matter, Renaissance and Enlightenment thinkers developed their conceptions from the model of mechanistic science : particles of matter moving according to fixed laws of motion. This model had proved fruitful in astronomy and physics; and they gave it wide use in their conceptual schemes.

Descartes, for instance, conceived of matter as subject to mechanistic laws of motion, and mind as analogously subject to fixed laws of thought which move us from one idea to another and organize our ideas into propositions and arguments and theories. The model is mechanistic : particles of thought (clear and distinct ideas) organized by "rules for the regulation of the mind." The mind is subject to different laws than the body and contains a different kind of ingredient; it is therefore separate and self-contained except for occasional interaction. My mind is isolated from any other mind by two bodies : my own and his. It is a Robinson Crusoe stranded in a body, but my body is an hospitable island whose shores are only occasionally touched by other men.

While Thomas Hobbes rejected Descartes' dualism of mind and body, he used the same mechanistic model in his own conceptual scheme. Words, for example, are the isolated atomic constituents of speech which are organized by the laws of grammar and syntax. Particles of speech are like particles of matter and the rules of language are like the laws of motion. The same model appears in Hobbes' social theory : isolated individuals drawn into an organized society out of mutual fear like Crusoe and Friday.

John Locke employed the same model and reaffirmed Descartes' mind-body dualism. According to Locke, experience is composed of simple ideas, the atomic particles of consciousness which are drawn together by laws of psychological association. And society is composed of isolated individuals each replete with natural rights, who unite under the rule of reason for the sake of life, liberty, and property. But by nature each man is an island . . . like Robinson Crusoe and his man Friday.

The Crusoe theory with its mechanistic model contains the following assertions :

(1) Man is by nature an isolated consciousness without direct encounter either with bodily things or with other minds.

(2) Man is by nature rational, his ideas clear and distinct and unalloyed by emotion. This kind of pure reason can relate a man effectively to the world and to other individuals.

(3) The individual is self-contained and self-sufficient, not needing others in order to be or to be himself.

(4) The natural state of man is one of isolation, not community. Love and friendship and community and the state are all posterior to individuality and arise from other needs than those essential to meaningful individuality.

The Crusoe theory, as we have labelled this concept of individuality, means that we achieve individuality alone by developing the rational resources which nature has provided. This means classical education; so Crusoe needed books. It means ruling nature; so he needed implements and weapons. It means shunning "enthusiasm," that emotional exuberance displayed by the unenlightened masses and in popular religion where reason does not rule. It means that an individual can by himself find fulfilment through developing the intellectual and cultural powers that enable him to rule nature and subdue savagery. Robinson Crusoe succeeded. Although just another sailor, he established an identity for himself and became an authentic individual . . . alone.

But Crusoe also believed in God. Locke and Descartes did too, not with unthinking enthusiasm, but as a rational conclusion to which the mind of man is logically drawn. Descartes' theism was part of his metaphysic and each depended on the other. Locke wrote *The Reasonableness of Christianity* to exhibit the logical character of what he believed. Faith, he declared, is the acceptance of propositions in addition to what can be known by reason alone. Other writers opted for reason alone : Thomas Paine in his *Age of Reason,* and Immanuel Kant in his *Religion within the Limits of Reason Alone.* But for all of them, with or without revelation, religion adds to the knowledge and strengthens the reasoning by which an individual guides his destiny, rules nature, relates to other people. Religion and society depend on what individuals know respectively of God and of each other.

2. *Epistemological problems*

I have placed the Crusoe theory in its metaphysical setting for two reasons, first to prepare the way for a contrasting model I wish to introduce later in conjunction with the phenomenological account, and second because the weaknesses of the mechanistic metaphysic become the weaknesses of its conception of individuality. The salient difficulty has to do with the isolation of the individual mind from its body, from its world, and from other minds. Berkeley and Hume raised the question : how can the isolated mind know that an external world exists, or other minds, or God?

Consider the problem of other minds. If I conceive myself to be a mind within a body, somehow interacting with but separate from my body, then I am separated from any other mind by both my body and his. I have "privileged access" to my own mind that I do not have to his.[2] In other words, how can one Crusoe on one island ever get to know another Crusoe on his?

It is suggested that I know other minds on the basis of their likeness to my own,[3] by analogical inference from the conjunction of behavior (B_1) and mental states (M_1) in my own case to their conjunction (B_2 and M_2) in other cases. Thus, $B_1 : M_1 : B_2 : M_2$. I know B_1 and M_1 and B_2. An analogical inference can therefore be drawn to M_2.

But Descartes and Locke must first make sure of their bodily self-knowledge. They know their bodies (B_1) indirectly through their mental states (M_1). All they know directly is their mental states, not bodies which inhabit another world external to their minds. The inference to other minds therefore depends on our prior knowledge of our bodies and of the external world, which Berkeley and Hume found problematic enough in itself.

Analogy, moreover, could only give insight into other minds in those cases where constant conjunctions between thought and bodily action occur, and where parallels exist between my own and other cases. Uniquely individual mannerisms, which in ordinary life are extremely revealing, would afford no logical clue at all. Moreover not all our bodily acts have corresponding mental states, and not all our thoughts are evident in our behavior. What analogical inference can tell us, therefore, is rather little, in fact considerably less than we actually seem to know in our normal relationships with other people. I know my wife far better than it seems I should. Perhaps we are not two Crusoes nor on different islands after all.

My point is this : if the Crusoe theory is true then something like the analogical inference theory must also be true and the individual can well find fulfilment alone. But the analogical inference theory is not wholly true, and similar objections have been raised to a more purely behavioristic approach to other minds, such as would be required by Hobbes' concept of man. Therefore the Crusoe theory is not wholly true either : it is a mistake to conceive the individual in a way that makes my mind a world apart from my body and from other minds and bodies too. The mistake comes from supposing that self-knowledge is a direct introspective awareness of mental states, rather than a more indirect awareness mediated through my words and actions in the world. I know myself whole, in being myself, bodily involved in the world rather than separated from it. As P. F. Strawson points out, the concept of a person is in ordinary language logically prior to separate concepts of mind and body.[4]

As a result we reject the Crusoe theory.

(1) A man is not an isolated consciousness without direct contact with either things or people.

(2) A man is not pure reason, his ideas already clearly and distinctly grasping the bodily world and other minds.

(3) An individual is not self-contained and self-sufficient, needing no others in order to be fully himself.

(4) The natural state of man is not one of isolation but of community.

This, moreover, is the conclusion reached by phenomenologists independently of the criticisms I have cited of the Crusoe theory.

II.   THE PHENOMENOLOGICAL ACCOUNT

1.   *A view of individuality* differing from the Crusoe theory can be traced in the history of thought. I shall start with three familiar examples.

The first is Socrates' art of dialog. Whereas the Sophists used rhetoric to control people as if people were objects, Socrates uses dialog, the language of interpersonal relations, to draw them out. He pictures himself as an intellectual midwife bringing to birth what others have in mind, and making clear to them what they really are. "Know thyself," he says, and helps them find themselves in dialog. But this requires friendship, for it is in friendship that one opens up to another, reveals his inmost thoughts, and discovers who he is. So in the *Lysis,* Socrates asserts that dialog is the natural occupation of lovers. For him, the knowledge of other persons is inherent in life and love and society rather than an inference from one Robinson Crusoe to another.

My second example is Aristotle's view of friendship. Man by nature is a social animal. Bodily existence and social relations are alike essential to his being. Without friends, Aristotle tells us, nobody would choose to live, but with friends a man is better able both to think and to act as he must. A friend is an *alter ego* who helps me both to be and to know what I am and what I can become.

My third example is Hegel's picture of the master and his servant. The master needs his servant; he establishes his own identity and gains status in the world through the role he imposes on his man. But the servant needs his master : if he is to identify himself he must contrast himself with others; so he finds meaning in life by serving his lord. Each achieves self-knowledge in relation to the other; no man is an island.

We may add that Marx and Sartre seize on Hegel's dialectic and argue that man is alienated from himself by the society in which he lives, that he finds himself and freedom in antithesis, in revolt, as he negates the others. On the other hand, people like Husserl, Heidegger, Buber, Merleau-Ponty and John Macmurray have explored the structure of I-Thou relations as the vehicle of self-knowledge apart from dialectical conflict. The point of view they expound stands in complete contrast to the mechanistic model.

Edmund Husserl puts it two ways.[5] First, my knowledge of other selves is gained along with my knowledge of objects generally. I live in a world of objects that I perceive and use, but not alone, for I find that others experience my objects as well. I myself am an object for them, along with my world, and so are they for me. Ours is a public world, not private, an interpersonal world of shared objects and overlapping horizons which other individuals enjoy as well as I.

Second, Husserl affirms that this knowledge of other selves is accordingly not an inference but a direct experience. The "we" is known not by inferring another self and adding that one to my one, so that the two make a "we," but directly in the experience of "togetherness." Nor is this a purely intellectual experience as the Crusoe theory supposed, but involves feelings and the bodily self-consciousness of a "shared" existence. This knowledge of others also involves an analogy between myself and them, but it is not an inference. Husserl calls it "analogizing transfer," an act of empathy : I recognize signs of sensing and feeling and thinking which awaken me to the presence of another subject like myself.

Martin Heidegger follows suit. I encounter others in my being-in-the-world. Their world intersects with mine. My "being-here" is not a state to which I have "privileged access," but a sharing of horizons and worlds with others. "Togetherness" (*Mitsein*) is an existential characteristic of my being-in-the-world, even when nobody else is at hand. When I am lonely, my aloneness is simply a lack. Togetherness rather than aloneness is fundamental to the lived character of my existence.[6]

Others have brought similar phenomena to our attention, and we may summarize the phenomenological evidence in contrast to the Crusoe theory.

(a) *I have direct experience of other persons,* not just an indirect knowledge drawn from the ideas in my isolated mind. We reveal ourselves to each other as persons in the dynamics of being, living, working, growing, and thinking together. I open up to you as you open yourself to me. My knowledge of other persons and of myself crystallizes out of this chemistry of personal relationships. As Buber says, "I-Thou" is as much a primary word denoting a basic form of consciousness as is "I-it." I experience "We-ness" directly : the "we" of our togetherness is not an intellectual construct or an inference at all, but a directly experienced thing.

The mechanical model suggests a theory of external relations. A vast number of isolated individuals relate to each other only incidentally and indirectly like marbles in a bag, so that personal relations resemble the rattling rebound of one marble from another. The phenomenological account, however, uses a theory of internal relations. You in a sense are a part of me and I of you. I place a chair beside a table and it makes no difference to either one. But I seat you at dinner beside a certain young lady, and the relationship that ensues can change your whole life and hers. She has entered your life. Personal relations are direct and intimate. We are one in community, in a way the furniture and a bag of marbles are not. The family, we say, is a "unit," and so indeed it is.

(b) *Personal relations involve emotion as well as intellect.* These two ingredients are inseparable in the unity of the self, and when I enter your life I do so as a whole. The more I involve myself with you, or you with your true love, the more emotion as well as thought appears. There are two poles to this that we shall call "rapport" and "distance," and both are important. I may feel close to you, we may have good rapport, yet we keep our distance and respect each others' rights. This is essential in the family, in education, in the community and in the nation. Rapport has been described in terms of empathy (by Edith Stein), of sympathy (by Max Scheler), of solicitude (by Heidegger) and of love (by Buber and Marcel).[7] Distance without rapport is alienation—a phenomenon which Sartre and Marx stress, and with which psychologists and sociologists are rightly concerned. It produces confrontations. It shows up when we dominate others or when we reject people along with their faults. Alienation inhibits growth; it destroys freedom; it perverts individuality. Distance there is and distance must remain, but rapport and love are needful as well. Personalities blossom in love as long as they are not "smothered"; we open up where there is rapport; we get to know one another and learn to be authentically ourselves.

(c) *I-Thou relations are possible because we share a common world. The world we share is first a world of bodily existence and common history.* It is our lived world, the life we live together. We share our bodily world. We work and play. We enjoy sunshine and success. A mother tickles the child she has just fed and wipes his tears. A man and his wife share bed and board, joy and sadness. And by these means persons, whole persons—not just bodies, nor isolated minds—are by nature united in a togetherness they each find fulfilling. It is in community that we discover who we are and become what we do.

The world we share is also *a world of language*. Some have said that language fails to capture the personalism of "I-Thou," because if words denote and sentences describe, then description loses the lived reality of our togetherness and reduces persons to objects rather than subjects. But this view of language belongs to the mechanistic model: words as linguistic particles associated by fixed laws of grammar, words that denote objects. There are other views of language. Not all words denote, not all sentences describe, and not all cognitive language objectifies like science. Personal language is different: it values and interprets what you are; it is an imaginative, creative exploration of persons and ideas. We have our language, you and I: the tale you tell that reveals what I am like, the drama we recount of life together, the reflections we enjoy of how good it all is and what it means, the paradoxical expression that captures the unique, the peculiarly creative words that poets and lovers use all the time. Personal language communicates well about persons—not exhaustively it is true, but well enough for you to dialog with me, and for me to introduce you to others you have never met. It reveals in a myriad of creative ways what each of us is like and what we find in one another. Personal language is the language of living literature. It is not the language of science but of the humanities, not of thoughtless emotion or emotionless thought but of life lived whole. To call it non-cognitive is to reduce art and literature and creative expression to nothing but emotive decoration.

On this phenomenological account in contrast to the Crusoe theory, interpersonal relations are the vehicle of individuality. A breakdown of relations can stunt growth and repress individuality, but interpersonal relations can also afford a depth of self-knowledge and fulfilment that is otherwise unknown.

2. *A metaphysical model* for conceptualizing this view of individuality and personal fulfilment may be drawn from certain idealists. This is no arbitrary device, for phenomenology grew on idealist soil; it shares with idealism its attention to human consciousness as distinct from the external physical world, and its methodical description of the structures of consciousness developed historically from Hegel's *Phenomenology of Mind*.

We approach the idealist model from the standpoint of the Greek model that has already been introduced. The Greek conception of individuals depended on the subordination of historical change to eternal and changeless forms, so that a particular man actualizes his human potential by participating in the universal nature of the species. As we have seen, this reduces individual differences to accidents unnecessary to the meaning of being human. The idealist, on the other hand, sees historical processes as themselves fulfilling ends and realizing values. In history, individual events and persons emerge with identities of their own. In place of particulars and universals, then, the idealist thinks of individuals in whom unrealized possibilities of the past come to concrete fulfilment.

Hegel's "world-historical individual" is the classic case: in a person like Napoleon past centuries converge and from him future horizons open up. He is the historical concretization of abstract possibilities, a universal ideal brought to realization, a "concrete universal."

This model is developed further by A. E. Taylor.[8] He points out that the self is one and individual in so far as its emotional interests and purposive attitudes express a central coherent interest and purpose. The self is a teleological unity organized around its values, not a particular unrelated to other particulars except by the rule of reason and law. Individuality is a purposeful, developing thing that grows in relation to surrounding persons and things. The freedom of the self therefore admits of degrees: I am free to the extent that I have abiding purposes that unify my life both now and into the future. A free man "knows his own mind." He is most fully free when he is unhampered in executing his central purposes by vacillating or conflicting interests of his own, by habits and addictions, or by the contrary influences of other people and things. Freedom, that is to say, is not indeterminism but self-determinism, the kind of creative harmony experienced by a mature and inner-directed individual. Such a person is "made-whole."

Such a man is no Robinson Crusoe, isolated from others and capable of finding fulfilment alone. Nor do his social relationships depend on mechanically imposed laws. On the contrary, the individual is immersed in social history: he both comes into existence and finds fulfilment by virtue of his internal relatedness to the community and the social processes in which he takes part. No man is an island.

We can now distinguish three metaphysical models. (a) The Greek model of form and matter, particulars and universals, separates our essential nature from temporal change and so has difficulty explaining and valuing individuality. (b) The mechanistic model with its atomic conception of Crusoe as an island of existence, somehow thrown together with Friday, has difficulty accounting for the intrinsically societal nature of our existence. (c) The "concrete universal" model with its appreciation of history conceives of the individual immersed in social processes through which he comes to know and be himself. This model obviously fits the phenomenological account.

According to (b) freedom is the natural endowment of an individual with his island to himself. But his island is constantly invaded by other islanders who force him to act against his will. So laws are imposed. According to (c) freedom is acquired gradually with individuality, and individuality means the pursuit of all-absorbing values in the historical situations in which we play a part.

According to both (a) and (b) religious faith is reasonable assent to true propositions, along with confidence in God. It is an act of will, but little is said of the bearing of faith on individual fulfilment. According to (c), however, faith is a more complete act, that unites the individual in a fulfilling creativity. We have now built a sufficient conceptual scheme to understand how this may be.

III. RELIGION AND INDIVIDUALITY

1. Human values and religious faith

The connection between religion and individual fulfilment is tacit in what has been said about model (c). It is our interests and purposes that unify the self and give coherence to a life. A person is a value-pursuing being; and religion contributes to our value pursuits in two ways; (1) It integrates a man's values around the God he believes to be supremely good and ultimately important; (2) the unifying power of the supreme good gives new direction and motivation to all life's interests and purposes.

I speak of "the God he believes to be supremely good and ultimately important" with two things in mind. First, traditional theism has regarded God as supremely good and declares the "supreme end of man" to be "to glorify God and to enjoy him forever." Second, the non-theist is "religious" in that he substitutes for the God of traditional theism some other "supreme good" that is of "ultimate importance" to him. Consequently what I have said about religion and human values may be applied to belief both in God and in a "God-substitute." The thing to bear in mind in both cases is that one has to accept some supreme good and involve oneself in the resultant "religious life" if religion is to contribute actively to the integration of values and thereby to individual fulfilment. Whether *any* good can be elevated to "supreme" status and achieve the same results, or whether it matters *what* we believe, is another question. At this juncture I can only suggest that if values have an objective status, rather than being the projections of individual desires, then *what* one believes to be supremely valuable does make a difference to our other values and to personal fulfilment.[9]

Differences between religions may be seen in this light. (1a) The concept of God varies between religions, so that resultant value-structures are likely to vary as well. (2a) The kind of religious activity expected of the believer varies with the religion, because religious activity is related to the values belief espouses. Religion-substitutes may be similarily perceived : (1b) an ideology like Marxism substitutes a humanistic supreme value for the divine, but it purports to integrate other values in similar fashion to religion; (2b) it expects of "believers" both "credal" affirmation and involvement in the "religious" life.

This relates in two ways to Maziarz' thesis concerning interpersonal relations. (1c) Religious experience in the Judaeo-Christian tradition is conceived as a man's relationship to a personal God. It is therefore described as the richest I-Thou experience available to man, one which underlies every other experience of men.[10] But since God is also the supreme good, religious experience also clarifies and unifies the values of the believer. (2c) The religious life is lived in a "community of faith," whether this be the Jewish community, the Christian church or the Communist party. It is not lived in Crusoe-like isolation. Worship, prayer, fellowship, religious instruction, and witness are all functions of the religious community, though performed by individuals. By these means the supreme value of God is communicated and clarified and reinforced in relation to other values, and the individual life finds ordered unity and fulfilment. The life of the community is also describable in "I-Thou" terms, that is, both in relation to the community of faith and in relation to God, religious experience is interpersonal.

2. *The phenomenology of faith*

Accounts of religious faith differ. One difference has to do with the evolution-revolution question : Does faith evolve gradually or does it burst out in revolutionary fashion? Is "conversion" a slow process or a sudden crisis?

Both points of view are evident in phenomenological writings. We shall look at Friedrich Schleiermacher, Paul Tillich and Sören Kierkegaard. Schleiermacher is an idealist; the other two are existentialists. Schleiermacher describes an evolutionary process; Kierkegaard experiences a revolutionary crisis.

According to Schleiermacher, the highest end of man is the development of individual self-consciousness in relation to the universe as a whole. We find two tendencies in ourselves, passive and active, that alienate us from full self-awareness and acceptance. On the one hand, we are beset by a "dread fear to stand alone against the Whole," while on the other, we need to establish ourselves as individuals. How can this tension be resolved and these two directions within the self be unified? The answer lies in religion conceived neither as a system of thought (the Medieval and Enlightenment models) nor as a set of duties (Kant) but as *piety*. On the one hand, a man becomes conscious of his absolute dependence on the Whole, gradually realizing that his unique structure as an individual is related, historically and organically, within the Whole; we express this feeling of dependence in our ideas of God. On the other hand, a man expresses the unique inner humanity he discovers in cultural activity in the world around. Thus the passive and the active sides of our being find fulfilment as they are drawn into one integrated being. Faith as piety is then a conscious dependence on the Whole and from it a free, unified existence grows.[11]

Coming to faith is a process. Schleiermacher sees the growth of faith as a natural evolution of self-awareness, stimulated by others who express their own religious feelings in the church. Conversion is therefore a process of liberating self-acceptance within the community of faith.

Tillich's phenomenology of faith develops Schleiermacher's ideas. Both of them speak of God as Being-in-Itself, the Ground of our being, the Whole. Both see God-concepts as symbolic expressions of religious experience rather than as literally or objectively true. Both see different religions as varying only in the degree to which they express the religious consciousness of man. And both see religious experience as the grounding of individuality by means of a newly integrated self-awareness.

Tillich lays more stress on the human predicament individually and culturally. He speaks of the "existential situation," and describes in Heideggerian terms the existential anxiety that results from the shock of realizing my finiteness and temporality and being-unto-death. Why is there something rather than nothing? I do not have to be. How then can I find the "courage to be" in the face of this nothingness?

Correlative to existential anxiety is my existential dependence known as faith. Tillich describes it as "ultimate concern," (1) because it transcends the flux of changing experiences with their passing concerns and grasps at the *ultimate* roots of my being. It heals the breach in human existence that created existential anxiety, by combining my act of trust as a *subject* with belief in an *object* of ultimate concern. (2) As ultimate *concern* it is unconditional. Faith catches up the whole self-conscious and unconscious, cognitive and emotive —and focusses all my creative energies on the ultimate. I rediscover myself as subject in relation to Being, and so I project myself as a subject into a world that treats men as objects. I find courage to be.

For Tillich, faith brings a revolutionary change. To be without ultimate concern is to be without a center—whether or not anyone is ever altogether "uncentered." Being with ultimate concern gives life new depth, new direction, new discipline. It gives to art and science and morality a creative *eros*, and unites them in a common aim. Thereby it exerts the healing power of faith and love in all of life.[12]

Kierkegaard too, laments the human situation. In *The Present Age* he complains that lack of passion makes us superficial and uncreative. Learning has become a game, an erudite hobby without purpose or personal involvement in depth. Religion has become a social propriety without either risk or demand on the adherent. A similar loss of ultimate values is evident in the uncertainty and anxiety of the times, the absence of real personal relations and of the kind of moral character that takes "inwardness." Authenticity is missing. This loss of values is therefore a loss of authentic individuality.

Kierkegaard's concern focusses on religion. Largely to blame is the concept of God and individuality we met in Schleiermacher. The immanent God of German Idealism allows us the comfortable luxury of a general sort of religious consciousness and a socially congenial faith. It is unduly quietening and far too optimistic about man and culture. Kierkegaard reaffirms the transcendence of a personal God. "God is in heaven and you are on earth," he cries, "so keep silence and listen!"

Faith in God is not self-acceptance in calm dependence on the Whole. Nor is it a reasoned assent to the objective truths argued by the Enlightenment theologian on philosophical or historical grounds. The objective path leaves us infinitely removed from God, because it reduces him to an object from whom we remain both personally and logically remote. God is rather a subject : he speaks; he acts. Faith is triggered therefore when the eternal God breaks into time; it is elicited most fully by the Incarnation. Faith is an *individual* response, passionate, in depth, with all the authenticity of a human subject.

This depth-response (Kierkegaard calls it "subjectivity") is what has been lacking. It restores a man's authenticity : it creates new love and new life. Ultimate values are restored, and the individual accordingly finds fulfilment through the act of God in Christ. This is a revolution.[13]

Do the similarities and differences between these three representatives clarify at all the evolution-revolution question in regard to the individual and his faith? They all speak of authentic individuality resulting from faith. But Kierkegaard sees faith in terms of (a) the passion of our being before God, and (b) the *revolutionary* change that results—be it gradual or sudden. Schleiermacher on the other hand sees an *evolving* self-awareness. Two questions therefore emerge to occupy the remaining pages. (i) Do differing conceptions of God make the difference between evolution and revolution in religious experience? (ii) Do different conceptions of the human predicament make the difference between evolution and revolution in religious experience?

3.   *Theology and the changing individual*

(a)   Does the traditional Christian conception of a transcedent God demand "revolution" in the individual in contrast to the evolution that results from Schleiermacher's immanentism? Does the supernatural intervention of a transcendent God in human history and consequently in the life of an individual necessarily have revolutionary effects? Traditional theology rejects any dichotomy of natural and supernatural. To believe in God as creator of heaven and earth is to believe in God's creativity in all of nature and history as well as the Incarnation. The difference between the Incarnation and the rest of history is not that God acted in one but not in the other, but that God acts differently in different cases. In other words, the God of traditional theology is not as utterly transcendent as Kierkegaard supposed, not as immanent as Schleiermacher's Whole.

An idealist might suppose Jesus to be the highest fulfilment of an evolutionary ideal made possible by the divine immanence in all of history. Kierkegaard regards his coming as a startling and unheard-of intrusion of the eternal into time. But in the historical context of first-century Judaism, the Advent of the Christ was the fulfilment of a growing expectation, the greatest in a long history of theophanies, unique indeed but by no means either as unheard-of or as natural a phenomenon as might be supposed.

It is to the idealist's credit that he recognizes with early Christian thought that, because all creation has meaning, both nature and history ultimately make sense. It is to the idealist's loss that he immerses the unique event in history's evolutionary unfolding. Kierkegaard rediscovered the unique, but he did so at the cost of forgetting his doctrine of creation and the resultant meaning of all nature and history as the work of God. Schleiermacher and Kierkegaard, then, need not be supposed to give us the only possible accounts of how one's God-concept affects one's faith.

At this point we reintroduce our models in order to conceptualize more clearly the divine activity that elicits faith. On the Greek model of particulars and universals, a unique act of God remains a total mystery because it does not fit into any natural species of events. On the mechanistic model it can be seen in some cause-effect relation to other events that are equally isolated if not so unique. But just what "cause" means in this context is difficult to know. The "concrete universal" model pictures events bringing past processes to unique fulfilment; the unique event both preserves what was immanent in the past as a whole and transcends the past as it moves uniquely towards a different future. If a unique historical event is conceived on the "concrete universal" model, then it is not the shocking intrusion that Kierkegaard describes nor need its effect on men be as revolutionary as he supposes. It exhibits discontinuity with the past, but continuity as well, and so may the religious experience that is coordinate with it.

(b)   What then differentiates revolution from evolution in religious experience, if it is not the act of God to which faith responds? We have spoken of "continuity" and "discontinuity." Let us define "evolution" as the process of completing an integration that has already begun. Then we may define "revolution" as the introduction of a new supreme good around which a new integration of values and life takes place.

Degrees of continuity and discontinuity seem to vary from one individual to another depending on the values that shape a life before one becomes a believer. Suppose that Kierkegaard had not lived in nineteenth-century Denmark, where religious and social values merged and where men could believe in God without passion or risk. Suppose he lived in a passionate, activist age, where people want above all else to relate meaningfully to one another, and where in-depth relationships evoke the repressed authenticity. What changes would faith then bring? The "new man" would evidently have more continuity with the "old," both in the values he conceives of and the integration he achieves. The "old" might be a life internally inhibited from achieving the life desired; the "new" might mean creative freedom to live life whole. One's conception of the Supreme Good and one's basic value-structure may not have changed except for clarity and reinforcement, although the realization of those values is another matter. Behaviorally it may or may not look like a revolution, for in actuality the freedom of faith has brought a previously struggling process closer to fulfilment.

Perhaps an analogy exists in lesser value changes than religious conversion. What happens to men when they fall in love? When they marry? When they become fathers? What happens when a man joins the school board or becomes a poetry-lover or a peace-marcher? There are too many variables to permit generalization. But in measure it depends whether our man was a lady's man before he fell in love, whether he was the domestic type before he married and settled down, whether he was really interested in education before he ran for the school board, whether he had aesthetic inclinations other than poetry, whether he was a social activist anyway before he marched. It also depends on whether our man is temperamentally given to passionate involvement or is the methodical, serene type. But if the degree of continuity and discontinuity depends on various things, is the evolution-revolution question answerable?

We have defined revolutionary discontinuity in relation to a *new* integrating good, and evolutionary continuity as the completion of an integration already begun. Thus the idealist sees continuity both in history and the individual. He looks for a teleology that brings past processes to their natural fulfilment. Reinhold Niebuhr finds this view far too optimistic about the nature and destiny of man.[14] Tillich and Kierkegaard too; for as existentialists they look at the dysteleology in history and so emphasize the discontinuity of past and future. The question of evolution or revolution does not depend only on what individual differences of values and temperament may exist; it depends on the predicament from which a man is restored. If the human situation is as Schleiermacher says, and all a man needs is to realize his dependence on the benign nature of Being as a whole, then an evolution may well lead that man to faith. But if being is more tragically ruptured, if the existential predicament haunts men regardless of differences in personality and culture, or if men are both alienated from one another and, as theologians have said in defining sin, alienated from the God who is the highest good, then a new integrating center appears needful, and religious conversion becomes revolutionary after all.[15]

To call conversion a revolution suggests that Robinson Crusoe suddenly encounters God in a whole community of faith . . . or rather that God encounters him. But we have argued that Crusoe is not on an island at all, but already in community. What if he lives with people who believe and, in a nominal way, holds their beliefs and values? What if these have been "growing on him" for some time? Then his conversion is an event that brings past processes to concrete fulfilment. After all, it is possible in Western culture to adopt traditional Judaeo-Christian values, to affirm that "the supreme end of man is to glorify God and enjoy him forever," even to try to integrate one's life around that supreme good, without interiorizing these values and beliefs with the total passion of an ultimate concern —that is to say, without becoming a believer. It is possible to pay lip-service to a life-fulfilling hope and no more. In such a case, faith may bring towards fulfilment a process previously if superficially begun. The fulfilling thing about religious conversion, after all, is not that a man has to throw out everything he believed and valued before (though he may have a lot of rethinking and re-evaluation to do), but that he is gripped with an ultimate concern that unifies his life and makes his beliefs and values wholly his own. He is made whole to an extent that previously eluded his experience. Even if the human predicament is as serious as has been suggested, the line between revolution and evolution in religious experience may still be extremely difficult to draw.

One final observation: religious experience of the sort we have discussed sometimes involves a transition from one religion or religion-substitute to another rather than finding an initial faith that has known no antecedent at all. In such cases we have to say either that a man's prior faith failed to give the kind of personal authenticity he later found, or that self-authentication is not the only function of a religion nor its only criterion of truth or reason for believing. Self-authentication may be a consequence of religious faith, but this does not mean it provides sufficient reason either for religious belief in general or for a particular religion. All we have provided, therefore, is a phenomenological description of religious faith, not an account of any other functions religion may have nor an argument for the truth of all religions or even of one.[16]

Notes and References

1. This suggestion is made by Karl Barth in his classic historical study, *From Rousseau to Ritschl* (London: SCM Press), chapter 1.

2. See Gilbert Ryle, *The Concept of Mind* (Hutchinson's University Library, 1949), chapter 6.

3. Rene Descartes, *Discourse on Method,* part v; John Locke, *Essay Concerning Human Understanding,* book IV, chapter iii, paragraphs 17, 27; George Berkeley, *Principles of Natural Knowledge,* sections 145–147; J. S. Mill, *An Examination of William Hamilton's Philosophy,* chapter 12. Among more recent writers see A. J. Ayer, *The Foundations of Empirical Knowledge* (New York: Macmillan, 1953), chapter 3; *The Problem of Knowledge* (New York: Macmillan, 1954), chapter 8. A still more recent defence of the theory appears in Alvin Plantinga, *God and Other Minds* (Ithaca, New York: Cornell University Press, 1967), part III.

4. *Individuals* (New York: Anchor Books, 1963), chapter III. Cf. G. F. Stout, *God and Nature* (Cambridge, England: University Press, 1952), chapter XVI.

5. *Cartesian Meditations* (New York: Humanities Press, 1960), fifth meditation.

6. *Being and Time* (New York: Harper and Row, 1962), pp. 153–163.

7. See Edith Stein, *On the Problem of Empathy* (New York: Humanities Press, 1964); Max Scheler, *The Nature of Sympathy* (New Haven: Yale University Press, 1954); Gabriel Marcel, *The Mystery of Being* (Chicago: Henry Regnery Company, 1951), volume I, chapter 9; Martin Buber, *I and Thou* (New York, Charles Scribner's Sons, 1958).

8. See his *Elements of Metaphysics* (New York: Barnes and Noble, University Paperbacks, 1961; first published 1903) Book IV, chapters iii and iv. On concrete universals, see F. H. Bradley, *Principles of Logic,* volume I. p. 188 passim; Temple, *Nature, Man and God* (New York: Saint Martin's Press, 1956), p. 100 ff. B. Blanshard, *The Nature of Thought* (New York: Macmillan, 1939), volume I, chapter xvi xvii. See also N. K. Smith, "The Nature of Universals," *Mind,* XXXVI (1927), 137, 265, 393; M. B. Foster, "The Concrete Universal," *Mind,* XL (1931), 1; and H. B. Acton, "The Theory of Concrete Universals," *Mind,* XLV (1936), 417 and LXVI (1937), 1. While this concept of individuality was developed by idealists, we shall see in section III that it need not be confined to the tradition.

9. See further A. F. Holmes, "Philosophy and Religious Belief," *Pacific Philosophy Forum,* V (1967), 4. 3–51. Also Basil Mitchell "The Justification of Religious Belief," p. 178 in Dallas M. High (ed.), *New Essays in Religious Language* (New York: Oxford University Press, 1969).

10. This is Buber's terminology, employed by Christian theologians like Emil Brunner in *Truth as Encounter* (Philadelphia: Westminster Press, 1964), chapters 8 and 9, and William Hordern in *Speaking of God* (New York: Macmillan, 1965).

11. Friedrich Schleiermacher, *On Religion* (New York: Harper Torchbooks, 1958), Speeches I and II.

12. Paul Tillich, *Dynamics of Faith* (New York: Harper Torchbooks, 1958), chapter I; *The Courage to Be* (New Haven: Yale University Press, 1952), especially chapter 5.

13. Sören Kierkegaard, *The Present Age, and Of The Difference Between a Genius and an Apostle,* (New York: Harper Torchbooks, 1962); and *Concluding Unscientific Postscript* (Princeton University Press, 1944), pp. pp. 23–55, 115–224.

14. See his *Nature and Destiny of Man,* (London: Nisbet and Company, 1941), volume I, chapters i–iv.

15. Traditional Christian doctrine therefore insists it does matter *what* one believes; and faith is more a revolutionary act than the idealist conceived. Kierkegaard undoubtedly had this in mind in conjunction with his reference to the Incarnation.

16. For an account of religious knowledge with regard to its historical, moral and metaphysical ingredients as well as the personal element, see A. F. Holmes, *Faith Seeks Understanding* (Grand Rapids: Eerdmans Publishing Company, 1971), chapter 6.

CONDITIONS OF ALIENATION
GEORGE SCHRADER
Yale University

1. I TAKE THE principal thesis to which I am to respond as asserting that: the human individual can be located on a map with coordinates which can be significantly designated as I-Thou, I-It, I-They, and I-We. The primary reference point is taken to be the individual self; but the important and valid claim is made that the self stands always in relation to the other coordinates. The understanding of man and his world turns on the way in which these coordinates are related or disrelated. Alienation is recognized both as a possibility and, to some extent, a necessity for the individual. Yet, the emphasis is upon the *continuity* of what is termed "evolution-revolution" and the prospect of overcoming current modes of alienation.

Although I sympathize with the basic orientation of the thesis, namely that man is an interpersonal being who can achieve an adequate form of existence only within an appropriated natural/ social context, there are both detailed and general aspects of the thesis which merit further examination and, perhaps, revision. The counter-thesis in this instance will consist more in a questioning of the thesis to the end of a clearer and more adequate understanding of the human situation than in an attempt to refute it. The presentation of the thesis is made in such a way as to permit and invite further questioning; or, if you like, it is self-referentially consistent in soliciting dialogic encounter. It is in this spirit that I wish to respond.

2. The term alienation has been so extensively used and misused in twentieth-century literature, both popular and academic, that it has accumulated an almost unmanageable array of meanings. To compound the conceptual ambiguity of the term, it evokes a variety of confused and ambivalent feelings. The burden rests clearly upon

329

anyone who uses the term alienation to explicate its conceptual meaning and show how it is applicable to the human situation. A major part of this essay will be devoted to that task.

In its most fundamental sense alienation suggests a state in which a being is separated either from another being or beings or some part or aspect of itself. Already the question arises whether alienation is a category which might apply to any and all beings irrespective of type or whether it applies only to persons. The answer to the latter question will doubtless turn on the consideration whether alienation requires the consciousness of being alienated. Where then do we start? From some mode of basic ontological alienation such that experienced alienation might be only a special case? Or must we start from the experience of alienation and explore the boundaries of the alienated situation?

I am highly skeptical of proceeding by first establishing alienation as a general ontological category and then trying to understand experienced forms of alienation in terms of it. In the first place, alienation, whatever it may mean precisely, has to do with otherness and thus carries with it a reference to the human self. We might, for example, inquire if any two beings are necessarily alienated from one another because they are separate. We might then deal with the problem of identity and negation so as to follow the guidelines of classical metaphysics. But, to do so, we would either have to suppress our cardinal reference point or make use of it without having brought it explicitly into the analytical context. To talk about any being as having an other is already to invoke categories which arise in the context of self awareness. In its most primitive meaning, an other is a non-self. It is an other only by not being the self to which it refers in its own otherness. This point could be elaborated but the exercise would be both tedious and unfruitful. If alienation involves otherness and a necessary reference to the reflexivity of the human self, we would only be engaging in an act of self-deception to commence with the presumed otherness of being to being. Nor would it help to regard identity and negation as more fundamental notions which might be predicated of entities in some primordial ontological context. The only valid starting-point is the situation of a subject which experiences alienation—or, by the same token, identity and negation.

It is actually Hegel to whom we are indebted for the philosophical concept of alienation. For Hegel, alienation is a category of spirit and refers to a condition in which the self confronts an unappropriated other. Since the other is an other *for* the self which is aware of it, the condition of otherness is relative rather than absolute and requires a positive mode of relatedness. Moreover, self and other presuppose a supportive context. For example, the child-parent relationship refers to and presupposes the family as an immediate context; the self-object relationship refers to and presupposes a region of nature. The self is mediated by and mediates its other; both forms of mediation involve the contextual background as a third term. Hegel was, of course, concerned about alienation on a grand scale. He employed it as an ontological category to account for the disequilibrium which motivates the dialectical movement of spirit. Within the Hegelian schema a guarantee is provided that alienation can never be insurmountable. However acute, alienation is concompassed by the identity of spirit which both establishes it through an act of self-diremption and promises always to transcend it through a further act of self-integration. If we were convinced Hegelians, alienation would give us little pause. Although we would be spared neither the pain of our own divided state nor the terror of history, we could console ourselves with the thought that the drama is meaningful in its most minute detail and considered in its entirety; it is nothing less than the self-unfolding through time of absolute spirit. Ironically, since Hegel the increasing sense of human alienation has been in direct ratio to our declining hope of surmounting it.

It is apparent from what has been said about Hegel's concept of alienation that it is far from a pejorative term. Although involving negativity and estrangement, alienation is, in the first instance, the development of a more immediate situation which is not only less explicit but less interesting. The task confronting the self is never to obliterate otherness or cancel out alienation but rather to incorporate it in a new project. What seems to have happened since Hegel is that the other has become an absolute other and alienation a state of pure estrangement. If that is in fact the situation in which we find ourselves, there may indeed be no alternative to perpetual unhappiness and despair.

Alienation involves otherness, but it is not apparent that otherness as such is sufficient to constitute a state of alienation. There is a *prima facie* reason for not regarding otherness as a sufficient condition of alienation, since to do so would trivialize the concept. One can imagine the skeptic, who regards all talk about alienation as pretentious breast-beating, responding to such an identification. If to say that someone is alienated from someone or something is only another way of saying that he is not that person or object, why not say what you mean in the first place? We would never have thought of denying that Alice was not the Red Queen, he might add, but neither would we have thought that to express the fact requires a special category or that the fact itself has any particular significance! So Alice is not the Red Queen! Alice is Alice and the Red Queen is the Red Queen; and they can speak or not speak or go their independent ways. If for the Red Queen to be other to Alice is simply for the Red Queen to be the Red Queen, we are dealing in the emptiest of tautologies while pretending to engage in philosophy. If I am alienated from all other men simply by virtue of the fact that I am not any one of them, this is true but trivial. There are more direct and less obscure ways of expressing such patent facts about the world!

3.   *Alienation and self-awareness.* Alienation cannot meaningfully be regarded, then, as a condition of otherness. Although the self and some not-self are essential to alienation, they do not suffice to constitute it. Our problem is thus to delineate those special features of otherness which are sufficient to constitute a state of alienation. We have noted already that the co-presence of two objects can provide neither the otherness nor the estrangement which are essential to alienation. If the human self were only a complex object coexisting with other objects, we would doubtless have separateness and relatedness, but not otherness. To say this may appear to beg the whole question by defining otherness in such a way as to imply a self. This is indeed our starting point and in a sense our ultimate pre-supposition. It is non-arbitrary only because we cannot bypass experience. Otherness is a term which is commonly if not universally employed to designate that which a self-conscious being is aware of as not itself—and in that quite literal sense an other. What needs elucidation is the possibility of the original experience of otherness which is foundational for all manifestations of it.

At the very least, to be aware of a being as other to oneself is to recognize (a) that it is another *being*, and (b) that the other being is given in a world as a copresence with oneself. Moreover, the other can be understood as other only in terms that permit one to apprehend oneself. To condense Hegel's elaborate and definitive analysis of this relatedness, the self sees itself reflected in the other and the other as a reflection of itself. This entails that the other as object is problematic for the self and, further, that the relatedness itself is problematic. As both Hegel and Sartre have made dramatically clear, the problem is even more complex if the other is another self.

Both Hegel and the perspective of dialogical involution expressed in the thesis are anti-Cartesian. For Descartes the other is necessarily an absolute other since consciousness is the absolute property of an individual ego. If only I can appear in my consciousness, then no other can legitimately appear. Either I must infer the presence of the other from clues in my own consciousness or regard myself as possibly the sole being in the world. Hegel took consciousness as a contextual field within which both self and other appear. No language, conceptualization, or genuine perception occurs save within the interpersonal context of shared consciousness. But a shared consciousness or experience is possible only because of the reflexivity of the individual self. The self can be conscious of itself as a self and thus be truly a self only through the mediation of an other and, ultimately, another self. There is an important difference on this point between Hegel and Sartre. Whereas for Hegel the self is a being-for-itself only through the mediation of an other, for Sartre it is intrinsically a being-for-itself and thus more independent ontologically. Sartre qualifies the ontological independence of the self by acknowledging that subjectivity requires the refused objectification by another consciousness. In *Being and Nothingness,* however, he tends to view each consciousness as relatively absolute and to permit no genuine interpenetration. It is for this reason that he must deny the possibility of a mutually satisfying love relationship. Sartre's analysis of the for-itself not only condemns inter-personal relationships to perpetual conflict and frustration but relegates non-human objects to a more or less meaningless status. They are simply there as a receptacle for our own projected meanings. In the last analysis, otherness for Sartre is relatively absolute whether the other is a person or a natural object.

If we define a self as a being which is capable of being aware of itself or self-conscious, then the self must be in some sense an object to itself. The question at issue is whether it is or can be an object to itself independently of its relationship to other beings. It is on this point that Hegel and Sartre differ. The difference is crucial, since it establishes the basis for our interpretation of the inter-personal. If the self can be reflexive and self-conscious independently of its relations to objects and persons, it is essentially a Cartesian ego. If, on the other hand, it requires the mediation of objects and persons for its own self-definition, the self is fundamentally inter-subjective.

Hegel is, I believe, clearly correct on this issue. Our most elementary self-awareness is the awareness of ourselves in relation to objects and persons. Both self and other appear within an encompassing context. In the most inclusive sense the world is the context, and consciousness is co-extensive with experience. The subject is revealed as an object in relation to other objects and as a subject in relation to other subjects. It is altogether fanciful to believe that any human person ever became aware of himself save in the context of the family or an equivalent human group.

To talk about consciousness is to suggest an artificial model of awareness as if our understanding of ourselves and others were exclusively or primarily reflective. In fact, our primary mode of understanding is in the form of practice and behavior. Although the term experience conveys something of this pragmatic quality, it shades off too easily into disembodied awareness. The hunter is aware of himself in relation to the animals he hunts and needs and with whom he identifies. The farmer understands himself as mediating the processes of natural growth and fruition. Each takes his meaning as much from the objects in his world as he projects meaning onto those objects. It is never a case simply of subject and objects, but of a context, whether pastoral, agricultural, or industrial, in which all have their function, status, and meaning.

Although Heidegger has made much of this point in his attempt to repudiate the Cartesian cogito and establish a mode of existential understanding, he greatly overstressed the contributory role of *Dasein*. In *Being and Time* Heidegger remains far too Cartesian to recognize that man is as much shaped by his practice, the tools he uses, the objects to which he refers and relates as he is the shaper of tools and instruments. Heidegger comes closest to recognizing the full importance of the inter-subjective context in his exploration of *Das Man* (the anonymous one or the "they"). (I wish to come back to this point since it is central to the argument of the thesis.) But Heidegger had serious difficulties, not unlike those of Sartre, in attempting to account for being-with-others (*Mit-Sein*). The world of *Dasein* is too self-contained to permit inter-subjectivity. Not insignificantly, for Heidegger authentic existence depends upon the adequacy with which each individual confronts the ultimately isolating phenomenon of his own death.

4. *Alienation from nature.* It is one of the merits of the thesis that it attempts to recover the positivity of man's relation to nature. In recent years, we have become increasingly aware of the importance of our relation to nature and of the manifold ways in which a dis-relationship to our natural environment alienates us from ourselves and other men. It is fairly easy to exhibit the reciprocity involved between man and nature within a pastoral environment. The man whose life is engaged in cultivating the soil or the caring for animals participates in the cycle of nature as a midwife to processes which he accepts as a condition of his existence. We have become increasingly aware of the ways in which man's relation to machines gives form to his personal and social existence. As man once interacted with the processes of nature as a creature of that same nature, he now interacts with the complex apparatus of automated technology. The limitation in Heidegger's interpretation is his virtually exclusive stress on man's projects and his understress on his response, initially to nature and subsequently to his material culture. If not dialogic, the relationship between man and nature or man and the machine is highly dialectical.

Marx was concerned about man's alienation from himself through his labor. He regarded such alienation as a function of the socio-economic system and, hence as amenable to correction through a basic reorientation of the social context. It is important to note that Marx did not regard work itself as alienating. Nor did he view industrialization as establishing a social context which made alienation from one's labor inevitable. In retrospect Marx appears to have been a utopian dreamer, for we are now less confident than he that a non-alienated existence is possible within a technologically developed society. A good many social prophets and a significant proportion of the younger generation despair over the prospect of a humane existence within a rationally ordered technical society. Some of them are presently attempting to reverse the tides of historical progress—as we have long considered it—by reviving earlier patterns of pastoral and village life. For them, the city has come to symbolize an environment which has been de-humanized by man's creation of an artificial but polluted habitat. The seriousness of this issue is indicated by the fact that at present no utopias are being promulgated.

The standard perceptual model of the object is inadequate for understanding otherness in that it posits the object merely as something looked at or observed. A more adequate model views the object as related to practice and behavior. Though work is not the only mode in which man is related to objects, it is singularly important in that through his work upon the object man identifies himself with it. He transcends himself toward the object and to some degree makes it his own. Thus human labor may be viewed in one sense as attempting to overcome man's alienation from his material environment rather than as constituting the alienation itself. So long as man views himself as a creature of nature there is nothing intrinsically alienating about his situation. If man is at home in the natural world and regards his needs and activities as essentially natural, he cannot be estranged from that world. There can be little doubt that at the present time man does experience a profound estrangement from nature and longs to recover his sense of belonging to it. This recurrent nostalgia for an intimate relationship with nature reflects an alienation from nature; it does not, however, invoke an original alienation in the natural situation itself. The question at issue is whether the continuity between nature and human culture asserted by the thesis is possible.

5.   Natural processes themselves consist both of cumulative growth and radical leaps. It seems to me altogether misleading to suggest that natural evolution has been uniformly gradual. If by revolutionary change we mean radical change which introduces new life forms, natural history is replete with examples of revolutionary change. By the same token, cultural changes exhibit both the gradual working out of consistent patterns and the introduction of radically new patterns. If by evolution we mean the development of the natural world, it is confusing to contrast it with revolution as the development of the human or cultural world. It is possible to distinguish between nature and history and search out developmental patterns in both spheres. But there is no *a priori* reason to assume that natural change has been any more gradual than cultural or historical change. If by evolution we mean gradual development or unfolding of patterns, and by revolution abrupt change in patterns, we can expect to find both types of change in nature and history. The evolution-revolution distinction as considered in the thesis not only fails to provide a helpful distinction between nature and human history, but obscures the meaning of revolutionary change.

We can, I take it, talk sensibly of the evolution of the horse. We can note how the form of the animal has changed over a considerable period of time to produce the animals we now recognize as members of that species. But we can speak with equally clear sense of the evolution of the motor car from its first versions to its contemporary manifestations. In both instances we are confronted with variations on a common theme. By contrast, the emergence of man as tool and language-using animal was a revolutionary development which introduced a radical gap in nature and even a transcendence of natural process itself. The emergence of the human species represents one of the most revolutionary changes in evolutionary development. Moreover it is a highly paradoxical development since it is an event within natural history which founds a new mode of time and a new form of existence. Man stands both altogether within nature and altogether beyond it. Man is the hyphen, as it were, between nature and history.

Whether gradual or abrupt, natural change presumably takes place without conscious direction. Historical change, on the other hand, involves human consciousness and some conception of the direction and goal of change. In view of the fact that man continues to be a natural being and lives out his life within the context of a natural environment, it would be misleading to treat the dichotomy between nature and history as an absolute fissure. The complex and paradoxical fact is that man carries nature within him. The relationship between nature and history must not be reduced to a simple continuity nor broken apart into a total disjunction. There is indeed a *hyphenated* relation between nature and history. It would be mistaken, however, to interpret this relation as a tension between man and nature since this would be to assign man to one side of the hyphen. The fact is that man himself is both natural and historical and, further, that nature is taken up within the context of human history. However nature may have been transformed by human inventiveness, it cannot be cancelled or completely superseded.

How, then, are we to interpret man's experience of alienation from nature? Is it that when man becomes fully aware of his power to establish ends for himself the human world is necessarily set in opposition to nature? If human freedom consists in the power to negate whatever is given and to substitute for it one's own meanings, alienation from nature is the necessary price for human culture. In his exaggerated representation of human freedom, it is significant that Sartre views being-itself (the natural world) as intrinsically meaningless. As Sartre interprets it, the freedom to create values depends upon the power to negate. In taking this position Sartre is not asserting an idiosyncratic view of freedom; on the contrary, he is only stating in unqualified form the conception of freedom which had become dominant in the western intellectual tradition. It was implicit already in the Cartesian cogito and quite explicit in the philosophy of Kant. If man's ability to legislate for himself requires that he legislate also for nature, the realm of freedom stands in opposition to the realm of nature. It is not the otherness of nature as such which constitutes the resulting alienation, but the fact that nature is fixated in a negative relation to human freedom. Man is freed only to the extent that he sets himself in opposition to nature for the sake of his own power. Nature is rendered meaningless in itself and man is elevated to the status of an unlimited legislator. But, alas, the meanings he creates he can also annul; the alienation from nature thus establishes the possibility for a further negation, namely the repudiation of human ends themselves.

It was suggested earlier that alienation does not consist in otherness as such but in an unappropriated or negated otherness. If man now finds himself alienated from nature it is not simply because he transcends nature through his own self-consciousness. It is rather because he has neither appropriated nature in relation to his freedom nor considered his own ends as integrally related to natural processes. If animals are viewed as meaningful only in so far as they satisfy human purposes, they may be destroyed at will. If the resources of nature exist only for human use, they may be limitlessly exploited. The present concern about the pollution of the environment is, to a considerable degree, the rediscovery of the inter-penetration of natural and human processes. After centuries of regarding it as a more or less accidental platform for human projects, man is rediscovering the earth as his natural habitat. History and culture are now seen as encompassed by natural space, and as dependent upon primordial natural processes. Freedom is alienated whenever it negates its own foundation; man is alienated whenever he repudiates his abode.

6.   In order to have his freedom as its own foundation man has not only cut himself apart from nature but established himself as God—albeit a finite one. His assertion of his supremacy over nature has resulted in the positing of nature as an inessential "other" to be overpowered and manipulated. The possibility of any dialogic relation to nature has thereby been annulled. Since he is irremediably

a creature of nature, man's assertion of radical autonomy has involved an alienation, also, from himself. If his freedom is without foundation, his existence is no less gratuitous than that of nature itself. What we have discovered is that man is, after all, a creature who eats and breathes. It is nothing short of incredible that this discovery should be so widely celebrated. The overcoming of this alienation will require both that the original negation of nature is acknowledged and man's natural habitat be reappropriated. The irremedial dependency on nature must be accepted and a reciprocal relationship re-established in which nature is regarded as a source of meaning and a foundation for human freedom rather than a surd. The positive relationship to nature is always there as witnessed by the fact that to exist man must breathe. It is the disrelationship which must be overcome if anything like a dialogic relationship is to be possible.

Even as the turn away from nature to create an independent and artificial environment eventually became radical and revolutionary so far as human existence was concerned, the reappropriation of nature can be simply turning back. If man now finds himself alienated from nature, it is by virtue of the artificial environment which he has created. How is modern man to live if not in the city or its immediate environs? And how is he to sustain himself if not through use of the complex mechanisms of industrial society? A handful of people might be able to return to a pre-industrial mode of existence, but even they are inevitably dependent upon money, transportation, communication, and sophisticated modes of fabrication. Increasing numbers of people have homes in the country where they lead a quasi-rustic existence. But their lifeline is tied to the city and the intricate processes of the industrial economy. The farms in New England, which can no longer provide revenues sufficient to pay taxes and keep the buildings in repair, furnish a retreat for those affluent enough to maintain them as non-productive units. But they provide no real alternative to urban existence or industrialized culture, since they are wholly dependent upon precisely those conditions for their sustenance.

It has been suggested that man is not alienated from nature so long as there is a genuine reciprocity and inter-dependency in his relation to it. It is only when he comes to regard himself as a power asserted against nature that the alienation occurs. Nor does the fabrication of goods and the division of labor it entails in and of itself constitute an alienating situation. The fact, for example, that one man makes boots and another pots does not mean that they are estranged from their own labors or from other men. On the contrary, men can and frequently do serve each other's needs in such a way as to have a genuine sense of community. To make shoes for another man does not alienate one's own labor; it may, indeed, make one's labor the more significant. If the organization of work and the exchange of goods take into account the primacy of human needs, there is no reason for them to be alienating. The needs of the shoemakers are, of course, as legitimate as those of the man for whom he makes shoes. If he becomes a skilled craftsman, his contribution to the community deserves to be rewarded. So long as shoes are valued for their use and their quality, there will be no reluctance to support the craftsman. Alienation occurs only when the reciprocity is broken and the exchange of goods becomes a means for individual profit. At that point men are no longer positively related through their work; instead of working to serve his own needs and the needs of others he works only for personal profit. The other person becomes a means toward that end rather than a legitimate need to be served. And one's own work has value only by virtue of the profit it returns rather than the quality of the object it shapes or the need it serves. There is a vast difference, for example, between a restaurant which exists only to return a profit to its owners and one that is operated primarily to serve the needs of its customers. In the former case the chairs may be made deliberately uncomfortable so that there will be maximum turnover of customers. One then wants the customer to be no more comfortable than necessary while taking up the minimum possible time to consume his meal. In such a case, the relationship between the consumer of food and the purveyor is based primarily on profit and secondarily on need. Each uses the other as a means for the satisfaction of his own needs. The consumer is required to forego more inclusive personal satisfactions he might have obtained through eating and, thus, to reduce his need to a perfunctory one. He feeds himself at the food bar much like he fuels his car at the service station. He is forced to treat himself as an instrumentality and thus is alienated from his own most basic needs as a human being. But he is related as an instrumentality, also, to the one who serves him, since he is primarily an excuse for the meal check. He may try to overcome that relationship by elaborate tipping and by dining in expensive restaurants which specialize in "service." But even then he knows that the special recognition he receives is a response to his ability to pay premium prices rather than acknowledgement of his needs as a person. In similar fashion, he is alienated from those with whom he eats. Since none counts for himself or others as persons in this situation, they can have only the most impersonal of relationships. One has only to observe people for a few minutes in a discount house to see what effect such a context has on their behavior toward one another.

Two important and familiar questions arise at this point : (1) Is the fact that work is for profit rather than to meet human needs in and of itself alienating? (2) Does the fact that individuals are related to others in highly indirect ways necessarily constitute a depersonalized situation in which little if any reciprocity is possible? These are distinct though related questions. The two considerations are interrelated to the degree that production for profit determines the system of production and distribution of goods. If, for example, it is more profitable to organize the production and distribution of goods on a large scale this affects the participatory role of the individuals involved in the enterprise as workers and consumers. A chain of restaurants is much more likely to regard its workers and customers impersonally than a locally owned and managed restaurant which has other concerns than that of achieving maximum profit. Occasionally one comes upon a marginal establishment of the latter kind in which the feeding of people is obviously more important than turning a profit; and one is troubled by the thought that it may not long remain in existence. In our society it is clearly a luxury to reverse these priorities.

There is a difference, of course, between working or producing goods only for profit and receiving payment in money rather than goods or services. Only in a fairly primitive barter economy would it be possible for everyone to be rewarded with goods in return for his work. Although barter is a far more direct medium of exchange and retains the element of reciprocal need more conspicuously in focus, it is clumsy and inefficient. Thus even pre-industrial societies usually have some medium of exchange which serves as money for that group, such as feathers for at least one New Guinea tribe. So long as money was tied to a strict gold standard, money was a measure of sorts which could be substituted for goods. But when money has only an arbitrary relationship to goods and services, it is no measure at all. If, as now tends to be the case, virtually everything that might be owned or exchanged is valued in monetary terms, and money value is either arbitrarily assigned or is dependent upon factors which have nothing to do with the intrinsic value of goods or services, monetary value is only accidentally related to the real or human value of goods. The irony involved in this development is that money, which was originally introduced as a convenience, a token in exchange for goods which were valued either in themselves or for their use, has come to be the end value. Inflation is but one sign of the tenuous relationship between the human value of goods and their monetary value or current price. If, for example, a stock or piece of real estate is valued only for the added money it might bring at some future date when sold, monetary value has clearly displaced other values which ownership might have. It is then understandable that to be rich is to own a quantity of dollars. If one then measures the value of the goods he buys with these dollars in terms of their cost or price, one is involved in a circular system in which there is no rational way to assess the real value of anything at all. The problem is not, therefore, simply a question of profit but of a system in which money has become the primary standard of value.

A monetary economy is alienating to the degree that monetary value is only externally related to the goods to which it is assigned. Intrinsic value is severed from extrinsic or monetary value and made subservient to it. The translation of value into monetary terms substitutes a quantitative for a qualitative measure. Moreover, it establishes an essentially impersonal standard of value. The economist and the fiscal expert would be quick to remind us that money is too useful an instrument to be given up. And they are doubtless correct. No one can seriously propose that we return to a barter economy to avoid this form of alienation. The difficulty is that the instrumentality has become the end with a resulting dehumanization of value. Like the machine which was designed to be the servant of man but which increasingly has become the master, money has tended to become the ultimate value in our society. It enables us to abstract from all other values and compute them on a common scale. It exempts us from the need to consider the value of things in and of themselves or in relation to primary human needs. Monetary value is more than a convenience; it represents an integral phase of man's attempt to achieve mastery over his world. It is a conspicuous way in which he legislates the meaning of all that he beholds. Man and man alone is the declarer of price; he and he alone is the possessor of money.

The alienation involved in a system of monetary value is not due to the instrumentality itself but the mistaken assumption that everything has its price. Only when it is recognized that money provides no measure at all for human values, will it be possible to reverse the priorities and reinstate a personal standard of value. Whatever else it may signify, to put a price on things is in no sense to establish their intrinsic or even their instrumental value. To know the value of anything is to understand and appreciate it in terms both of its intrinsic qualities and its significance for human existence. To the extent that considerations of profit are predicated on a quantitative monetary standard, they represent only a special case with respect to the more general problem of value. If, on the other hand, profit can be measured in other terms, it becomes a question of appropriate reward for human effort and response to human need. The point I wish to make is that a monetary value system, because it is abstract and quantitative, cannot accommodate personal values. If profit is measured in monetary terms, it is alienating for all parties concerned. The system itself depersonalizes values and hence makes it impossible to adjudicate questions of legitimacy and exploitation. The most fundamental issue about the way in which goods are produced and distributed pertains to the persons involved. Are they instrumentalities in a fundamentally inhumane process which provides personal satisfaction for no one involved? Or do they find personal and communal satisfaction through their participation in the economic system? Today large numbers of men find little personal satisfaction in their work not simply because they are exploited for profit but for the reason that the entire operation is measured in quantitative and impersonal terms. It is the prevailing value standard which is at the root of the alienation.

The thesis has, echoing Heidegger, considered man's alienation from himself and others through the mode of the anonymous one or the "they." Another way of characterizing this type of alienation is as the mode of the impersonal. There is nothing intrinsically bad about the impersonal as such; it becomes destructive only when it contributes to the depersonalization of individual men. If men are related through intricate social structures which make direct personal relationships impossible and, further, which remove the significance even of indirect relationships, participation in such structures is fundamentally depersonalizing. If connections cannot be traced out, imagined graphically, or understood between one's own role and the way that role affects and is affected by others, the roles are deprived of personal and interpersonal significance. For example, if a worker on an assembly line takes pains with his particular part of an operation but finds that no one cares, but, on the contrary, is exclusively interested in efficiency in the productive process, his personal concern about his work and his indirect concern about other persons who might benefit from his concern is invalidated. This is precisely the way in which many people feel in contemporary society, namely that their personal concerns and efforts are invalidated by the system which provides work and disposes of the work done.

Heidegger's answer to this dilemma is for each individual to find an authentic existence for himself in his role. But how is that possible (a) if his work involves the fabrication of a socially useless product, or (b) if he is unable to connect his work with the needs and concerns of other persons? Heidegger bypasses the social dimension of this problem, but even so his recommendation is inadequate. To accept a depersonalizing role with personal passion and resoluteness does not change the objective of the role. The most common attitude toward the depersonalized roles people are required to play in our society is cynical resignation. The role is accepted with fatalistic submission and rendered as indifferent and inessential as possible. In other words one surrenders a major portion of the self to the impersonal mode of the "they" and attempts to recover a personal core of meaning outside of one's social role. Heidegger suggests that authenticity can be achieved through a modification of such everyday and impersonal modes. But if this modification is taken seriously, it must involve more than an inner and subjective change within the person. Could it, in effect, become a matter of life and death for a man how he attaches a part to what he knows will be a shoddily made automobile? Can he make a socially useless and futile role personally meaningful by his own resoluteness? He can, perhaps, in two senses. First, he can takes what he does with personal seriousness, regardless of the way it is regarded and ultimately disposed of by others. It becomes, then, a matter of personal pride to attach a part as if one were making something designed to be excellent and to be used and enjoyed by another human being capable of appreciating it as such. Such an endeavor is admirable in many ways—but pathetic! It is admirable in that it involves taking personal responsibility for one's work and the refusal to surrender one's own standards to others. But it is pathetic in that it is condemned to regard what may be socially useless work as if it were ultimately important. In other words, it must proceed *as if* what one is doing and the way that one does it actually makes a significant difference to other persons. Hence, this attempt to overcome the alienation of the "they" cannot succeed. It isolates the individual even further from other men and, in effect, requires him to abrogate the interpersonal dimension of his world altogether.

But there is a second possibility which consists simply in carrying the first move a step further. If one takes full responsibility for one's role and one's work, that could motivate action to transform the system which maintains and requires the depersonalization. Instead of pretending that one's work is useful or socially significant, one might face the fact that it is not and either refuse to perform it or attempt to change the system which renders it personally and interpersonally trivial. It is at this point that the question of revolution must be seriously faced. As I see it, the thesis has not actually confronted this issue. Like Heidegger, the proponent of the thesis has assumed that an authentic existence which transforms the "They" into a "We" or an I-Thou is readily possible. Because the thesis regards all human programs as "revolutionary" because they are consciously conceived, it denatures the concept of "revolutionary change" and thus manages to dissolve the issue altogether. Neither Heidegger nor the author of the thesis take man's alienation from himself and others in a depersonalized society very seriously. They do not appear to recognize that a mere change of heart or of existential attitude might only render the alienation more acute. If, for example, an individual who no longer flees from his participatory role in the social system becomes concerned about the fact that the performance of his role requires him to live in bad faith with himself and others, he may find the continued acceptance of that role intolerable. Personal conversion cannot suffice if radical change in the social order is called for. To paraphrase an expression of Rheinhold Niebuhr's, the individual cannot achieve an authentic existence in a depersonalizing society. It is by no means obvious how anything like a "dialogic relationship" can be established within impersonal institutions, or, for that matter, with co-participants within those institutions. Attempts at humanizing such institutions as, for example, by providing sensitivity groups for staff and workers or clubs and social activities within the organization, serve only to carry the depersonalizing process a step further. It fosters the pretence that a mode of relatedness which treats persons as means is actually concerned about persons. One is encouraged to develop a new set of skills so that one's capacity for personal friendship and satisfaction can be sacrificed to the requirements of the institution. Instead of overcoming the alienation between the "I" and the "They" it only accentuates it and renders it more pernicious by denying the "I-They" distinction altogether. It may well be far healthier to recognize that one simply does not care personally and is not personally involved with one's co-workers than to pretend that one stands in a personally significant relationship to them.

Lest I be misunderstood, let me hasten to add that I do not mean to say : (a) that one's co-participants are not persons and to be regarded as such, or (b) that it is impossible to have a personally meaningful relationship with a co-worker. What I do mean to urge is that the attempt to overcome the alienation of the "I-They" by converting it to a pretended "I-We"—where the basis for being together remains unchanged, is dishonest. Is it sensible to expect the convict to overcome his alienated situation by making friends with the turnkey? He might do that, but even if he did their relationship as turnkey-inmate would remain a problem to be faced by the two of them. As a matter of fact it is only because both of them are able to transcend their roles as turnkey-inmate that they can have a personal relationship at all.

I wish to argue, therefore, that it is vain to hope that the "I-they" relationship can be directly converted into an "I-we" relationship. On the contrary, an "I-we" relationship can be established only if the depersonalized "I-they" relationship is recognized, refused, and transcended. It must be remembered that it is the personal which founds the impersonal and, hence, that underlying the impersonal means-means relationships which obtain within the organization, are the persons involved. They can establish only a false "We" relationship if they attempt to consolidate an interpersonal relationship around depersonalizing goals and activities.

The question here is just what alternatives are available. Can a serial and disjunctive impersonal relationship be transformed into an interpersonal "we" relationship by an inner transformation? Is it reasonable to expect that it can result either (a) from an existential conversion of the individuals involved so that they play their roles differently, or (b) from a new mode of group interaction? Heidegger proposes the former alternative whereas encounter and sensitivity groups aspire toward the second. As I see it, both alternatives are inadequate because they divorce form from content, feeling from behavior. The emphasis on inner meaning and the affective level of communication is crucially important; these factors constitute a necessary though not a sufficient condition of dialogic exchange. The sales manager might, if existentially converted and group-sensitive, play his role somewhat differently, so as both to gain more personal satisfaction from his work and relate more humanely to his associates. Still, if his work and that of others is essentially predicated on seduction and exploitation of others and, further, if there is little or no genuine autonomy permitted within the objective order, his personal conversion can provide no fundamental answer to the question either of personal meaning or social alienation. It is no less alienating to regard the objective order as unessential to the persons involved than to regard subjectivity as unessential. It is doubtless a useful corrective to stress the importance of the subjective dimension; but, once that is accomplished, the problem of outward form must be confronted. The pure heart and the noble intention are inevitably corrupted by a depersonalizing system which makes final disposition of them. If personal warmth and sensitivity is socially valid only for merchandising shoddy products, the subjective authenticity is subverted and becomes the instrument of the "bad" social order it set out to oppose and transform.

Orthodox Marxists have stressed the objective order as the crucial factor in constituting man's alienation from himself and others; existentialists and, more recently, specialists in group dynamics have emphasized the subjective dimension. It is no more realistic, however, to believe that a change in the objective order will dispose of the problem of human alienation than to believe that interpersonal conversion will accomplish that result. Though man is a subject, he lives and must live in an objective order. And even though he lives in an objective order, he can never be exhaustively identified with his objective social role.

The basic position taken by the thesis is, I believe, correct. Man does come to self-consciousness as an individual within an interpersonal context which both generates him and provides the initial meaning-matrices. Most immediately, it is the family which provides the decisive context. But the family itself is located within the objective social order and, to a considerable extent, is defined and limited by that order. It is now widely recognized that the family as an institution is in serious trouble. It is not simply that it has become nuclear rather than extended and, hence, cannot provide a sufficiently rich interpersonal context for its members. Even worse, it is without a nucleus. The family is no longer an economic unit, save perhaps as a special nexus of consumers. It is certainly not a significant productive unit. The economic and much of the social and educative function of the family has been taken over by larger social agencies. The family is plugged into the social network in such a way that the outward orientation and centrifugal force is far more significant than the inner core. Virtually the sole remaining function of the family is the nurture of the young. But in view of the fact that the young are not needed as long-term participants in the family venture, the family loses its *raison d'être* with the maturation of the children. It is literally true that the contemporary family is oriented toward its own dissolution.

The contemporary American family is expected to be able to move across the country lock, stock, and barrel one day and be totally plugged in and ready for operation the next. The family is little different from the individual in this respect. Both are expected to be rootless and adaptable for an instant plug-in like the T.V. set or the electric toaster. One wonders, therefore, if one can accurately view the contemporary family as a "We," a genuine nuclear unit, and the remainder of society as a "They?" If the individual and the family can function only when plugged into the supervening social system, there is as much a problem of the "We" as of the "I." If day-care centers should become the institution of the future, this trend will have worked itself out to its logical consequence. Children will then presumably be divested of any residual traces of identification within an intimate group; indeed, intimacy itself will presumably have been successfully dissolved altogether. I hasten to point out, however, that there is nothing radical or revolutionary about such a prospective development. It represents nothing more than a further stage in a process of social change which has been taking place on a wide scale for several decades. Insofar as it is revolutionary, it is but a phase of a larger development which has altered the foundations of human life. Revolutionary change at this point would require a calling into question of the large-scale processes which have generated and virtually require this further step. The issue regarding the family is not an isolated issue to be met by devising *ad hoc* measures for holding the family together, but part of the larger question about the way in which our whole society is organized. We must either face this crisis or be swept along by current tides.

7.   To persist in an alienated state requires that one cultivate either a cynical or a despairing consciousness. Since the cynical consciousness is itself a form of despair, in the last analysis there is but one alternative. The cynical consciousness recognizes the forms of alienation but accepts them with fatalistic resignation. One sees through the sham, as it were, but retains a modicum of freedom through one's lucidity. In other words one accepts *bad faith* as the only viable alternative and cheats on the system. One is then dishonestly honest and honestly dishonest. One withholds and withdraws even as he participates. It is a despairing posture because one gives up all hope of an integration of the subjective/objective, the I-They, etc.A more acute and overt form of despair relinquishes the hope of preserving one's freedom through compromise. Whereas the cynical consciousness gives rise to resentment and apathy, the overtly despairing consciousness wills either its own death or the annihilation of the tyrannical other. It posits either suicide or violence. The only positive alternative is to refuse the alienated state as a necessary condition of human existence and to attempt with others to overcome it. At the present time we are much clearer in our diagnosis of the situation than about prospective cures. Yet this is the perennial challenge confronting man, namely, to find scope for his personal freedom in what must remain always a partially alien context. It is the pain of alienation which provokes both personal and social growth.

Albert Camus has argued that man cannot be free to assert his freedom save insofar as he rebels metaphysically, that is, until he confronts and refuses to resign himself to death and finitude. His answer is thus rather different from that of Heidegger, who urges that we live our death in an anticipatory fashion and thus fully accept our finitude. The truth, as I see it, lies somewhere in between these two points of view. Death and finitude are the ultimate boundaries which must be acknowledged. As both Heidegger and Camus emphasize, man must remain in important respects ever a stranger in the world. Human dread and anxiety are ineradicable. Man can never expect to be transparent to himself any more than he can expect to achieve perfect understanding of his world. Nor can he realistically expect to achieve complete mastery over himself, other persons, or nature. Yet, to live in anticipatory resignation toward death is to capitulate to fatalism as if one could identify with one's own fate; whereas to live in a state of revolt against death is to exist in a posture of defiance. In a certain sense death is the supremely alienating factor which both individuates and isolates man. Coming to terms with death is a way of acknowledging one's finitude and thus liberates the individual for the exercise of his freedom. Camus is correct in holding that an evasion of this fact condemns man to live in fear and requires that he accept whatever circumstances will remove or distance the threat of death. There remains a question, however, whether death is so final and absolute a power as both Heidegger and Camus suggest. Only if the individual is at least relatively absolute, can death be an absolute power. If man is the sole source of meaning, then death must be viewed as a final negation. But if nature is considered as the source of personal existence, death simply returns life to its origin. It is unreasonable to expect that man himself should be able to overcome the alienation signified by his finitude. For death to be positively meaningful requires that there be a source of meaning beyond human existence which encompasses both birth and death.

By common consent the most pervasive and acute form of alienation confronting man today is his inescapable involvement in a depersonalizing material and social culture. This culture is predicated on a negation from the natural world. But the cultural world which has gradually supplanted man's natural habitat is one in which he finds it increasingly difficult to be at home as a person with others in a community. If he is to cope constructively with this estrangement he must, first, become fully aware of his predicament and second, he must trace it back to its origins. He must, in other words, understand the fundamental motivations which have been operative in the depersonalizing historical/cultural process. He can have a genuine option only insofar as he apprehends the full scope of his condition and, in so doing, calls it radically into question. He cannot expect to undo his own history or return to a pristine state of nature, but he can insist upon making a radical turn at this juncture in history. Though he may feel utterly solitary and impotent, he is never really alone; his is a shared fate and a universal human predicament. Although radical and even revolutionary change may be called for, he cannot accomplish it by himself. He needs the reinforcement of others in an enterprise based upon common concern. And if he is to be at home in the world he must revise his conception of his own freedom. He is now at the point of questioning whether nature can legitimately be viewed as an alien force to be subordinated to human projects. If man really were only what he makes himself and has made himself to be, there would be little basis for hope that he could surmount the predicament in which he finds himself. Human history is full of surprises precisely because man is not his own foundation and can never fully appropriate his own potentialities. The answer to his distress may depend less upon new rationally conceived projects to revise the old order than upon his ability to listen and recover his ties with his own being. Perhaps, as Heidegger has suggested, he needs more to let himself be than to make himself be. And this applies, also, to his being with others and with nature. What the author of the thesis has termed dialogic involution is a valid ideal, but the achievement of it demands a revolutionary turn—including both personal conversion and fundamental alteration in our social consciousness. The first state in this turn is the full experience of the estrangement. The next and subsequent stages remain to be formulated.